S0-CGR-019

Western Lands and Waters
XIV

The Don Pedro Dam and Powerhouse.

LAND, WATER AND POWER

A History of the Turlock Irrigation District 1887-1987

by
Alan M. Paterson

THE ARTHUR H. CLARK COMPANY
Glendale, California 1987

LIBRARY OF CONGRESS CATALOG CARD NUMBER 87-070947
1SBN 0-87062-177-7

Contents

Illustrations

Maps and Charts

Preface

The scene repeats itself early every morning from March to October. A white pickup drives westward along the bank of the main canal, its driver following the water down from the reservoir, setting the boards and gates that regulate the flow to match the volume of each day's releases. Rising above the canal bank are the steel towers of the main transmission line from the generators at the Don Pedro powerhouse, their wires catching the sun as it rises from behind the Sierra. There, in one place, can be found the symbols of the present-day Turlock Irrigation District; the lines of water and power flowing down from the mountains to the farms and cities below.

In the Central Valley of California, as in many parts of the American West, the irrigation ditch is among the most significant of man's works. Aridity typifies the West, and in an arid region agriculture is impossible without putting more water on the soil than nature provides. Irrigation has transformed deserts into gardens, and has shaped institutions, livelihoods and communities where it is practiced.

Irrigation is as complex as it is fundamental, and perhaps no more so than in its institutional structure. All but the simplest canals require some sort of organization to finance, build and operate them. Pioneer irrigation systems were mostly small affairs, built either with private capital, often as part of land sales schemes, or by cooperative companies owned by the irrigators themselves. Later came public irrigation districts, which, like the mutual companies, gave the irrigators a voice in the management of an asset as vital as the land itself. These were all independent, local institutions and that is an important point to remember because twentieth century irrigation development has been dominated by the federal reclamation program. However, despite the obvious significance of federal projects, local irrigation agencies have had a continuing importance in Western water management.

An understanding of grassroots water agencies is made no simpler by the fact that they are a disparate lot. Differences in their size, organizational framework and the extent of their responsibilities often stand out more clearly than their similarities. They do have a great deal in common, however, in terms of their basic functions.

For example, they are all responsible, in one way or another, for establishing a water supply for the lands they serve. That may involve no more than contracting with the Bureau of Reclamation for the delivery of water from federal aqueducts, or it may mean operating the agency's own dams and canals, or something in between. Regardless of where the water originates, most local irrigation agencies maintain and operate the canals or pipelines that carry water to farms and manage its distribution, determining when, how often and in what quantities irrigators are entitled to receive water. And last, but by no means least, they represent and speak for their irrigators in all those matters of negotiation and lobbying that can affect the welfare of the local irrigation community. Thus even though the federal government, and in California, the state government, have assumed responsibility for major water projects, local agencies still play a vital role in the development, distribution and use of water throughout the West.

Although they might seem to be primarily of local interest, these mutual irrigation companies, irrigation districts, water storage districts and others are such an integral part of the history of American irrigation, and of the arid West in general, that they deserve more attention than they usually get. This book tells the story of one such agency, the Turlock Irrigation District. The Turlock District is the oldest example of an important type of irrigation institution, the irrigation district. It is not necessarily a typical irrigation district. The TID and its partner agencies are among the few to have built and operated large dams and powerhouses without any substantial federal or state assistance. A major contributing factor to its independence is the TID's unusual role as an electric utility, retailing the power it generates or purchases directly to consumers within the district. In common with local irrigation agencies throughout the West, however, the Turlock Irrigation District faced problems of water rights, canal management, and internal politics. And the impact of irrigation on agriculture and community life in the Turlock region mirrored its impact, in magnitude if not in detail, elsewhere.

This book is the culmination of a research project commissioned in 1981 by the Board of Directors of the Turlock Irrigation District.

For the district, the most immediate reason for the project was the approaching centennial of its founding. Perhaps of equal importance was the need to examine those elements of the district's past that may help shape its future. Historic practices, contracts and agreements provide the framework within which the district, like any organization, must operate. For district personnel and for district residents old and new, there is considerable value in knowing how and why the district developed as it did.

It has always been my hope that local readers, students of irrigation history and the district itself would all find something of value in this telling of the district's history. However with the needs and interests of three distinct audiences to keep in mind, the finished product involves some unavoidable compromises. Professional historians might prefer a more comparative approach, while other readers may wish for more of the lore of local people and places and less detail on the unexciting but important matters of finance and engineering. While I can readily sympathize with such desires, my basic goal had to be a balanced, and above all, comprehensive account of the district's history. It is not, however, the final word on the TID's complex and fascinating past. Many themes that could only be touched on briefly here could be explored in greater depth, and of course, the final outcome of the most recent events will not be known for years to come.

Regrettably, it is impossible to express my appreciation by name to all those who played a part in my research. The TID opened its records to me without limit or exception, and I am deeply indebted to the many employees who not only went out of their way to help me locate nearly forgotten documents, but made the district office a pleasant place to work. It is important to note that at no time was there any attempt to influence my research or conclusions and I wish to thank the Board of Directors and management for insuring that I had that freedom.

I spent considerable time at the library of the California State University, Stanislaus, and at the Stanislaus County Library, and used the University of California, Davis, library as well as the California Water Resources Center Archives and the Bancroft Library, both on the campus of the University of California, Berkeley. Any researcher knows how important the cooperation of library staffs can be, and I certainly found the people at these institutions to be friendly and helpful. A special thanks should go to the many district residents and past and present TID employees who allowed me to interview them. Without exception, every interview yielded information unavailable from other sources and gave me a feeling for the way people viewed the TID and its role in the community. I met many wonderful people during the course of this project, and they contributed immeasurably to the result. I am also grateful to Donald J. Pisani of Texas A & M University for providing me with an advance manuscript of his book *From Family Farm to Agribusiness: The Irrigation Crusade in California and the West, 1850-1931,* and for reading a draft of my work. My debt to my former professor and mentor, W. Turrentine Jackson, must also be acknowledged, for it was at his suggestion that I first turned my attention to water resource history. Of course, any errors or omissions are entirely my own responsibility.

It is not unusual for an author to express his appreciation to members of his own family for their assistance or moral support during the often taxing tasks of research and writing. I have a special debt, however, to my father, Grant Paterson, a native of the Turlock District and a lifelong farmer. Through him I learned what I know of practical irrigation and irrigated farming. I would like to think that my own insights into irrigation and the Turlock Irrigation District were enhanced by the hours spent watching the water spread across his orchards. Finally, I would like to thank my wife Patty and our daughters, Cindy and Susie, both born during work on this book, for the aid and comfort they gave at even the most frustrating moments. For them especially, my gratitude is beyond words.

Land, Water and Power

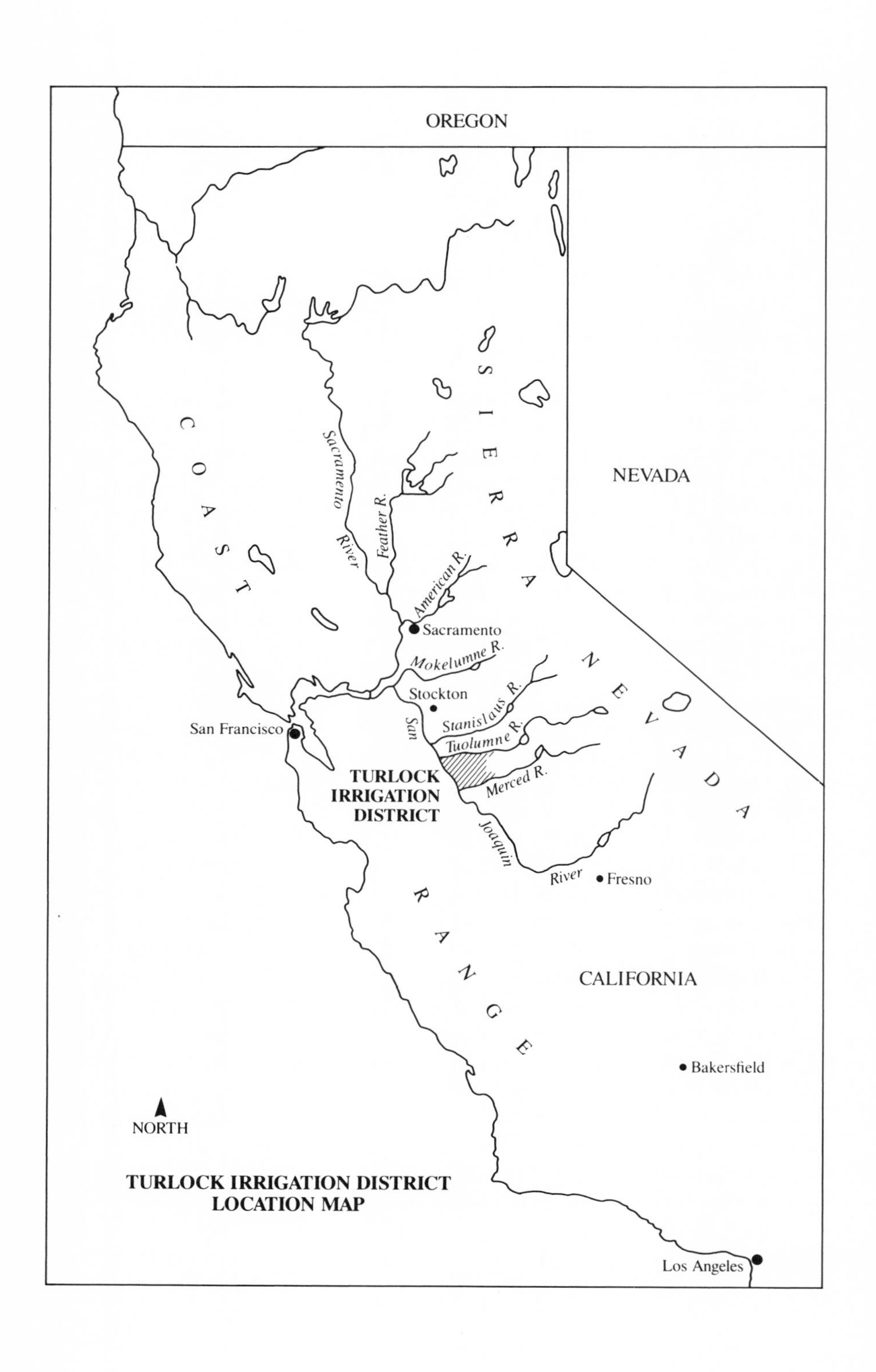
OREGON
NEVADA
CALIFORNIA
COAST RANGE
SIERRA NEVADA
Sacramento River
Feather R.
American R.
Sacramento
Mokelumne R.
Stockton
Stanislaus R.
Tuolumne R.
Merced R.
San Joaquin River
San Francisco
TURLOCK IRRIGATION DISTRICT
Fresno
Bakersfield
Los Angeles
NORTH
TURLOCK IRRIGATION DISTRICT
LOCATION MAP

1

Antecedents: Natural and Historical Background to Irrigation

The story begins with the natural elements of land, weather and water. In the development of irrigation anywhere, fertile soils and a climate that lacks only enough natural moisture during the growing season are the first and most fundamental prerequisites. The land that became the Turlock Irrigation District lies near the heart of California's four-hundred-mile-long Central Valley. It is bordered by three of the rivers that flow out of the Sierra Nevada Mountains — the San Joaquin and two of its tributaries, the Tuolumne and the Merced. On the valley floor the landscape between these streams now seems monotonously flat, whatever slight variation it once had having been nearly erased by decades of farming and land-levelling. In fact it rises gradually in elevation from the swampy lowlands along the San Joaquin River eastward toward the first gently rolling ranges of the Sierra foothills that border the district. Thousands of years ago, the mountains gave birth to these plains as little bits of sand and soil eroded from the peaks and washed into the valley or the shallow waters that once covered it. The alluvial soils of the Turlock region are predominantly sandy, reflecting their heritage of decomposing mountain granite. Although they are generally light and well drained, there is often considerable diversity in soil texture and structure even within relatively small areas. Soil classification experts have defined and mapped these minor variations in composition and given them names like Tujunga loamy sand, Ripperdan sandy loam and Delhi sand. For the most part they are fertile, easily tilled soils, but they are not all easy to manage. From near Turlock south to the Merced River, the sands are noticeably coarser. At Delhi the wind once piled the loose sand into low dunes and throughout the area early settlers remembered the sandstorms that accompanied the spring winds. Severely alkaline soils are found only in limited areas near the San Joaquin River, but there is a substantial area of slightly saline soil west of Turlock. Hardpan, a compacted, usually impervious subsoil layer found at varying depth, is a more common nuisance that can restrict water penetration and root growth. Flat and sandy, the Turlock region offered an adaptable, if unexciting, landscape that would in the course of a few decades go from grasslands to grain fields to irrigated farms.[1]

After the landscape, weather is the chief architect of the valley environment. Before irrigation, life on these sandy plains depended entirely on the seasonal cycles of sun and rain in a climate characterized by cool, moist winters and hot, dry summers. The rainy season typically begins in October or November when the weather fronts that march eastward across the northern Pacific Ocean gather strength and begin to slip further and further south toward central California. Rainfall averages between nine and a half and fourteen inches annually, almost all of it coming in the winter and early spring months.[2] Snow is a rarity in the valley and freezing temperatures generally occur only at night. Dense ground fog, known as tule fog, forms in the calm night air following a winter rain, often lifting to form a low, dark overcast that can persist for days or even weeks. Winter is short, however, and usually by February temperatures are noticeably milder. As spring progresses storms become weaker and less frequent, leading to a summertime pattern of cloudless skies and hot days. The heat that characterizes valley summers can arrive as early as May, with temperatures climbing into the nineties and often over one-hundred degrees Fahrenheit. Fortunately for human comfort the hot valley air is relatively dry. Relief from the heat comes from the afternoon and evening breezes which, after several days of high temperatures, draw cooler coastal air into the valley. On rare occasions unpredictable summer storms might dampen the ground briefly, often coming, when they come at all, from the remnants of errant tropical storms moving northward from Mexico. Usually, though, there is no rain until fall, when the cycle starts over once more.

Before white men came to the valley, it was a vast open prairie. With the first autumn rains the native grasslands began to turn green. Perennial bunchgrasses like purple needlegrass and nodding needlegrass dominated the Central Valley prairie, along with associated species like blue wild rye, pine bluegrass and deer grass.[3] Although riparian forests lined the rivers, only widely scattered oaks or occasional groves broke the monotony of the plains. By February and March the warming days of spring brought the grasslands to life in a brief period of ample moisture and sunshine. Explorer John C. Fremont, travelling through the San Joaquin Valley in early 1846, recorded these detailed impressions of springtime on the valley plains:

> At the end of January, the river bottoms in many places, were thickly covered with luxuriant grass, more than half a foot high. The California poppy, *(Eschscholtzia Californica),* the characteristic plant of the California spring; *nemophilia insignis,* one of the earliest flowers, growing in beautiful fields of delicate blue, and *erodium cicutarium,* were beginning to show a scattered bloom. Wild horses were fat, and a grisly bear, killed on the 2nd February, had four inches thickness of fat on his back and belly, and was estimated to weigh a thousand pounds. Salmon were first obtained on the 4th February in the To-wal-um-ne river, which, according to the Indians, is the most southerly stream in the valley in which this fish is found. By the middle of March, the whole valley of the San Joaquin was in the full glory of spring; the evergreen oaks were in flower, *geranium cicutarium* were generally in bloom, occupying the place of the grass, and making all the uplands a close *sward.* The higher prairies between the rivers presented unbroken fields of yellow and orange colored flowers, varieties of *Layia* and *Eschscholtzia Californica,* and large bouquets of the blue flowering nemophlia nearer the streams. These made the prevailing bloom, and the sunny hill slopes to the river bottoms showed a varied growth of luxuriant flowers. The white oaks were not yet in bloom.[4]

Even in 1846 the plains had been influenced by the distant settlements of Spanish-Mexican California. Wild horses competed with the native tule elk and pronghorn antelope, and non-native plants, such as the *Erodium cicutarium* or red-stemmed filaree noted by Fremont, had begun their conquest of the original prairie grasses. By May the brilliant tapestry of spring was fading and the store of moisture from the rainy season was quickly being exhausted. The annual grasses, which had sent out new shoots up to two feet high, turned brown above their stems, conserving moisture for the long dry season ahead. In some climates winter is the time of dormancy but on the plains of the San Joaquin Valley the native grasslands approached that condition in the baking heat of summer and early fall.[5]

Given fertile soils and a suitable climate, the final material necessity for irrigation is a water supply. Although surrounded on three sides by rivers, the history of the Turlock Irrigation District centers on only one stream, the Tuolumne River, the district's northern boundary and the source of its water and power. The Tuolumne rises high in the alpine granite of the Sierra crest. Forks originating on the slopes of Mount Lyell and Mount Dana, both over 13,000 feet in elevation, combine with other little snow-fed creeks at Tuolumne Meadows, in the high country of Yosemite National Park, to form the river's main stem. Plunging through the steep Grand Canyon of the Tuolumne, the river drops 5,000 feet in about 15 miles into Hetch Hetchy, once a deep, sheer-walled valley, now a reservoir for San Francisco's municipal water supply. Below Hetch Hetchy the Tuolumne soon leaves the granite cliffs behind, passing in a deep, V-shaped canyon down the western slope of the mountains. Tributary forks and creeks that originate in the pine and fir forests of the 1,880 square mile watershed join the river as it continues a rapid, tumbling descent toward the valley. Gradually the hills above the river become lower and more rounded, and their covering changes from dense conifers to scrubby digger pines and low-growing chaparral and finally to the oaks and grasses of the lower foothills. Before its innundation by reservoirs the lower canyon held the Tuolumne through the foothills to a final fall or rapids just east of the town of La Grange. Although the river is still surrounded by the last rolling foothills for a few miles west of La Grange, it has slowed to become a valley stream. Cut deeply into the valley floor the river and its narrow flood plain are separated from the surounding plateau by steep bluffs that rise on either side. Finally, it empties into the San Joaquin River, which carries its water into the Delta where it joins the other great river of the Central Valley, the Sacramento, for a final surge to the sea through San Francisco Bay.

California's dry summers and moist winters give the Tuolumne River a distinct seasonal pattern. The winter storms that bring rain to the valley floor blanket the high mountains with snow. Storm after storm, it accumulates on the high country while rainfall on the lower watershed swells the river with each passing storm. By April of most years the storms have subsided, leaving snow piled to a depth of seventy-five to one hundred inches on the upper watershed. The snowpack is the natural reservoir of the Sierra, where moisture is stored in the form of snow until its release at the onset of the dry season. At first the spring sunshine melts only a bit of the crust each day before the pack refreezes at night, but soon the whole mass begins to soften and then liquify. Rivulets from

The upper Tuolumne watershed at Slide Canyon in the Yosemite high country. Photo by Ned Peterson.

each snow bank feed the creeks and streams of the watershed and they in turn fill the Tuolumne with snowmelt. Snow on the lower mountains melts rapidly, but around the high peaks the process takes longer, insuring heavy runoff through June and into July in most years. The river's most spectacular floods are the result of torrential winter rains, but they cannot compare with the sustained volume of the spring runoff from the snowpack. Perhaps two-thirds of the water the Tuolumne will carry for a year flows down from the mountains during a four month period between the beginning of April and the end of July. With the snowpack exhausted the river slows to a trickle during the summer, finding its way around the rocks it easily covered a short time before. Towering clouds build up over the high country on warm summer afternoons but their thunderstorms do little to replenish the river. Like life on the valley plains, the river, too, must wait for the fall rains for its rebirth.

A river can be measured in many ways – by the size of its watershed or by how much water it carries. The most common yardstricks are those of volume, expressed in acre-feet (the amount of water needed to cover one acre one foot deep), and the rate of flow, as measured in the number of cubic feet per second passing a particular point (commonly referred to as "second-feet"). The Tuolumne is the largest tributary of the San Joaquin River, averaging about 1,800,000 acre feet per year. The average is deceiving, however, because there is seldom an average year. Since accurate record-keeping began in 1897, precipitation and runoff have fluctuated over a wide range from year to year, from a high of 4,631,000 acre-feet in 1983 to a low of 383,000 acre-feet in 1977. The seasonal variations in flow are somewhat more predictable, but sudden storms in the winter and the rate of temperature change in the spring, as well as the overall size of the snowpack, can cause each year and each season to vary considerably from what is termed "normal". The Tuolumne's ample resources, and its vagaries, were as significant as the soil or the climate in shaping the history of the Turlock Irrigation District.[6]

From the start the river played an important role in the area's history. The first settlements in the Turlock region were along the rivers rather than on the open plains. They owed their birth to the tide of Gold Rush trade and travel that swept across the Central Valley to the mining camps of the Sierra. Ferries were established for overland travellers and shallow-draft steamboats pushed up

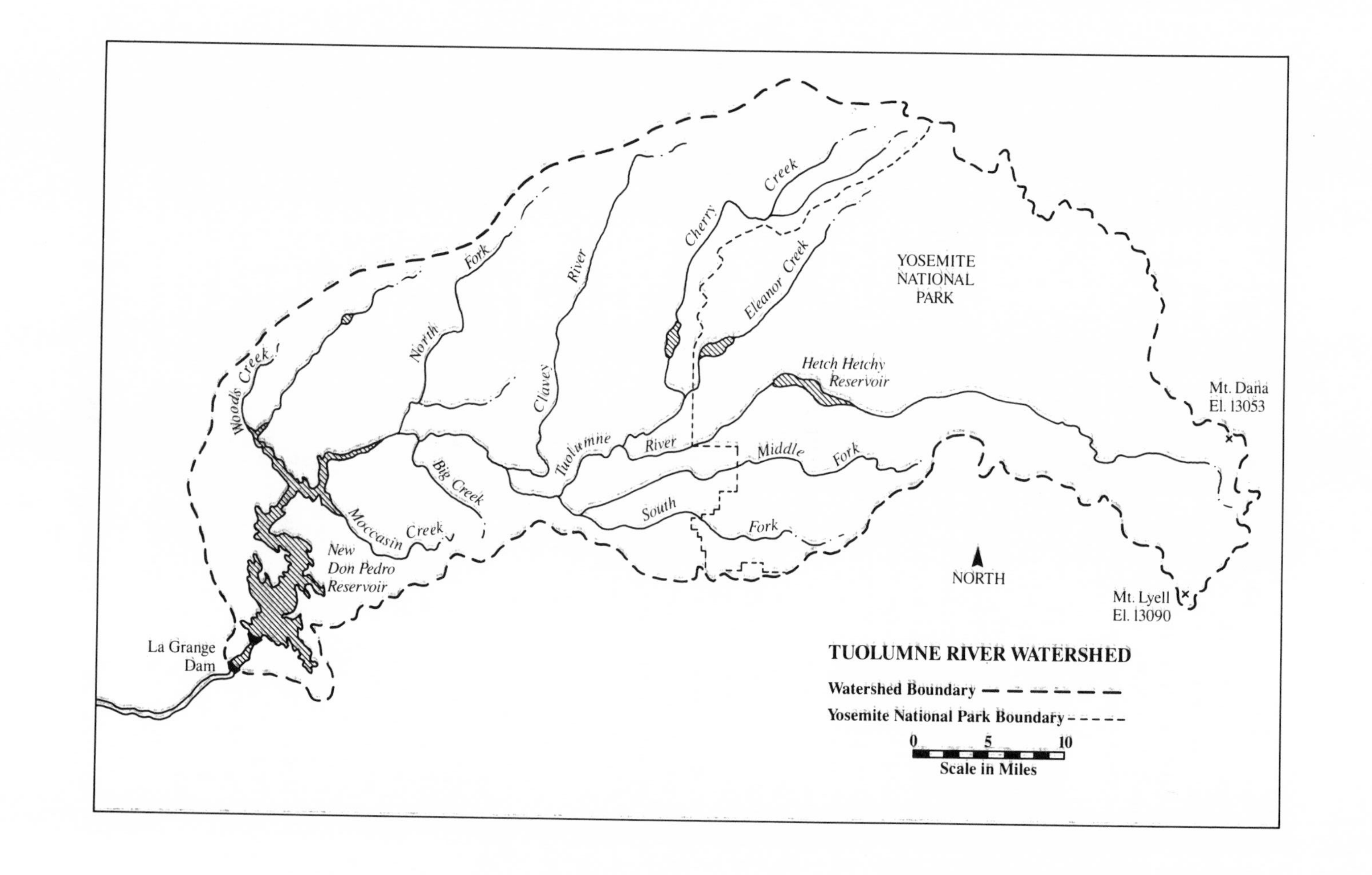

TUOLUMNE RIVER WATERSHED
Watershed Boundary
Yosemite National Park Boundary
0
5
10
Scale in Miles
NORTH
YOSEMITE
NATIONAL
PARK
Mt. Dana
El. 13053
Mt. Lyell
El. 13090
La Grange
Dam
New
Don Pedro
Reservoir
Woods Creek
North
Fork
Clavey
River
Cherry
Creek
Eleanor Creek
Hetch Hetchy
Reservoir
Tuolumne
River
Middle
Fork
South
Fork
Big Creek
Moccasin
Creek

the rivers with cargos destined for the Mother Lode. At ferry crossings or where the boats could go no further, towns sprang up to serve travellers and shippers. One of these, Tuolumne City, was located west of present-day Modesto on the north bank of the Tuolumne River. It was described in April of 1850 as a "real thriving busy little town" whose location at what was thought by some to be the head of navigation on the river would "in a short space of time make it one of the first among the cities of the interior."[7] Unfortunatley, high hopes were not a firm foundation for permanance. Low water resulting from the dry winter of 1850-1851 halted navigation and effectively shut down the budding metropolis of Tuolumne City, as well as Empire City, a rival head of navigation even further upsteam.[8]

While the little villages along the Tuolumne went through a rapid cycle of boom and bust, a livestock industry was being established on the plains. Even before the Gold Rush wild horses and cattle, the descendent of strays from the great coastal herds, had wanderd into the Central Valley, where they shared the grasslands with the native antelope and elk. The demand for meat to feed California's new population suddenly made these cattle valuable, and still more were brought in from as far away as Texas. While cattle roamed the plains in more or less wild bands that were handled only during an annual roundup, the early-day cattlemen made their homes along the rivers, where the availability of water and timber created a more congenial environment than the dry and nearly treeless prairies.

Cattle covered the plains until a combination of flood and drought decimated the herds. From mid-November 1861 through January 1862 torrential rains fell on California. The Tuolumne County mining town of Sonora, for example, recorded an unbelievable eight and half *feet* of rain during that period. Rivers everywhere went over their banks, turning the valleys into wide, shallow lakes. In the Sacramento Valley riverboats cruised across once dry fields in search of survivors, and the situation was probably no better along the San Joaquin River.[9] Some cattle were drowned outright by the runaway rivers while others died hopelessly mired in the mud left by the receding waters. A devastating drought that lasted until 1864 followed the floods. William Brewer of the California Geological Survey crossed the San Joaquin Valley south of the Turlock region in May and June 1864 but his observations were probably representative of the situation throughout the valley. Of

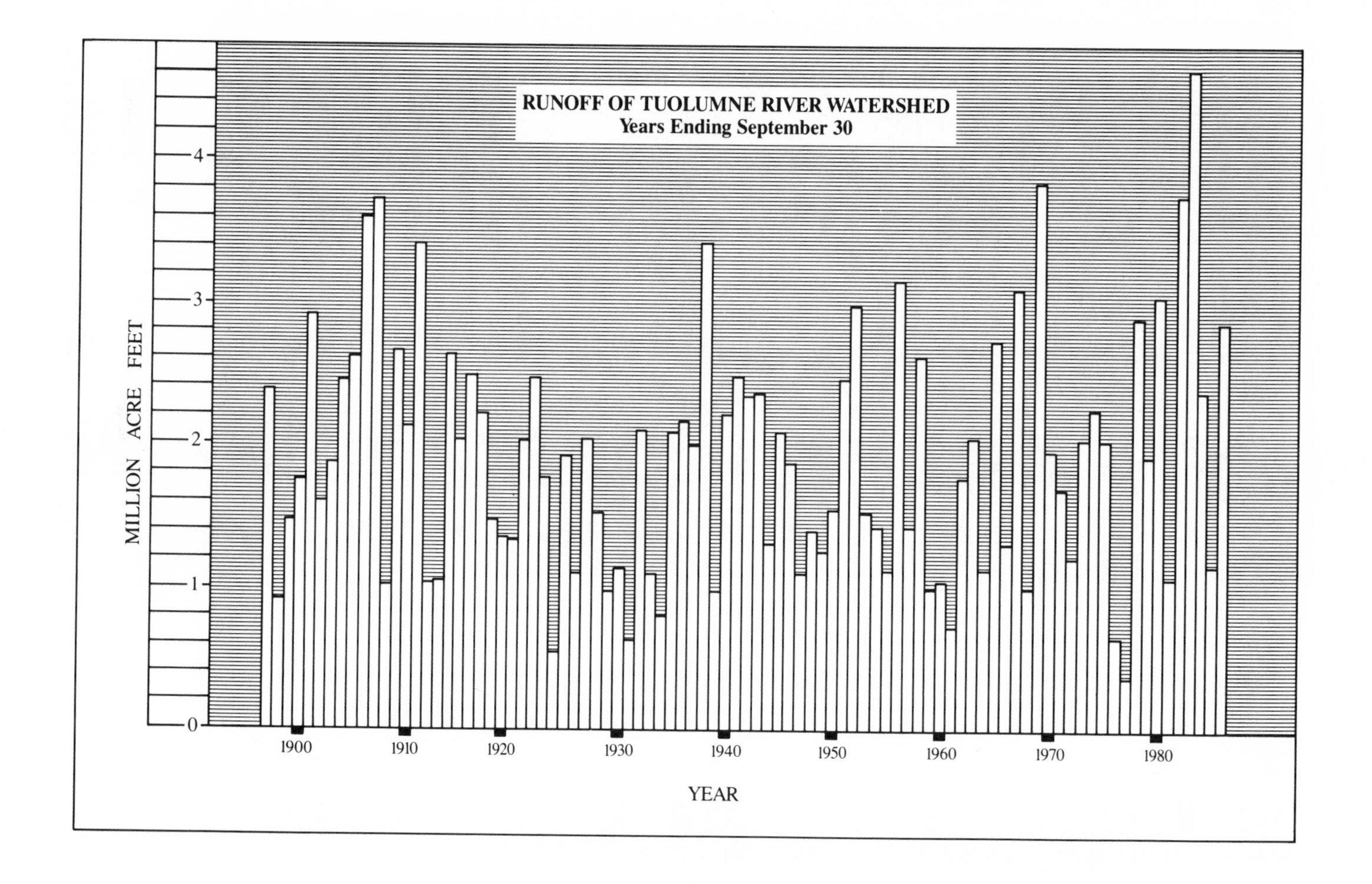
RUNOFF OF TUOLUMNE RIVER WATERSHED
Years Ending September 30
MILLION ACRE FEET
0
1
2
3
4
1900
1910
1920
1930
1940
1950
1960
1970
1980
YEAR

his journey from Pacheco Pass to Firebaughs Ferry he wrote: "The ride was over the plain, which is utterly bare of herbage. No green thing greets the eye, and clouds of dust fill the air. Here and there are carcasses of cattle, but we see few living ones — not twenty during the day, where nearly as many thousands could have been seen two years ago."[10] The cattle industry continued to decline even when the rains finally returned, and by 1870 the state had only half a million beeves where perhaps three million had roamed in 1862. Following the drought sheep became an increasingly popular alternative to cattle. They needed less water and could be more easily herded to fresh pastures when necessary.[11] But even as sheep began to replace cattle on the valley ranges, larger changes were in the making; changes that would transform the plains from grazing land to farm land.

The crop that conquered the plains of the Central Valley was wheat. It was California's second great bonanza – a source of agricultural wealth to rival the product of the mines. Wheat was well adapted to the state's climate, producing a desirable hard, white grain, and by 1858 a modest surplus was available for export. During the nineteenth century wheat was shipped from California primarily to the British port of Liverpool, where it competed in the international grain market with crops from the eastern United States, Europe, Russia, India and later Canada and Australia. Although the export market held promise, events in the early 1860s postponed major foreign shipments until 1867. The flood of 1862 and the subsequent dry years reduced the size of the crop, the Civil War disrupted commerce and the silver mining towns of Nevada's Comstock Lode offered a readily accessible market for California growers. The large crop of 1867 was the first to be exported in significant quantities and the timing proved excellent, for world grain stocks were low and prices were high. When it became apparent that wheat grown for the world market could yield handsome profits, efforts were quickly begun to bring new lands, including those of the Turlock region, into production.[12]

By 1867 wheat was already being planted in the section known as Paradise Valley, just north of the Tuolumne River. The rapid growth of the local wheat industry was reflected in statistics on the cultivated acreage in Stanislaus County, which rose from about 30,000 acres in 1866 to over 200,000 acres the following years.[13] In 1867, land sales, an important precursor to agriculture, finally began on the plains south of the Tuolumne.[14] Not only wheat but the promise of a railroad down the valley prompted farmers, and speculators, to take a sudden interest in the San Joaquin Valley.[15] Although the public domain in this part of the Central Valley had been surveyed and offered for sale by the federal government in the 1850s, there had been no takers. The cattle and sheep ranchers had acquired title to some lands close to the rivers, but they had no compelling reason to purchase the grasslands where their livestock grazed. Farming, however, was a different matter. Crops and homes demanded the security of property ownership, so the first rush onto the plains was in reality a rush to the U.S. Land Office.

The manner in which land was acquired was the first indication of the character of the wheat business and the course that the area's economy would take in the next few decades. Full sections of 640 acres were commonly sold as a single unit, and a few men of large means were able to accumulate huge holdings. John W. Mitchell was easily the largest landowner in the vicinity, his 100,000 acres stretching from north of Turlock southward beyond the Merced River. Near the Sierra foothills just south of the Tuolumne, L.M. Hickman amassed 16,000 acres, and Daniel Whitmore bought 9,000 acres near what would soon become Ceres. Even though it offered an opportunity for free land, homesteading on 160 acre tracts was comparatively rare in the Turlock area, confined primarily to the western portion of the district and nearer the rivers.[16] Most buyers wanted much more than 160 acres and many were clever enough to know how to acquire title for substantially less than the government's minimum price of $1.25 per acre. For example, "greenbacks," the depreciated paper currency of the Civil War era, were only worth about seventy cents on the dollar in the late 1860s, but were accepted by the Land Office at face value.[17] Land could also be obtained cheaply by the use of land scrip originally issued to certain states and to some military veterans. Military bounty warrants were rarely used in the Turlock area, but agricultural college scrip authorized by the Morrill Act of 1862 was commonly traded for land. With more vacant public lands than cash, Congress had provided financial support for state agricultural colleges by granting land to each state, which could in turn be sold to pay for the schools. For those eastern states without any unsold land, scrip was substituted, allowing them to go into the western public domain to select lands to support their colleges. Rather than choosing and selling lands themselves, most of the states entitled to scrip simply sold it for whatever it would bring. As scrip flooded the market after the Civil War, its value, and consequently the real value of the land acquired with it, declined. In 1868 more public land was purchased in California with agricultural college scrip than by all other means combined. John Mitchell was reported to have gotten about one-third of his 100,000 acres in this manner. Congress placed limits on the use of scrip at the end of 1868, but by then most of the land in the Turlock region had passed into private ownership and the creation of a new wheat farming area was underway.[18]

The movement of agriculture onto the plains briefly rejuvenated commerce on the rivers and with it the river towns. Following its collapse in 1851, Tuolumne City had been reduced to no more than a ferry boat and a saloon, but by 1867 it once again warranted a post office and the following year 10,000 tons of wheat were shipped by riverboat and barge from its docks.[19]

While Tuolumne City became the most important town in the area, other river towns also flourished, as steamers loaded grain at least as far up the Tuolumne as Empire City when the river was high enough. Inland navigation, however, could be a risky proposition, depending as it did on having enough water in the river. The problem of reliable transportation was solved by the arrival of the railroad, and with its coming the center of community life moved permanently away from the rivers and onto the plains. The Central Pacific reached the Tuolumne River from the north in the fall of 1870. A new town, Modesto, was laid out beside the tracks on the north side of the river, and like a magnet it soon drew the citizens, businesses and even many of the buildings away from the river towns. The construction of a bridge over the Tuolumne took until July 1871, and that summer the line was extended across the Turlock region to the Merced River.[20] In August Modesto's newspaper commented that, "We understand that a switch and depot will be established on the railroad near the county line, on Mr. Henderson's ranch, which is some thirteen miles south of this place."[21] No buildings were erected at Henderson's station and the depot site was soon moved about a mile north to the place on John Mitchell's land where Turlock grew.[22] By the end of 1871 the town had been established and lots were being sold. About nine miles up the tracks to the north, the town of Ceres was founded by Daniel Whitmore, and other sidings and warehouses were built at Keys Switch and on Mitchell's property at Delhi, just north of the Merced River.

The founding of towns along the newly completed railroad marked the completion of the first great transformation of the Turlock region. Settlement had finally come to the plains and as fast as the plows could do their work, agriculture was replacing grazing. In the short space of half a dozen years the patterns of economic and community life that would last until the coming of irrigation three decades later had been established.

Wheat was the cornerstone of the era. Writer Charles Nordhoff toured the San Joaquin Valley in 1872 and found it little more than a gigantic grain field.

> Between Stockton and Merced lie about six hundred square miles of wheat. The railroad train runs through what appears to be an interminable wheat-field, with small houses and barns at great distances apart . . .
>
> Wheat, wheat, wheat, and nothing but wheat, is what you see on your journey, as far as the eye can reach over the plain in every direction. Fields of two, three and four thousand acres make but small farms; here is a man who 'has in' 20,000 acres; here one with 40,000 acres, and another with some still more preposterous amount – all in wheat.[23]

In that year, 1872, Stanislaus County led the state in wheat production with a five million bushel crop.[24]

Nordhoff's description pointed out two of the most outstanding features of Central Valley agriculture in the grain era; reliance on a single crop and the breathtaking size of some of the ranches. At a time when most American farms were small, family operated and diversified, the wheat ranches of California were just the opposite. They grew virtually nothing but wheat year after year, and although they were unexcelled in the use of labor-saving equipment, they still depended on large numbers of hired seasonal workers.

Many of the farms were also larger than those found in the rest of the country but it is easy to exaggerate their size. The few ranches of truly baronial proportions, encompassing many thousands of acres, attracted the attention of visitors like Nordhoff, who made it appear as though most wheat growers operated on the same immense scale. Some idea of farm size in the Turlock region can be gained from an examination of the early assessment rolls of the Turlock Irrigation District. The first tax assessment was made in 1888, and in that year 182 taxpayers owned almost all the agricultural land between the Tuolumne and Merced rivers as far east as the Sierra foothills. The remaining fifty-four taxpayers included the railroad, the owners of town lots in Turlock and Ceres and a handful of property owners with less than eighty acres apiece. Only four men owned more than 3,000 acres. John W. Mitchell had nearly 43,000 acres in the district, L.M. Hickman had over 6,700 acres in the TID and more than that outside the district boundaries, Daniel Whitmore owned more than 8,600 acres and Hiram Hughson had nearly 3,400 acres. Nine taxpayers owned between 2,000 and 3,000 acres, while another twenty-eight had from 1,000 to 2,000 acres. Thirty-two property owners held from 640 acres – the equivalent of one square mile – to 1,000 acres. The seventy-three taxpayers with 640 acres or more owned over 80 percent of the land in the district. Forty-two landowners had between 300 and 639 acres, and of the remaining sixty-seven owners of agricultural property over half had 160 acres or more. Many of the smaller parcels appear to have been located near the San Joaquin and Merced rivers rather than out on the plains.[25] These statistics do not by themselves tell the full story of farm acreages in the wheat era. Where members of the same family owned adjoining parcels, they were probably operated as a single unit. At the other end of the spectrum, the largest landowners commonly used tenants to farm their land. Daniel Whitmore is said to have rented out tracts of 800 to 1,000 acres, and Turlock pioneer C.F. Lander recalled that, "Any man who could get hold of 6 horses could start farming. Mr. Mitchell would build him a house and barn, furnish plows, seed, feed for stock. He could go to almost any store and buy harness, household supplies, and groceries on one years time."[26] Tenants usually supplied only the teams and men in exchange for half the crop.

The land was plowed and planted each year after the first fall rains had softened the soil. Plowing the wheat fields was not a matter of walking behind a plow turning one or two furrows at a time, but of riding a California-developed device known as a gang plow. Gang plows consisted of a three-wheel wooden frame to which four to eight small plowshares were attached. They did not cut deeply, usually only three inches, and Nordhoff commented that in the Central Valley "if a man ploughs five inches deep he thinks himself a hero."[27] In some cases seed boxes were attached to the plow frame, scattering the seed just ahead of the shares so that the land was turned and seeded in a single operation, while other ranchers preferred mechanical broadcast seeders mounted on wagons or carts. Wheat germinated and grew during the cool winter months and by March it might stand a foot high. The success of each crop depended on having just enough rain in the fall for timely planting, and additional amounts into the spring to sustain the growing plants. By June the weather was dry and the ripened grain was ready for harvest. In harvesting as in plowing, California in the 1870s and 1880s was a world apart from the rest of American agriculture. On the Great Plains reapers cut and bound the grain, which was then stacked in shocks before being hauled to the thresher.[28] In California a variation of the reaper known as the header was used. Pushed through the fields by six or more horses, headers cut only the heads and perhaps a foot of the stalk, conveying them up a belt and into a specially constructed header wagon driven alongside. The wagon carried the grain directly to the thresher or to large stacks where it was stored until the thresher arrived. In time, giant combines drawn by thirty-six horses or mules replaced the headers and threshers, and the mechanization of the wheat fields was complete.[29]

The great unfenced fields of a single crop stretching as far as the eye could see or the sight of a squadron of gang plows moving over the land in echelon could stir the imagination, but wheat farming was an economic system and a way of life that stirred profound criticism. For example, Charles Nordhoff deplored what he considered the profligate ways of the wheat growers.

> There are hundreds of farmers in California, men who would be thought wealthy in any farming community in the East, who own several thousand acres, and who do not raise even a potato for their families. Wheat, wheat, wheat, is their only crop, and for this every thing else is neglected. Their families live on canned fruits and vegetables; all their house supplies are bought in the nearest town, of the groceryman; in a good season they sell their wheat for a large sum, and either buy more land or spend the money in high living; and when a dry year comes they fall into debt, with interest at one per cent. a month; and when the next dry year comes it brings the sheriff.[30]

When Nordhoff reproached one of the farmers for this lack of diversification and self-sufficiency, he was told bluntly, "We don't go a cent on anything but wheat in this country; we all want to get rich in two years."[31] The same sentiment was expressed still more powerfully by Frank Norris in his 1901 novel, *The Octopus.* Norris gave the fictional wheat grower Magnus Derrick the traits that many thought epitomized the wheat era in the San Joaquin Valley.

An 1880-era harvest scene on the Daniel Gallagher ranch.
Three miles west of Turlock. Source: Elliot and Moore, 1881.

> At the very bottom, when all was said and done, Magnus remained the forty-niner . . . the gambler, willing to play for colossal stakes, to hazard a fortune on the chance of winning a million. It was the true California spirit that found expression through him, the spirit of the West, unwilling to occupy itself with details, refusing to wait, to be patient, to achieve by legitimate plodding; the miner's instinct of wealth acquired in a single night prevailed, in spite of it all. It was in this frame of mind that Magnus, and the multitude of other ranchers of whom he was a type, farmed their ranches. They had no love for their land. They were not attached to the soil. They worked their ranches as a quarter of a century before they had worked their mines. To husband the resources of their marvelous San Joaquin, they considered niggardly, petty, Hebraic. To get all there was out of the land, to squeeze it dry, to exhaust it, seemed their policy. When at last the land, worn out, would refuse to yield, they would invest their money in something else; by then they would all have made fortunes.[32]

Seen through the eyes of Nordhoff or Norris, the wheat growers were a species of speculators, masquerading as farmers.

The indictment was a damning one, but was it accurate? Charges that they mined their land, growing wheat season after season without doing anything to restore the soil's fertility were substantially true, but the practice was not without justification. Wheat and other grains were all that could be grown in the San Joaquin Valley without irrigation. Diversification or soil-building crop rotation was ruled out by the climate as much as by the attitudes of the landowners.

The monomania for wheat also contributed to the growers' reputation as gamblers. Not only were they unable to spread the risks inherent in agriculture among more diverse crops, but their rejection of traditional rural self-sufficiency left them tied tightly to the cash economy of the marketplace. Central Valley grain farmers were a prototype for the equally specialized, market-dependent growers who became the norm in the twentieth century, but in their own time they were criticized for risking their annual livelihoods on one crop and one far distant market. The significance of the international market was obvious to the Modesto editor who commented in 1885 that:

> A good rain and a war between England and Russia would no doubt be very acceptable news to the most of our farmers. It would probably tend to the spreading of English civilization and at the

Combine harvester on the N. L. Tomlinson ranch east of Hughson.
Source: Margaret Sturtevant.

> same time increase the price of wheat . . . All Californians, if not Americans, are financially interested in seeing the old lady get angry enough to fight.[33]

Closer to home the farmers faced other threats to their prosperity. Those who speculated in the growing of grain had to deal with men who speculated in other aspects of the wheat trade – the ship brokers and sack merchants – whose periodic attempts to corner their respective markets added to the risk and instability of the business and often cost the ranchers dearly. Of course, the critique of the wheat growers implied that their gamble was of another sort – that they would stay in the wheat game, as would any shrewd card player, only until they had won their fortune. That was not how they actually behaved, at least in the Turlock region. They stayed on their land through good years and bad until in time their chances of making a fortune disappeared. They built finer homes, planted shade and fruit trees, and raised hogs and chickens and a few garden vegetables. All this is not to say that the critical observations of Nordhoff and others were entirely without merit. They probably contain more than a little truth about the character of men who did revel in their mile-wide fields, and who may very well have arrived as opportunists more than farmers. In the end, though, they generally became farmers who adapted as well as they could to the land and the climate of the San Joaquin Valley.

The farmers were not the only ones who gambled on the weather and the Liverpool grain market; everyone else in the communities dependent solely on wheat took the same chances. C. F. Lander put it simply when he said,

> If the crop was good the farmer could pay his debts – if the crop failed the farmer went under and the merchant went with him. Under these conditions it was impossible for the town to be really prosperous. That is the reason that Turlock remained nothing but a village for the first 30 years. Being a pioneer of Turlock was no snap.[34]

Lander recalled that Turlock in 1876 had half a dozen warehouses, two hotels, two meat markets, three blacksmith shops, a tin shop, a harness and shoe shop, two saloons and a population of about 200.[35] In 1880 Turlock's population was offically recorded at 192, and about ten years later the town was described as "consisting of three hotels, one restaurant, three general merchandise stores, one drug store, one tin shop, one boot and shoe shop, three livery stables, two blacksmith shops, a butcher shop, five warehouses, and sixteen saloons."[36] From these accounts, it seems that the principal growth in nearly twenty years of existence was in the number of saloons.

Decades after the close of the wheat era, Modestan Sol Elias reminisced that, "A beautiful place was Old Stanislaus in these bygone wheat-growing days – beautiful in the simple vocation of its citizenry, in the contentment and prosperity of its inhabitants and in the social intercourse of its people."[37] Time may have mellowed the scene for Elias. Prosperity was really only intermittant but life and the social order did have a certain stability and simplicity due largely to the fact the whole community depended on a single "simple vocation": the growing of wheat.

Prosperity became more elusive with the passing years as the soil's natural fertility was worn away. In the early years the newly broken plains were so fertile that seed lost in harvesting would produce a satisfactory volunteer crop with little additional work or expense. In time the land became so weak that it could no longer support a crop every year and farmers had to resort to fallowing, a practice of resting the land for a year between crops. Not only were yields declining but ample world grain supplies led to a downward trend in wheat prices after 1883.[38] By the 1890s farmers were forced to turn increasingly to other, less valuable grains – barley, oats and rye – rather than wheat. With the earning potential of each acre reduced, only the largest landowners could hope to make any profit at all. It was later said that, "The big ranchers carried a buckskin pouch of eagles and the little ranchers carried a mortgage." [39]

"It is not a pleasant system of agriculture," wrote Charles Nordhoff of the wheat era, "nor one which can be permanent." [40] Nordhoff made his prophecy in 1872, only a few years after wheat was first grown in much of the San Joaquin Valley. The wheat farmers did manage to achieve a measure of permanence and the history of irrigation would show that some struggled to maintain their way of life. For the most part, though, Nordhoff was proven correct. By exhausting the soil and tieing the area's economy to the shaky foundation of a single crop, the grain era sowed the seeds of its own destruction and was replaced by an altogether different system of irrigated agriculture.

CHAPTER 1 – FOOTNOTES

[1] Univ. of Calif., Agricultural Experiment Station, *Soils of Eastern Stanislaus County, California* by Rodney J. Arkley, Soil Survey No. 13 (Jan. 1959), pp. 1-2,13, 16-17, 118-119; Univ. of Calif., Agric. Experi., Sta., *Soils of Eastern Merced County, California* by Arkley, Soil Survey No. 11 (Nov. 1954), pp. 58-59; U.S. Dept. of Agric., *Soil Survey of the Modesto-Turlock Area, California* by A.T. Sweet, J.F. Warner and L.C. Holmes (1909), pp. 16-19, 41-48.

[2] "Climatological Summary," Exhibit T & M 22, Before the Federal Power Commission, *In Re Application of the Turlock Irrigation District and Modesto Irrigation District for License for New Don Pedro Project.*

[3] L.T. Burcham, *California Range Land: An Historico-Ecological Study of the Range Resource of California,* (Sacramento, 1957), pp. 90, 105.

[4] John C. Fremont, *Geographical Memoir upon Upper California . . .* (San Francisco, 1964), p. 18.

[5] Elna Bakker, *An Island Called California,* (Berkeley, 1972), pp. 148, 155-157.

[6] Turlock Irrigation District, "Monthly Average Unimpaired Flows of Tuolumne River at La Grange." Tabulations show flow in acre-feet and in second-feet. Figures used are for water years running from October 1 to September 30.

[7] *Daily Alta California,* Apr. 27, 1850, quoted in Sol Elias, *Stories of Stanislaus,* (Modesto, Calif., 1924), pp. 266-267. General information about the early history of the Turlock region can be found in Elliot and Moore, publishers, *History of Stanislaus County, California,* (San Francisco, 1881), and in Helen Hohenthal, "A History of the Turlock Region," M.A. Thesis, Univ. of Calif., Berkeley, 1930.

[8] I.N. Brotherton, *Annals of Stanislaus County: River Towns and Ferries,* (Santa Cruz, 1982), pp. 113, 139-142.

[9] William Brewer, *Up and Down California in 1860-1864,* (Berkeley, 1966), pp. 241-244.

[10] *Ibid,* pp. 509-510.

[11] Raymond F. Dasmann, *The Destruction of California,* (Collier Books edition, New York, 1966), pp. 67-68.

[12] Rodman W. Paul, "The Wheat Trade Between California and the United Kingdom," *Mississippi Valley Hist. Rev.,* XLV, (Dec. 1958), pp. 391-397.

[13] Brotherton, p. 114; Elias, p. 16.

[14] Richard Allen Eigenheer, "Early Perceptions of Agricultural Resources in the Central Valley of California," Ph.D. disser., Univ. of Calif., Davis (1976), pp. 273-274.

[15] Khaled Bloom, "Pioneer Land Speculation in California's San Joaquin Valley," *Agricultural History,* 57 (July 1983), p. 298.

[16] Eigenheer, pp. 281-331.

[17] Bloom, p. 298.

[18] Eigenheer, pp. 281, 329-335.

[19] Brotherton, pp. 113-114.

[20] *Stanislaus County News,* June 30 and July 21, 1871.

[21] *Ibid.,* Aug. 18, 1871.

[22] Horace Crane interview in Hohenthal thesis, Supplementary Material.

[23] Charles Nordhoff, *California for Health, Pleasure and Residence . . .* (New York, 1873), p. 182.

[24] Elliot and Moore, p. 11.

[25] Turlock Irrigation District, Assessment Book, 1888.

[26] George H. Tinkham, *A History of Stanislaus County,* (Los Angeles, 1921), p. 235; C.F. Lander to H. Whipple, Sept. 1, 1921, extra sheets in Whipple Letters, Turlock Public Library.

[27] Nordhoff, p. 179, also Nordhoff p. 184; Horace Crane interview in Hohenthal; and Reynold M. Wik, "Some Interpretations of the Mechanization of Agriculture in the Far West, "*Agricultural Hist.,* 49 (Jan. 1975), p. 78.

[28] Gilbert Fite, *The Farmer's Frontier, 1865-1900,* (N. Y., 1966), pp. 79-85.

[29] See Elliot and Moore; Wik; and Wallace Smith, *Garden of the Sun,* (Fresno, Calif., 1939), pp. 233-234.

[30] Nordhoff, p. 131

[31] *Ibid.*, p. 188

[32] Frank Norris, *The Octopus,* (New American Library edition, New York, 1964), pp. 211-212.

[33] *Modesto Daily Evening News,* Apr. 4, 1885.

[34] Lander to Whipple, Sept. 1, 1921

[35] *Idem.*

[36] Helen Alma Hohenthal and others, John Edwards Caswell, editor, *Streams in a Thirsty Land: A History of the Turlock Region,* (Turlock, 1972), p. 58. [Hereafter cited as *Streams in a Thirsty Land*]

[37] Elias, p. 19.

[38] Paul, pp. 410-411.

[39] John T. Bramhall, *The Story of Stanislaus,* (Modesto, Calif., 1914), p. 7.

[40] Nordhoff, p. 187.

2

Genesis of Irrigation

Silas Wilcox, first County Surveyor for the newly formed county of Stainslaus, submitted his annual report to the Surveyor General of California in November 1854. In it he wrote:

> The plains of this county could be irrigated by taking the water from the rivers running through it at the foot of the mountains by means of canals. It is not expedient at present for it would be attended with great expense and have but few consumers. We have good reason to believe from the situation of the arable land of this county that artesian wells could be sunk successfully; if so,it would be more convenient than any other mode of irrigation.[1]

Wilcox's vision was remarkable. Not only was the new county barely populated and its plains vacant except for roving bands of cattle, but irrigation was an idea that only a short time before had been entirely alien to the American experience. Until the 1840s the American agricultural frontier had been confined to the forested, well-watered lands east of the Great Plains. A few pioneer farmers had gone further, passing beyond the unfamiliar plains all the way to the Oregon country in search of the kind of environment they knew and understood. However, attention soon turned dramatically to the drier climates of the Southwest, where irrigation as a way of life predated the arrival of Europeans in America. The Spanish missionaries, soldiers and settlers who came to the northern borderlands that stretched from Texas to California brought with them their own long heritage of irrigated agriculture. They built community ditches and administered the use of water according to a complex set of rules adapted to the conditions they found.[2] In 1846 the Mexican War sent Americans into this region, and into contact with irrigation for the first time. It was in 1846, too, that the Mormon migration to Utah began, and with it the beginning of an American irrigation experiment where whole communities were planned around ditches and irrigated farms. How much Silas Wilcox may have known about Spanish-Mexican or Mormon irrigation practices is a matter of speculation, but he readily understood the prospective value of irrigation in the dry San Joaquin Valley and the potential for its development from the rivers that flowed out of the Sierra Nevada.

Just how far ahead of his time Silas Wilcox was may be seen in the slow growth of irrigation in California. During the 1850s most of the state's energy and resources went into mining, but the foundation for agricultural growth was being laid, without irrigation, in the valleys of northern California. Rainfall there was usually sufficient and land titles were often too insecure to permit investment in ditches. The most important irrigation developments in the 1850s and 1860s took place in southern California, where the need for irrigation and the heritage of Hispanic community ditches were most obvious. Among the best known were a Mormon colony established briefly near San Bernardino, and a group of Germans, incorporated as the Los Angeles Vineyard Society, who settled Anaheim as a cooperative irrigation colony. These colonies were the precursors of those that would one day cover coastal southern California. Elsewhere, irrigation was a rarity, confined mainly to a few small ditches, although several grandiose and splendidly impractical proposals for grand canals in the Sacramento and San Joaquin Valleys and a lake in the Salton Sink gave publicity, if not practical encouragement, to the idea of irrigation.[3] The idea needed whatever encouragement it could get because there was still considerable doubt about the necessity of irrigation, especially on the part of promoters eager to convince potential immigrants that such an expensive, and even un-American, practice was not needed in bountiful California.

When agriculture finally replaced grazing on the plains of Stanislaus County there was no immediate thought given to irrigation. It was assumed that there was enough rain in most years to mature a wheat crop before the summer dry season. However, farmers had no sooner moved onto the plains south of the Tuolumne than they faced a two and a half year drought that lasted until the rains of December 1871.[4] The rapid development of agriculture on the plains and the prolonged drought helped produce the first of several proposals to bring water to the lands along the Tuolumne – proposals that were especially interesting because they came at a time when irrigation in California, and especially in the San Joaquin Valley, was still in its infancy.

On May 1, 1871, Michael Kelly, the owner of water rights on the

Tuolumne River near La Grange dating back to 1854, filed a notice of his intention to build a dam at the falls a mile and a half above the town of La Grange. The notice also described his plan to construct canals along either side of the river "for the purpose of running water out of the Tuolumne River into said Canals or Ditches for to sell and dispose, for mining, irrigating, and propelling machinery and for general use."[5] Shortly thereafter Kelly sold his claim to a partnership of J. M. Thompson, Charles Elliott and M. A. Wheaton, who formed the Tuolumne Water Company. In June 1871 they offered a proposal, subject to negotiation and revision, under which they would build irrigation canals and supply water to farmers on both sides of the river. They suggested a forty-nine year contract at an annual rate of $1.25 per acre, to be paid whether water was used or not, with a lien on the land to guarantee payment. In exchange, the company would promise to supply one and a quarter acre-feet of water to each acre of land, delivered during the first six months of the year. During the remainder of the year they promised to distribute any available water as fairly as possible. The drought-stricken farmers were interested in irrigation but unenthusiastic about the terms presented by the company.[6] A committee appointed at a farmer's meeting recommended reducing the contracts to five years and the cost to seventy - five cents an acre. The company promptly rejected any reduction in rates but did consent to shorten the life of the contracts to ten years.[7] No further offers were exchanged but the company still went forward on construction of its $24,000 dam. When completed it stood thirty feet high and solidly constructed of a quarter million board feet of sawed lumber and sixteen tons of bolts.[8] A first step, at least, had been taken in the direction of irrigation.

The cost of the dam was only a small down-payment on the price of a complete irrigation system. The long-term contracts sought by the company would have guaranteed a market for its water and might have helped convince investors to put up whatever additional capital was needed. When it became clear that local farmers were not willing to sign such contracts, financing became more difficult. The company therefore conceived a novel scheme to raise money to build canals from its dam. They drafted legislation to permit Stanislaus County to sell $300,000 worth of county bonds to subsidize the company's work.[9] Although the money would have to be paid back with interest, the company contended that it needed the loan to build its ditches, notwithstanding the fact that the amount would probably pay for no more than the difficult foothill stretches of the two canals. At the same time they let it be known that water contracts running only two or three years were now acceptable.[10] At first the idea seems to have received at least a measure of support. Stanislaus County's state senator, T. J. Keys, a member of the farmers' committee that had dealt with the company the previous year, introduced a bill in early 1872 authorizing the plan, although the maximum amount of the subsidy was cut to $150,000. If enacted the bill would have required an election within ninety days on whether or not county bonds should be issued. If they were approved, the company was given ten years to repay the loan at 10 percent interest and the county Board of Supervisors would be given authority to regulate water rates above a minimun of $1.25 per acre.[11]

Local opposition was soon evident. At issue was not so much the merits of irrigation as the use of a public subsidy for the benefit of a private company. Not many years before, money had been spent by state and local governments, in addition to large federal subsidies, to aid the construction of the transcontinental railroad, only to find that the railroad brought California the blight of monopoly and hard times instead of its promised blessings. Widespread disillusionment with railroad subsidies fostered opposition to any further public financing of private businesses. In December 1871 the first suggestion of an irrigation subsidy brought only a muted protest from the normally anti-subsidy *Stainslaus County News.* As dangerous as subsidies might be, the lure of irrigation was so strong that the paper merely urged local voters to think carefully about what they were doing.[12] By March of the following year the *News* had overcome its timidity and was actively opposed to the bill. While Senator Keys had first been able to present a petition with 130 signatures endorsing the subsidy, a petition opposing the measure carried some 400 names.[13] Local resistance became so strong that Keys himself and other Stanislaus County legislators finally voted against the bill. Despite the stand taken against it by the area it was supposed to benefit, the bill passed the legislature but died on the governor's desk.[14]

Irrigation institutions could come in three basic forms: corporate, cooperative and public. The Tuolumne Water Company was an unsuccessful example of the first type, which depended on outside capital and involved the ownership of water independent of the land. It was not the kind of institution suited to circumstances along the Tuolumne. When corporate irrigation did succeed, it was usually because the developers could sell both the water and the land it irrigated. In the Fresno area, for example, land speculators acquired immense acreages at the same time, and in the same manner, as land passed into private hands in the Turlock region. During the 1870s they established canal companies to bring water from the Kings River to their lands and then subdivided their holdings into colonies of small irrigated farms. The land sold for a higher price because it was irrigated and the buyers received, by contract, a proportionate share of the water right held by the land or ditch company serving the tract. The farmers' only on-going expense was as annual charge for the operation and maintenance of the irrigation system. Profits came not from running water but from selling irrigated land. That fact was confirmed in 1877 when the Bank of Nevada took over a canal in the Fresno area and hired William Hammond Hall, one of the state's best known irrigation engineers, to run it. Only five months of operation convinced the bank that there was no money to be made from a canal alone, and they resold the property to one of its former owners, a land developer.[15] The minor capitalists who formed the Tuolumne Water Company had no land to sell and no assurance that landowners would buy water once the canals were in operation since there was usually enough rainfall to grow grain. Their attempt to secure a county subsidy meant that private lenders, even in that speculative age, doubted the project could ever be profitable.

At the same time that the subsidy bill was being debated, Senator Keys introduced an alternative, or perhaps complementary, irrigation plan. Key's bill, which he submitted at the request of some of his constitutents, authorized the formation of public irrigation districts in California. The concept of special units of local government for the financing and control of irrigation development did not originate in 1872, though that was the first time the term "irrigation district" was used. A similar idea was adopted as early as 1854 when the legislature allowed the residents of any township in specified agricultural counties to elect three water commissioners to plan an irrigation system and supervise construction using citizen labor.[16] A Utah statute of 1865 authorized irrigation districts in that territory but defects in the law kept it from having much effect. Keys' bill, too, had serious defects. The districts it authorized would have been mere appendages of county government rather than independent agencies, and without the right to issue bonds it would have been difficult for them to finance canal construction. In order to protect the interests of private irrigation developers, districts would not have been allowed to condemn any existing water rights, and the law did not apply at all in the counties of Kern, Tulare, Fresno and Yolo, where most irrigation in northern California was then practiced. The motives behind the bill remain obscure. A district organized under its provisions in Stanislaus County could not have forcibly displaced the Tuolumne Water Company. On the other hand the bill might have been an alternative means of aiding the company in case the subsidy bill failed, but it would have been an unwieldy instrument for that purpose. In any event the bill became law, but no irrigation district was ever formed under its terms.[17]

Corporate and public irrigation plans were suggested in 1871 and 1872 without success, and in 1873 the other institutional format, cooperative, may have been given a brief try south of the Tuolumne. Cooperative irrigation institutions are essentially those in which the irrigators themselves own and operate the irrigation system, in most cases through a corporation in which stock is held in proportion to each user's acreage or entitlement to water. Farmers could set up a cooperative organization from scratch, assessing each share of stock to raise enough capital for the project. On the other hand a cooperative company could take over an already completed system from a private developer and operate it. The latter plan was adopted by George Chaffey for his colonies at Etiwanda and Ontario in southern California and was widely copied. Each purchaser of land was given shares in the tract's water company in relation to his acreage so that as the colony was sold, ownership in its water supply passed to the irrigators.

A cooperative institution of some kind may have been the goal of several farmers living between the Tuolumne and Merced rivers, who in August 1873 filed articles of incorporation for the Merced and San Joaquin Irrigation Canal Company.[18] Old-timers knew that when the Merced River was full, some water escaped through a slough on the north side of the river, upstream from the Central Pacific Railroad bridge. It spread out over the plains to the northwest, finally draining into the San Joaquin and Tuolumne rivers near their confluence. The general plan of the company was to dam the Merced and force water into canals that would follow this natural drainage route.[19] An engineer, George H. Perrine, was hired and in September 1873, he made his report. A canal beginning at a four-foot-high dam about three and a half miles above the railroad bridge would follow the river until it reached the level of the plains at a point west of the railroad. From there it would flow north past Turlock and then recross the railroad near Ceres before emptying into the Tuolumne River. Perrine estimated that the canal would carry enough water to irrigate over 27,000 acres per month, but by raising the dam only two feet and making other minor changes, its capacity could be increased 20 percent.[20] Following the engineer's report the Merced and San Joaquin Irrigation Canal Company simply vanished. Perhaps the water supply or the extent of the acreage to be served by the canal prompted second thoughts or, more likely, the potential cost of the project discouraged the farmers who hoped to build it. Since the scheme collapsed so quickly there is no way to tell whether its promoters intended it as a private canal which would sell water or as a cooperative project.

Despite concern over Senator Keys' irrigation bills in 1872 and the brief flurry of activity south of the Tuolumne in 1873, interest in irrigation faded after 1871 in the face of adequate rain and good crops. Meanwhile, the holdings of the Tuolumne Water Company passed into the ownership of one of the partners, San Francisco attorney Milton A. Wheaton. Without money to build canals to the valley, Wheaton's dam served no useful purpose other than the diversion of a trickle of water into a short canal to land near the town of La Grange. In June 1876 Wheaton visited the office of the *Stainslaus County News* in Modesto in an effort to rekindle an interest in irrigation. He announced that he was now willing to sell his dam and adjoining property to the local landowners. Later that month he presented a sweeping proposal covering not only the sale of the dam but the construction of the canals that would give it value. Wheaton suggested the organization of a corporation made up exclusively of landowners who would subscribe one dollar per share, with each share representing an acre of land. Besides their dollar an acre investment, the shareholders would have to pay whatever additional assessments had to be levied to complete the canal system and put it into operation. What Wheaton was proposing was a classic cooperative scheme – a mutual water company through which the irrigators would pay for and control their water supply. There were contemporary precedents for that sort of development. In the early and middle 1870s, for example, a number of mutual ditch companies were organized by farmers in the area around Hanford and Lemoore in the southern San Joaquin Valley.[21] To assist a similar venture on the Tuolumne, Wheaton promised to sell his own property to the corporation "at just what it cost me, which is not a tithe of its actual value."[22] The offer, and the proposal for a corporation, elicited no immediate reaction and might have been forgotten entirely had it not been for the onset of another drought.

Rain failed to come in the fall of 1876 and it was not until February 1877 that there was enough moisture for the farmers to plow and plant. Dry weather and the threat of a crop failure worked as nothing else could to focus attention on irrigation. On April 4, 1877, a meeting of fifty to sixty farmers from north of the Tuolumne convened at the Grange Hall in Modesto to discuss the situation and take the first steps toward a solution. A five-man committee was appointed to investigate canal routes and hire a consulting engineer.[23] At a meeting a week later J.M. Henderson, a resident of the Turlock area, said that farmers on his side of the river were also becoming interested in irrigation and he got the group to adopt a resolution in favor of considering a joint undertaking with farmers from south of the Tuolumne.[24] The logic of a project serving both sides of the river was underscored by the engineering report presented at a meeting on April 21, 1877. The committee had hired William Hammond Hall, then chief engineer of the ill-fated West Side Irrigation District and soon to become California's first State Engineer, to make a general survey of the situation. Hall told the gathering that the Tuolumne River was the natural water source for the land on either side of it, and he urged that any plans to irrigate one side should be in conjunction with those to water the other side. He described the Wheaton dam as a good, well-located structure, particularly for irrigation north of the river. The cost of an irrigation project was estimated at three to four dollars per acre for the main ditch and perhaps six to eight dollars per acre for the complete system.[25]

After Hall had completed his own report, he read to the meeting a letter written a day earlier by M.A. Wheaton. In the letter Wheaton once again offered to sell his land and dam "for what they cost me in money" and also urged the adoption of the plan he had presented the year before.[26] As an added inducement to the more active group from north of the river, he pointed out that if they purchased his property themselves they would have something valuable "to sell which the south side will soon want for its use, and can sell on its own terms."[27] Even without the reminder that he was ready to do business, Wheaton's earlier suggestion of a landowners' corporation had apparently not been forgotten, for at the same meeting the committee on organization proposed a very similar plan. Under the committee's revision each acre enrolled would be represented by a single share of stock but at a much higher price of four dollars per acre. The report envisioned capital stock amounting to $600,000, which would require the subscription of 150,000 acres. Realizing that the high cost of stock might prove a deterrent to participation, the committee offered an alternative under which a landowner could convey his property to the corporation in lieu of paying cash for stock. The corporation could then mortgage the land to raise money. The farmer could recover his property at any time within ten years by paying off the cost of the stock and assessments plus interest at 10 percent. A sign-up sheet circulated at the meeting showed thirty-two landowners and 13,000 acres at least tentatively pledged to the plan.[28] A little quick arithmetic shows that the subscribers averaged just over 400 acres apiece. Although some might have signed up only a portion of their holdings, it seems probable that the plan failed to attract the more substantial farmers who controlled most of the acreage. Two weeks later it was decided that detailed surveys were needed, but before going forward with the work at least 40,000 acres had to be definitely committed to the project.[29] This time fewer than 5,000 acres were subscribed and the enterprise collapsed.[30]

It is doubtful that the proposed corporation could have succeeded. Its sponsors had promised to limit additional assessments to no more that ten cents a share, thereby making it impossible to raise over $4.10 per acre, well below William Hammond Hall's estimate of six to eight dollars per acre for a complete system.[31] Hidden in these figures was one possible reason, a physical one, why a cooperative institution could not succeed in irrigating the Turlock-Modesto area. Unlike the Kings River and many others, the Tuolumne was cut deeply into the valley floor. To bring water to the plains by gravity, it had to be intercepted in the foothills, as Wheaton planned to do, and carried by canals for miles through rugged terrain. Farmers could not use their own teams and scrapers to make the cuts, fills and tunnels that would be needed, and no cheap brush or cobble diversion dam would suffice either. Getting water out of the Tuolumne was going to be expensive, and there was simply not enough capital in the farming community to finance that kind of costly, long-term irrigation project.

So for as can be ascertained, local enthusiasm for irrigation waned follwoing the failure of the proposed corporation and nearly disappeared when the drought ended in January 1878. Despite an evident lack of interest, the legislature passed a bill in early 1878 creating something called the Modesto Irrigation District. There was no groundswell of popular support for the measure and what little notice was taken of it by the local press seems to have been negative.[32] Who the originators and supporters of the legislation were is a mystery. Steadfast local irrigationists unwilling to give up when their proposed corporation failed to materialize may have been behind the act, which was an attempt to combine a landowners' corporation with another public subsidy. The law applied only to land between the Tuolumne and Stanislaus rivers and thus did not directly involve the Turlock region. However, the Modesto Irrigation District Act merits at least passing notice because it did represent another step in the search for an irrigation institution suited to conditons in Stanislaus County.

It was essentially a cooperative plan, a landowners' corporation, but with public financing for its capital expenses. Under the act, any five or more landowners within the district could form a corporation, issuing one share of stock for each acre subscribed. Those joining the corporation did not have to pay anything for their stock and were subject only to annual assessments of no more than $1.50 per acre ($1.00 for irrigated grain) to cover the maintenance and operation of the proposed canal. Money to build the works would come from bonds sold by Stanislaus County for that purpose. When 50,000 acres had been subscribed to the corporation the county was to provide $25,000 in bonds to begin construction, issuing additional bonds up to a total of $500,000 as work progressed. The bonds were to be repaid exclusively out of the increased state and local tax revenues that were confidently expected to result from higher land values once the canal was completed.[33] The act became law on March 30, 1878, and less than two weeks later a corporation was formed in accordance with its terms.[34] Despite the offer of free stock, the minimum acreage was not subscribed and the whole scheme died as quietly as it had been born. Even with more local support it seems unlikely that the plan would have proven workable. The grant of public money was accompanied by only the vaguest of public controls over how it was to be spent. And since neither the state, the county nor the landowners were responsible for repayment of the bonds beyond whatever additional taxes irrigation might generate, they would have been virtually impossible to sell to investors.

The various efforts to bring irrigation to the plains adjacent to the Tuolumne River between 1871 and 1878 reveal several things about the nature, and limitations, of the irrigation institutions that had been proposed. And it was clearly an institutional problem that the would-be irrigation developers faced. There was plenty of water in the river and the basic engineering concepts for extracting it were fully understood. From the start it was known that a dam in the La Grange vicinity could supply canals on each side of the river. The problem was not what to build, but how to pay for it. The principal options, and in fact the only practical ones in the 1870s when no viable public institutions existed, were corporate or cooperative. The former meant that outside investors, like M.A. Wheaton, would put up the money to build the system in the expectation of profiting from its operation. Unfortunately, there was little opportunity for a profit from water development alone. Private canals were a good investment only in combination with land sales, which was impossible along the Tuolumne, where the land and water were already in separate hands. Cooperative irrigation institutions, on the other hand, were really non-profit associations of farmers who combined their own resources to build the canals they wanted, or in the case of some mutual companies, to operate the ditches put in by private land developers. When farmers decided to build their own systems they had to provide the necessary capital or do the work themselves, and it was questionable whether the grain farmers of Stanislaus County could have scraped together enough cash to tame the Tuolumne. A cooperative enterprise also requires a general commitment to the purposes it wishes to achieve, and it seems obvious that there was no widespread agreement on the need for irrigation in the Turlock-Modesto area in the 1870s. In fact, except for the Merced River canal survey in 1873, the irrigation plans considered in the 1870s all sprang from a combination of dry weather and the persistent efforts of M.A. Wheaton to turn a profit on the dam he and his partners had built near La Grange. In other words most of the initiative came from outside the local community, which remained interested but cautious and noncommital. If neither corporate nor cooperative means were suited to the area, the only alternative was some sort of public institution.

Only one truly public institution, the ineffective district scheme authorized in 1872, was proposed; the other plans presented were hybrids that added only a public subsidy to what were essentially corporate or cooperative concepts. As the 1870s ended, irrigation along the Tuolumne was caught in an institutional stalemate.

Meanwhile, irrigation was expanding rapidly elsewhere in California. Canals and colonies proliferated in the Fresno area, where small farms with orchards and vineyards dramatically broke the monotonous pattern of wheat. Just to the south, the cooperative ditch companies of the Mussel Slough country were active, and at the southern end of the Central Valley, the Kern County Land and Water Company vastly expanded the system of ditches drawing water from the Kern River. In southern California irrigation progress was especially rapid in the boom years of the 1880s as promoters set up new communities and irrigation colonies in the emerging citrus belt. And, in 1878, California got its first State Engineer, William Hammond Hall, who was, among other things, instructed to study the state's water resources and offer advice on irrigation. In the same year John Wesley Powell, famous as an explorer of the Colorado River and a scientist engaged in federal surveys of the West, submitted his *Report on the Lands of the Arid Region of the United States.* Powell pointed out that aridity was the most fundamental fact of life in the West and called for reforms in land laws to suit the needs of irrigation and the community control of water. Powell and Hall, who also advocated a more systematic, comprehensive approach to water development than the helter-skelter pattern of private and cooperative ditches, were ahead of their time. Powell's long-term effect, however, was substantial; the 1878 report began a chain of events that led eventually to a national reclamation movement, federal reservoir surveys and finally to the U. S. Reclamation Service. On all sides the interest in irrigation was intensifying and its practical value became more apparent in the prosperity of the places where it was practiced. In that atmosphere it would be only a matter of time before talk of canals would be heard again along the Tuolumne.[35]

The first sign of renewed interest came in May 1885 when an announcement appeared inviting "all landowners between Turlock, Ceres, West Point and the San Joaquin River" to meet on May 25 to form a canal company.[36] On that day some forty people, including twenty-five landowners, met at McDonald's store in Ceres to discuss irrigation between the Tuolumne and Merced rivers.

Just what prompted the meeting is unknown, but those in attendence wasted no time in beginning work. A committee consisting of J. B. Brichman, Joseph Vincent and C. N. Whitmore was appointed to collect subscriptions to pay for a survey of potential canal routes.[37] Preliminary surveys were made, but by the time a second meeting convened at Rogers Hall in Modesto on July 18 the committee had failed to raise enough money. The surveyor, General Bost, refused to make his report until his fee was paid in full, and it took "considerable drumming" of the crowd of a hundred people before the sum was raised.[38] With the financial formalities out of the way, Bost presented two reports. The first was similar to the 1873 survey of a canal from the Merced River starting a few miles upstream from the railroad bridge and capable of irrigating the land west of the tracks. The other report covered a diversion from the Tuolumne River at a point two miles below La Grange near Thompson's Ferry, which by a curious coincidence was exactly where J. W. Bost, J. A. Worthington and S. W. Geis had recorded a water right in May 1885 just before the canal survey began.[39] The Tuolumne canal's upper works would cost twice as much as the Merced River alternative, but by entering the plains at a higher elevation it could irrigate a much greater area.

The Tuolumne's larger water supply and the prospect of irrigating more land persuaded those at the meeting to favor the development of that river rather than the Merced. A joint stock company open to any investor was immediately suggested and some fifty shares at $100 each were subscribed at the meeting.[40] Another meeting was held August 15, 1885, at which the formal organization of the Tuolumne River Irrigation Company was approved and directors were elected.[41] Articles of incorporation were filed with the Stanislaus County Clerk in October, which showed that the company had broadened its scope to include the area north of the Tuolumne River as well. At that time the incorporators reported that only $22,400 of the projected $200,000 capital had been paid up.[42] After that no more was ever heard of the Tuolumne River Irrigation Company. It shared the same fate as all other proposals for private financing and undoubtedly for the same reasons – the costs were more than local farmers were willing to shoulder and the potential for profit was too remote to attract outside investors.

Despite the failure of the Tuolumne River Irrigation Company the events of 1885 were repeated in 1886. In August of that year a group of farmers from south of the Tuolumne hired George

Manuel, the engineer for the Fresno Canal and Irrigation Company, to survey a ditch running north from the Merced River.[43] At least some of those behind the new surveys – men like J. V. Davies, Levi Carter and George Perley – were veterans of the previous year's ill-fated venture. Manuel soon presented them with plans for diverting the Merced near Merced Falls, much further upstream than any previous survey. To save money the plan envisioned using portions of Dry Creek and Mustang Creek to carry water down to the plains, which would be reached about four miles northeast of Turlock. The *News* commented that "the farmers upon the route of the proposed canal have been indominitable in their efforts . . . The money for the survey was easily raised by the liberal contributions of the farmers in the vicinity to be benefited."[44] Even with the modifications suggested by Manuel's survey the Merced River was a less appealing choice than the Tuolumne. Since the land between the two rivers sloped downhill from the northeast to the southwest, a canal from the Merced River would run counter to the lay of the land. With this in mind, the same meeting that heard the formal report of the Merced River route ordered surveys made for a Tuolumne River canal.[45]

George Manuel was at work on the new survey in November and gave his report to a meeting at Rogers Hall in Modesto on December 16, 1886. Beginning at Wheaton dam, which he proposed to double in height, Manuel had mapped out a route that utilized sections of Granville and Peaslee creeks along with at least two tunnels to traverse the rough country near La Grange. He estimated that a flow of 1,200 second-feet, which he believed was sufficient to irrigate 192,000 acres, could be carried to the edge of the plains near Hickman for $95,000.[46] With the report in hand, the meeting quickly took up the matter of financing the project. It was decided that $150,000 should be raised to cover the main canal outlined in Manuel's report and extend it across the plains. A committee was named to confer with landowners and mortgagees regarding the possibility of bonding the land for from one to three dollars per acre to pay for the work.[47] Nothing came of the plan to raise money from the land to be irrigated and the matter was not vigorously pursued because by 1886 a new proposal for public irrigation districts was being readied – one that would hopefully solve the institutional dilemma facing the Turlock-Modesto region.

Accompanying the second round of surveys in as many years was an outbreak of irrigation fever in Stanislaus County. The first symptoms appeared in the newspapers in early 1886. Despite reservations about specific proposals, the local press had always been sympathetic to irrigation, but in 1886 the reasons why they were in favor of it appeared to change. A sampling of editorial opinion in 1885, when the Tuolumne River Irrigation Company was searching in vain for investors, illustrates what had been the long-standing attitude toward irrigation. For example, on August 15, 1885, the *Modesto Daily Evening News* asserted that so long as public money was not involved, a scheme like the proposed canal south of the Tuolumne,

> . . . addresses itself solely to the land owners, or those who may think it their interest to take stock in the enterprise. The benefits of irrigation have been so clearly proven by practical experience in Spain, Italy, France and India, as well as California, . . . that it would be presuming upon the intelligence of our people to say they needed more light on that subject.[48]

A September 5, 1885, editorial in the same paper concluded that, "Each farmer or land owner is as competent as the *News* editor in figuring as to the value it would be to each of the dry and parched acres of his fields."[49] Irrigation was, in this view, an antidote to drought and nothing more. If that were true, it was of interest only to those who would use the water.

Within a matter of months that attitude had changed. In reporting on a visit to the Rock River Ranch near Roberts Ferry in early 1886, a journalist told of a pleesent ride toward the foothills but added:

> . . . there is still something wanting to make it more pleasing to the eye of the lover of prosperity and civilization. The want consists of a system of irrigation, the cutting up of these large ranches into smaller farms; the creation of a greater number of farm houses; the planting of orchards, and a hundred other things that might be mentioned.[50]

In the same issue a letter signed only "Irrigation" hammered away at the same theme. "(A) canal," it said, "means smaller farms, a varied production, an increased population, more customers for the store keeper, for the black smith, the hotel keeper, the liveryman, the professional man, in fact, it means prosperity for everybody dependent upon public patronage."[51] Two days later an editorial exclaimed, "Give us water, and the money will come," and the following months witnessed a virtual barrage of enthusiastic rhetoric about irrigation.[52] In the view of the *News* at least, irrigation suddenly became more than a way to spread water on the "parched acres" adjoining the Tuolumne; it became a powerful instrument of economic and social reform.

The local irrigation crusade, and similar state and national irrigation movements, were rooted in a series of related assumptions – assumptions that to their adherents took on the certainty of religious convictions. Foremost was the idea that irrigation was synonymous with small, family farms and a diversity of crops. Of course that was not necessarily true, but the best known irrigation developments up to that time had involved small farms. Certainly, irrigated land was more expensive, so most settlers could afford to buy only relatively small acreages. Irrigation gave farmers a greater choice of potential crops, many of which, like orchards, demanded the kind of intensive cultivation suited to small holdings. Then, too, higher valued irrigated crops returned more income per acre, allowing a small-scale irrigator to earn as much, and more reliably, as the owner of a much larger dry farm. If it resulted in the anticipated community of small farms, irrigation was expected to have definite social benefits as well. It meant a dense rural population and close neighbors in contrast to the depressing loneliness of scattered homesteads and ranches. Market towns and villages flourished under these circumstances too, providing increased access to education and cultural opportunities rare in rural areas. This is what turn-of-the-century irrigation publicist William E. Smythe called the "blessing of aridity" – the advantages that could make life in irrigated areas more civilized as well as profitable.[53] These were articles of faith for the advocates of irrigation, who expected it to transform the local landscape and community just as it had in Fresno and southern California. When the rapid growth of those places was contrasted with the nearly stagnant economic condition of Stanislaus County, the lesson seemed obvious – irrigation was the key to prosperity.

The new irrigation crusade was a call to action, and the action that was needed was first of all political. No matter how brigeality without money, and for the lands along the Tuolumne River that meant some form of public financing. The brief history of the Tuolumne River Irrigation Company had been another reminder that, just as in the 1870s, private capital and private or cooperative institutions were unable to build canals to Turlock or Modesto. Efforts to merge public money with private construction had twice failed, leaving a thoroughly public irrigation agency as perhaps the only worthwhile alternative. A governmental institution, however, depended on more than the support of farmers alone; it required at least a measure of approval by the community as a whole. Once they had been infected by irrigation fever, the newspapers carried the message that everyone — merchants and townspeople as well as farmers —would benefit from irrigation. It was an argument that helped justify the creation of a public institution to manage and pay for irrigation development. By what may or may not have been a coincidence, the irrigation crusade suddenly sprang to life in Stanislaus County in the spring of 1886 just as the process of selecting candidates for the coming fall's general election was beginning. The stage was being set, locally at least, for a new public policy for the encouragement of irrigation.

C. C. Wright, the young Modesto attorney elected to the Assembly in 1886, repeatedly said he was chosen "for the express purpose of advocating some measure providing for the municipal control of water for irrigation."[54] Considering local interest in irrigation at the time and his subsequent performance in the legislature, Wright's statement that irrigation was the dominant issue in the election must be taken at face value. Unfortunately, the meagre reports of his campaign make it impossible to tell just when he drafted his program. Shortly after his nomination in the Democratic Party primary of May 1886, he was characterized only as "anti-Chinese – riparian — railroad," although in the parlance of the times the label "anti-riparian" was the equivalent of "pro-irrigation."[55] Even the staunchly Democratic and rabidly irrigationist *News* paid surprisingly little attention to Wright's campaign or his ideas on irrigation. The report of a rally in Oakdale said that his remarks on irrigation "won him many votes," without revealing anything about what those remarks were.[56] A summary of Wright's speech to a Democratic meeting in Modesto only a few days before the election shows that he gave considerable time to non-irrigation issues but also outlined a proposal for locally managed irrigation districts. The account said,

> His plan would be to establish irrigation districts throughout the state, apportioned with reference to the convenience of supply; each district to have a common source from which to obtain its supply of water; the people of each district to elect officers to control said district, repairs and necessary expenses to be obtained by assessments from those using the water; the farmers themselves to own the water and thus be masters of the situation.[57]

Wright undoubtedly did much more to explain his plans to the voters than the newspaper reports indicated, and they apparently liked what they heard. On November 2, 1886, Wright was elected by a better than two-to-one margin over Republican C.M. Stetson of Ceres.[58]

Following the election Wright began putting his ideas into writing and by the end of the year he had drafted the bill he would introduce when the legislature convened in January 1887. The bill would allow fifty or more landowners living in an area that could be irrigated by a single system to petition the Board of Supervisors of the county where the largest part of the lands were located requesting the formation of an irrigation district. The supervisors were required to hold hearings on the proposal, modify the district's boundaries if necessary, and then call an election at which all eligible voters in the area affected, not just landowners, would be allowed to vote. If the voters approved by a two-thirds majority the district was to be declared organized. A five-man Board of Directors, each member representing one of five equally sized divisions, would be elected to manage the affairs of the district, direct the construction of canals and be responsible for the operation of the system. A district could use the power of condemnation to acquire needed property or water rights. Financing was provided for in two ways. To fund major construction work, the district's voters could approve the issuance of twenty-year bonds, with no payments on the principal to be made for ten years. Normal operations, and the payment of any bonded debt, were to be covered by a tax on all land and improvements within the district, though the cost of operating a completed irrigation system could, if the directors wished, be paid for by charges on water actually used.[59]

Some years later Wright explained the rationale behind his bill in these terms:

> In devising plans it occurred to me that the best method whereby the people could organize, and when organized would have power to assess and collect money for the purpose of constructing works by the ordinary process of levy and collection was the proper thing, and that a local government based on the familiar lines of county government and officers, differing in no respect . . . except in the object to be obtained, would accomplish our purpose without chance of failure. . . There was nothing new or noval in that plan. The district when effected is the same as a county.[60]

C. C. Wright, attorney and State Assemblyman.
Source: McHenry Museum.

Wright's approach thoroughly embodied the concept then so much in vogue in Stanislaus County that irrigation would benefit not just the irrigator but everyone in the community. Since the district would be a unit of local government, the bill made urban residents eligible to vote and responsible for the payment of district taxes even though the benefits they received from irrigation would be indirect. It also required all landowners within a district to pay irrigation taxes regardless of whether or not they wanted to irrigate. In this way a majority of townsmen and smaller farmers could, in theory, overcome the opposition of even the largest property owners. Though the concept of irrigation districts was not new, Wright's bill for the first time provided for the formation of what could be viable, adequately-financed districts. It was also tailor-made for conditions in Stanislaus County. After reading the proposed legislation, a Modesto editor endorsed it with the comment that, "we are fully confident that its success will put new life into the business interests in this section of the State, and sincerely hope it may become a law."[61]

Based on the record of previous legislative sessions, Wright's bill or any other irrigation legislation might have stood little chance for success. Irrigation was a subject that had bedeviled the California legislature for years, and during 1885 and 1886 it had proven particularly troublesome. The perennial problem of finding practical ways to encourage irrigation had then been overshadowed by a bitter water rights dispute. In theory, the question was whether riparian rights, which were the exclusive rights of streamside landowners to the water in a stream, or rights acquired by appropriation, which meant the taking and use of water at any location on a first come, first served basis, would be supreme. California had inadvertently adopted both doctrines early in its history, creating the West's most tangled and uncertain system of water rights. In 1884, and again in 1886, the state Supreme Court ruled in favor of the riparian principle in the case of *Lux v. Haggin,* which had pitted cattle barons Henry Miller and Charles Lux, the owners of riparian lands, against land and irrigation developers James B. Haggin, Lloyd Tevis and William B. Carr for control of the Kern River. The Court's decision did not eliminate the appropriation doctrine and it did not necessarily imperil irrigation, which tended to rely on appropriative rather than riparian rights. Nonetheless, it was widely viewed as a serious threat to irrigation in California and the losers in the legal struggle were quick to capitalize on that notion. Haggin and his allies whipped irrigators throughout the state into an anti-riparian frenzy with the cry that their water was in jeopardy. The clamor, and the behind-the-scenes work of William B. Carr, convinced a sympathetic governor that a crisis was at hand and that the legislature should be called into special session to deal with it. When the special session convened in the summer of 1886, Haggin and Carr were ready with a package of bills that would effectively overturn the *Lux v. Haggin* decision, including a plan to pack the Supreme Court in their favor. Press reports claimed that bribery was commonplace, and in the end Haggin and company controlled the Assembly and Miller and Lux, the Senate. The result was stalemate, and the special session accomplished nothing. Obviously, if the deadlock continued into the regular session beginning in January 1887, Wright's irrigation district bill would probably have little chance of success.[62]

Fortunately, the stalemate did not endure. Following the special session a compromise was worked out between the Kern River antagonists to peacefully divide the river. With that one issue settled, the whole anti-riparian furor, which had been so well orchestrated by Haggin and Carr, began to fade, clearing the way for constructive action on irrigation. Even before the legislature convened it was widely known that Wright would introduce an irrigation bill. The *San Francisco Alta* commented that Wright's views, whatever they were, would command attention and added that he "comes from a county whose future is greatly involved with irrigation."[63] Wright's bill was not the only irrigation measure proposed that year. One bill sponsored by Assemblyman Brierly and supported by the governor and the state engineer would have authorized a two-year study by a commission of experts to recommend water rights legislation. Its passage would have effectively delayed any other action but it died in the Senate.[64]

Wright's proposal was well received and encountered no major opposition.[65] The committee in charge of the canal project south of the Tuolumne called a public meeting in Modesto to marshal local support for the bill, and as a result Wright was soon able to present petitions bearing hundreds of signatures in favor of his proposal. Modesto businessman George Perley and a number of landowners went to Sacramento to do whatever they could to aid Wright.[66] The *Stockton Independent* reported that, "Several of the large land owners in Merced and Stanislaus are here (in Sacramento) for the purpose of hastening irrigation legislation. They represent that the present outlook is a threatening one for the farmers, and unless some law is passed under which capital will operate in building canals and ditches many of them will be compelled to give up their farms which are heavily mortgaged."[67] The bill seemed to lead a charmed life, as it sailed through Assembly committees unscathed and was passed by that house without a dissenting vote. The Senate took up the bill immediately and made only a few minor amendments before it, too, approved the measure unanimously. The next day the Assembly voted its approval of the Senate's changes and the bill went to Governor Washington Bartlett, who signed it into law on March 7, 1887. When word reached the Assembly that the governor had signed the bill, its members burst into applause. [68]

Why had the Wright bill succeeded so easily where other efforts to create new irrigation institutions had failed? Certainly the relative absence of divisive issues, like the water rights controversy, made it easier to pass any substantial irrigation legislation. More importantly, though, the bill had a broad appeal of its own. The kind of strictly local and independent district that Wright had proposed was, after all, not the only kind of public institution that could have been established. For years efforts had been made in California to enact a uniform state-wide system of irrigation, more or less controlled by expert state officials. The plans varied considerably in detail, but it was apparent that such a paternalistic, and potentially expensive, approach was unlikely to win approval at that time. Supporters of the state-control position therefore fell in somewhat reluctantly behind the Wright bill as the best that could be done. On the other hand those who would have preferred to see no public irrigation development at all, including some cattlemen, large grain growers and land speculators, had more to fear from a strong, centralized state system than from the independent districts Wright proposed, which could perhaps be thwarted or manipulated. For anyone who shared that view the Wright bill was acceptable as the lesser evil.[69] C.C. Wright may have drafted his legislation with an eye to conditions in Stanislaus County, but the result was at least an acceptable compromise to almost everyone.

The day after the governor signed the Wright Act, plans were announced for a triumphant welcome when Wright returned to Modesto.[70] As the train carrying the assemblyman pulled into the station on Saturday, March 12, 1887, the Turlock Band was playing and an old signal cannon was booming. An enthusiastic crowd escorted Wright to Rogers Hall where he promised that the act would bring water to the plains at the cost of its development.[71] Other speakers extolled Wright's accomplishment and prophesied a new beginning for Stanislaus County, but Horace Crane remembered that before the meeting was over a row broke out between the anti-irrigation "dry" farmers and the "wets".[72] If true, the story of the fight was as accurate an indication of the future course of events as any of the optimistic rhetoric that came from the podium.

CHAPTER 2 – FOOTNOTES

[1] Elias, *Stories of Stanislaus,* p. 12.

[2] For a discussion of Spanish-Mexican irrigation institutions see Michael C. Meyer, *Water in the Hispanic Southwest: A Social and Legal History, 1550-1850,* (Tucson, 1984).

[3] See Donald J. Pisani, *From Family Farm to Agribusiness: The Irrigation Crusade in California and the West, 1850-1931,* (Berkeley, 1984), especially Chapter 4.

[4] Elias, p. 18.

[5] Notice of Michael Kelly, May 1, 1871 (Stanislaus County Records, Misc. Records, Vol. 1, p. 500) and "Testimony of William R. Gianelli," before the Federal Power Commission (1962), p. 6.

[6] *Stanislaus County News,* June 19, 1871.

[7] *Ibid.,* June 16, 1871.

[8] U.S. Geological Survey, *Irrigation Near Merced, California* by Carl E. Grunsky, Water Supply and Irrigation Paper No. 19 (1899), p. 44.

[9] *Stanislaus County News,* Dec. 8, 1871.

[10] *Ibid.,* Dec. 22, 1871.

[11] Thomas E. Malone, "The California Irrigation Crisis of 1886: Origins of the Wright Act," (Ph.D. disser., Stanford Univ; 1964), pp. 44-45.

[12] *Stanislaus County News,* Dec. 8, 1871.

[13] Benjamin F. Rhodes, Jr., "Thirsty Land: The Modesto Irrigation District: A Case Study of Irrigation Under the Wright Act," (Ph.D. disser., Univ. of Calif. 1943), p. 27.

[14] Malone, pp. 45-46.

[15] Arthur Maass and Raymond L. Anderson, . . . *and the Desert Shall Rejoice: Conflict, Growth and Justice in Arid Environments,* (Cambridge, Mass., 1978), p. 160.

[16] Malone, p. 25.

[17] Pisani, pp. 135-136, suggests that the Keys' bill may have been intended as covert assistance to the Tuolumne Water Company. See also Malone, p. 112, and *Cal.Stats.,* 1872, p. 945.

[18] *Stanislaus County News,* Aug. 1, 1873.

[19] *Ibid.,* Aug. 8, 1873.

[20] *Ibid.* Sept. 5, 1873.

[21] Maass and Anderson, pp. 189-191.

[22] Rhodes, p. 30.

[23] *Modesto Herald,* Apr. 5, 1877.

[24] *Ibid.,* Apr. 12, 1877.

[25] *Ibid.,* Apr. 26, 1877.

[26] *Idem.*

[27] *Idem.*

[28] *Idem.*

[29] *Ibid.*, May 10, 1877.
[30] Rhodes, p. 32.
[31] *Modesto Herald,* May 10, 1877.
[32] Rhodes, pp. 35-36.
[33] Cal. Stats., 1877-78, pp. 820-825.
[34] *Modesto Herald,* Apr. 18, 1878.
[35] On William Hammond Hall's views, see Pisani, pp. 175-179 Wallace Stegner, *Beyond the Hundredth Meridian: John Wesley Powell and the Second Opening of the West,* (Boston, 1954) describes Powell's career and his attitudes on irrigation and Western development.
[36] *Modesto Daily Evening News,* May 18, 1885.
[37] *Ibid.*, May 26, 1885.
[38] *Ibid.*, July 18, 1885.
[39] "Water Right Locations on Tuolumne River above La Grange." This document is in a series of loose-leaf volumes arranged by former TID Chief Engineer Roy V. Meikle. Hereafter this collection will be referred to as "Meikle files" and documents identified by volume and item number. The present reference is in Meikle files, vol. 3, item 21.
[40] *Modesto Daily Evening News,* July 20, 1885.
[41] *Ibid.*, Aug. 17, 1885.
[42] *Ibid.*, Oct. 30, 1885.
[43] *Ibid.*, Aug. 28, 1886.
[44] *Ibid.*, Sep. 21, Oct. 11, 1886. The quotation is from Sept. 21.
[45] *Ibid.*, Oct. 9 and Oct. 11, 1886.
[46] *Ibid.*, Dec. 16, 1886; *Stanislaus County Weekly News,* Jan. 14, 1887.
[47] *Modesto Daily Evening News,* Dec. 17, 1886.
[48] *Ibid.*, Aug. 15, 1885.
[49] *Ibid.*, Sept. 5, 1885.
[50] *Ibid.*, Mar. 29, 1886.
[51] *Idem.*
[52] *Ibid.*, Mar. 31, 1886.
[53] See William Smythe, *The Conquest of Arid America,* (New York, 1907) for an example of irrigation promotion.
[54] Richard J. Hinton, *A Report on Irrigation,* Senate Exec. Doc. 41, Part I, 52nd Cong., 1st Sess; (Washington, 1893), p. 95.
[55] *Modesto Daily Evening News,* May 31, 1886.
[56] *Ibid.*, Oct. 25, 1886.
[57] *Ibid.*, Nov. 1, 1886.
[58] *Ibid.*, Nov. 10, 1886.
[59] Cal. Stats., 1887, p. 29.
[60] Hinton, pp. 95-96.
[61] *Modesto Daily Evening News,* Dec. 29, 1886.
[62] The history of *Lux v. Haggin* and the special session are covered by Malone, pp. 131-181.
[63] *Modesto Daily Evening News,* Jan. 4, 1887.
[64] Malone, p. 196.
[65] *Modesto Daily Evening News,* Jan. 22 and 25, 1887.
[66] *Ibid.*, Jan. 29, Feb. 4 and 19, 1887; Malone, p. 198.
[67] Quoted in Malone, p. 198.
[68] *Modesto Daily Evening News*, Feb. 26, 1887; Rhodes, pp. 50-52.
[69] Malone, pp. 200-201.
[70] *Modesto Daily Evening News*, Mar. 8, 1887.
[71] *Ibid.*, Mar. 14, 1887.
[72] Horace Crane interview in Hohenthal, Suplementary Material.

3

Implementing the Wright Act: Birth of the District

The dust had barely settled from the celebration of C.C. Wright's return before efforts were begun to put the new irrigation district law into effect south of the Tuolumne. A great deal of preliminary work, including surveys, had already been done there and interest in irrigation had been aroused. By early 1887 only the means of financing a canal remained in doubt and the Wright Act at last seemed to offer a practical solution to that problem. A mass meeting on irrigation was called for the end of March, but then cancelled when it appeared it would accomplish little. Instead, A.S. Fulkerth and T.E. Price were appointed to circulate petitions calling for the formation of an irrigation district. Even before leaving Modesto for the Turlock region, Fulkerth collected fifteen to twenty signatures from farmers who were in Modesto on business.[1] He spent about two days travelling through the proposed district and reportedly encountered only two opponents to irrigation. Meanwhile, Price was also canvassing Turlock area landowners and meeting with equal success.[2] In less than a week, the petitions carried the names of seventy-seven landowners and were ready to be submitted to the Stanislaus County Board of Supervisors. About one-third of the district's landowners had signed and it was said that many more signatures could have been secured, but in order to have the petition heard at the supervisors' April meeting, it was necessary to act quickly since the law required publication of the petition at least two weeks prior to its consideration.[3] The Board of Supervisors took up the irrigation petition on April 11, 1887. Engineer George Manuel, A.S. Fulkerth, A.P. Boyd and others testified on the proposed project. The supervisors continued the hearings for five more days between April 13 and April 18, and on April 23 the board ordered an election for May 28, 1887, to determine whether or not the Turlock Irrigation District should be established.[4]

Opposition to the proposed district, and to the enormous changes irrigation portended, was all too evident. The *Stockton Independent* observed that, "Our neighboring county is cursed with a few men who are satisfied with the old order of things, and who are opposed to all kinds of innovations and all sorts of improvements. These men are, of course, arrayed against the proposition to form an irrigation district."[5] Writing years later, Elwood Mead may have come as close as anyone to explaining the psychology of some of the die-hard opponents to irrigation in the Central Valley when he compared the wheat growers to the cattlemen of the Great Plains.

> The owner of a range herd was more than a money-maker, he was practically monarch of all he surveyed. The cowboy on horseback was an aristocrat; the irrigator on foot, working through the hot summer day in the mud of irrigated fields, was a groveling wretch. In cowboy land the irrigation ditch has always been regarded with disfavor because it is the badge and symbol of a despised occupation. The same feeling, but in a lesser degree, has prevailed in the wheat-growing districts of California, and for much the same reason.[6]

Obstinate conservatism or a cowboy mentality aside, those opposed to an irrigation district may simply have been no good reason to change their crops or habits, or they may have been skeptical of the cost estimates for irrigation and fearful of higher and higher taxes. They might also have resented the coercive features of a law that compelled taxation of every acre, whether its owner wished to irrigate or not.[7] Anxiety over the possible strength of the opposition led some pro-irrigationists to discuss forming a syndicate to buy out disaffected farmers if they decided to sell their land rather than remain within an irrigation district. However, such talk was probably little more than a ploy to convince critics that they had nothing to fear from irrigation.[8]

At the upcoming election voters would decide not only whether to form an irrigation district but, if so, who would run it. In four of the five divisions meetings were held to nominate candidates for the Board of Directors, and in the process discuss the whole irrigation question. A combined meeting, picnic and dance was called for Saturday, May 21, in Turlock to nominate candidates for the district-wide offices of assessor, collector and treasurer. It was said to be the largest gathering the town had ever seen. Speakers made the air "redolent with irrigation eloquence," and C. C. Wright's remarks were said to have succeeded in "bringing over many converts" to the support of the proposed district.[9] Others who addressed the crowd tried to exhort the believers and woo the uncommitted with the usual predictions of irrigated prosperity. They also went to considerable trouble to refute claims that irrigation would prove too costly. George Manuel's estimate that water could be brought to the edge of the plains for less than $100,000 was recalled and from there it was assumed that the farmers themselves could build the remaining ditches.[10]

Eloquence and arguments failed to extinguish the opposition. Right up to the day of the election there were fears that "the most inexplicable opposition of some of the prominent land holders" would be enough to defeat the district.[11] The results proved otherwise. The district was approved by a vote of 291 to 73, a margin of about four-to-one and comfortably in excess of the two-thirds majority demanded by the Wright Act. A closer examination of the vote totals showed that the strongest support came from the Turlock and Westport precincts (Divisions 3 and 4), while in the Merced County precinct (Division 5) the vote was fifteen to fourteen against the formation of the district.[12] On June 6, 1887, the Board of Supervisors reviewed the election results and declared the district to be formally organized, certified the election of its directors and established its boundaries.[13] Only three months after the Wright Act became law the Turlock Irrigation District became the first to organize under its provisions, and across the river, the Modesto Irrigation District was following close behind. After years of effort the irrigation era seemed to be at hand.

The Board of Directors met for the first time on June 15, 1887, in the chambers of the Board of Supervisors in Modesto. The members — E. V. Cogswell of Hickman, R.M. Williams of Ceres, E. B. Clark of Westport, W. L. Fulkerth of Turlock and J. T. Dunn from Merced County — were all farmers, whose holdings ranged from Cogswell's 2,200 acres down to Clark's 160 acres. They went to work immediately, methodically planning to claim water rights and hire an engineer. Fulkerth and Cogswell went to Fresno to examine the irrigation systems in use there and learn what they could of engineers who had worked in that area. As a result of their inquiries, and undoubtedly on the basis of his previous service, George Manuel was quickly chosen to conduct the surveys for the canal.[14] Manuel arrived in the district within a few days of his formal appointment and went straight to work in the La Grange area.[15] The board was optimistic that actual construction would soon begin and it was even hoped that work on a diversion dam and on the main canal through the foothills could be completed that year.[16]

That kind of optimism vastly underestimated what had to be done. Manuel's surveys were not completed until September, and the results held some surprises. The canal route itself, beginning near the Wheaton Dam, was generally similar to the plan Manuel had developed almost a year earlier, though this time he extended the survey along the east side of the district down to the Delhi area near the Merced River. In July, however, he had announced that the canal should be designed to carry from 1,500 to 1,800 second-feet instead of the 1,200 second-feet he had deemed sufficient only six months earlier.[17] Increasing the capacity of the canal as well as extending it into the district and building the smaller lateral canals used to distribute water could certainly be expected to push the price well beyond the $95,000 Manuel had arrived at the year before. The new estimate, however, totalled $467,544.62 for a complete system of works, along with bridges and other accessories, needed to take water throughout the district. Under the circumstances it seems remarkable that the steeply revised estimate elicited no outraged response whatsoever, especially since opponents to the district had always contended that it would cost more than its advocates had dared think. The board simply accepted the estimate and resolved to ask the voters to approve $600,000 in bonds to finance construction.[18] This time, there was little debate and on October 8, 1887, the bonds passed nearly unanimously on a vote of 176 to 12.[19]

The directors seem to have understood from the start that irrigation district bonds, a new and unproven type of security, could prove difficult to sell unless it could be shown that they were a safe investment. Despite the unanimous passage of the Wright Act, and the optimism of many irrigationists, the law all too clearly had a host of implacable and often powerful foes. It was generally assumed that these opponents of the Wright Act would soon attack the constitutionality of the law and the legitimacy of the districts formed under it. Besides threatening the very existence of the state's irrigation districts, the legal challenge would also cast doubt on the validity and safety of their bonds, making them harder to sell. For that reason it was considered important to confirm the constitutionality of the irrigation district law. The Turlock board acted quickly to bring the constitutionality issue to a head itself, ordering on November 1, 1887, that "In the matter of

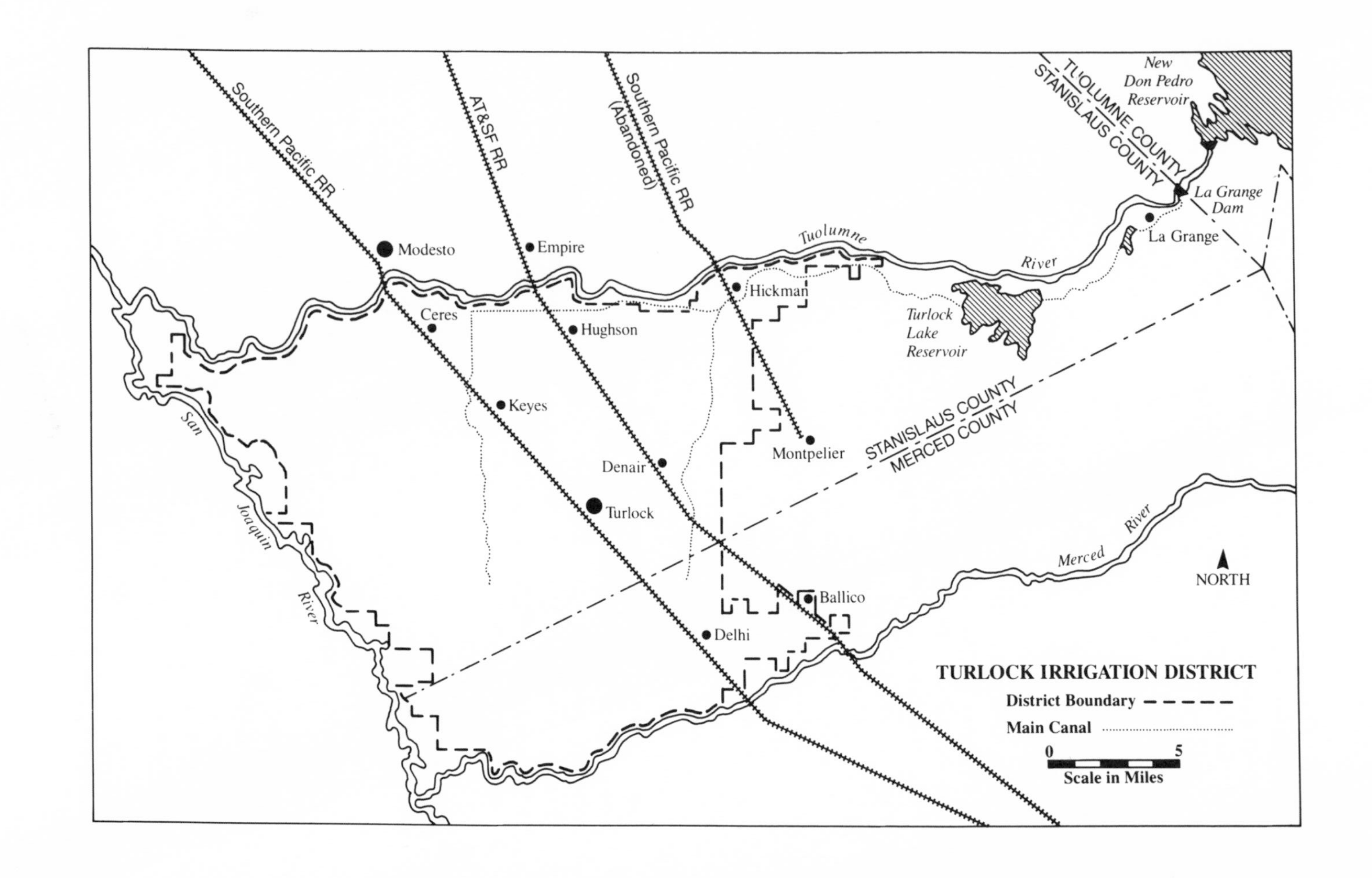

Southern Pacific RR
AT&SF RR
Southern Pacific RR
(Abandoned)
TUOLUMNE COUNTY
STANISLAUS COUNTY
New
Don Pedro
Reservoir
La Grange
Dam
La Grange
Tuolumne
River
Modesto
Empire
Ceres
Hughson
Hickman
Turlock
Lake
Reservoir
Keyes
San
Joaquin
River
Denair
Montpelier
STANISLAUS COUNTY
MERCED COUNTY
Turlock
Merced
River
NORTH
Ballico
Delhi
TURLOCK IRRIGATION DISTRICT
District Boundary
Main Canal
0
5
Scale in Miles

the constitutionality of the law . . . the attorneys be and they are hereby instructed to take such legal measures as in their judgement is necessary in order to insure the sale of the bonds of the district."[20] Fifty thousand dollars worth of the bonds had already been advertised for sale, and these were sold November 7 to Modesto banker Robert McHenry, who bid $45,000, or 90 percent of their face value, which was the minimum price allowed by law.[21] McHenry, who was also President of the Board of the Modesto Irrigation District, conditioned his bid on the resale of the bonds by him to other investors.[22] The bonds were soon printed, but in order to test the law, TID secretary Roger Williams refused to sign them. The district's lawyers were then able to request the state Supreme Court to issue a writ of mandamus compelling Williams to sign the bonds. It was planned that Williams and the attorneys representing him would offer every conceivable objection to the constitutionality of the Wright Act, to the formation of the Turlock District and to the validity of the bonds so that the Court could deal with all these issues at once. C.C. Wright was also involved and favored the action.[23] Arguments were heard in December, but months passed without a decision. While the matter was before the Court, surveys and other work came to a standstill.

On May 31, 1888, the Supreme Court finally ruled on the case of *Turlock Irrigation District v. Williams,* finding that the Wright Act was constitutional. In response to assertions by Williams' attorneys that irrigation districts were in reality private corporations for the benefit only of certain landowners, the court responded that, "This is not a law passed to accomplish exclusive or selfish private gain; it is an extensive and far reaching plan by which the general public may be vastly benefitted. And the legislature acted with good judgement in enacting it."[24] Likewise, objections that the system of taxation was unconstitutional were rejected. In fact the court ruled in favor of the Wright Act on every point. News of the decision set off celebrations in the irrigation districts. Modesto's reaction may have been typical.

> Huge bonfires illuminated the city and cannons peeled forth in thunderous tones the deep joy experienced by our people over the result. One hundred guns were fired. Everybody seemed to feel an uncontrollable desire to shake hands with every other body. Grave and dignified citizens were seen to stop on the street and seize their hats with a motion as if to throw them in the air and then as though remembering their dignity replace their head covering with a sheepish

look, conscious of having betrayed the buoyancy of their feelings. It was a great day for Modesto. It was a memorable event for Stanislaus County.[25]

With the basic legal issues thought to be settled, the board ordered the surveyors back into the field and planned to begin purchasing the right-of-way for the canal.[26] Optimism returned and there was even talk of building a dam before the beginning of the next rainy season.[27]

The board resolved in August 1888 to proceed with the sale of the remaining $550,000 in bonds.[28] However, when it came time to open the bids, only one had been received. L.M. Hickman, one of the most prominent Turlock area landowners, offered to take the entire amount of bonds at 90 percent of face value, but with the provision that the purchase price would be paid to the district only "on delivery of said bonds to me, but not before I can sell the same at not less than said price for cash."[29] Hickman also agreed to turn over to the district anything over ninety cents on the dollar that he might receive, less only the expenses of the sale. The sale of the bonds to Hickman and the previous sale of $50,000 worth to Robert McHenry, were not really sales at all. No cash whatsoever had been received for the bonds "sold," and in fact no one had ever offered to actually pay for the district's bonds. The so-called sales were only a formality that allowed the district to negotiate the placement of its bonds with bona fide investors without the necessity of readvertising them for bids. At that time no California irrigation district had sold bonds for cash and the financial community was still hesitant to risk any money on them.

The Turlock District tried to interest investors in San Francisco, the east coast and Europe in its bonds, but found it difficult to overcome a reluctance to invest in an enterprise that might still face local opposition and had yet to prove its ability to manage its finances or repay its debts. Finally, in early December 1888 representatives of a San Francisco syndicate considering purchase of the bonds toured the district and a few days later Judge James A. Waymire of San Francisco informed T. E. B. Rice, one of the district's negotiators, that the group would definitely take the bonds. Board members Clark and Cogswell left immediately for San Francisco accompanied by TID attorney P. J. Hazen, L. M. Hickman and Rice, to complete the arrangements.[30] On December 10, 1888, a contract between Hickman and the buyers was completed and countersigned by E.B. Clark on behalf of the district. P. J. Hazen telegraphed the news that, "Contract of the sale of the bonds of the Turlock Irrigation District fully executed. First payment on deposit."[31] The sale of half a million dollars in bonds for ninety cents on the dollar was the first sale of any bonds authorized by the Wright Act. Only two days after the bond sale was finalized the district passed another important milestone when John W. Mitchell, by far the largest landowner in the TID, became the first to pay his irrigation taxes, handing $1,581 to Collector A. N. Crow.[32] Collection of the remainder of the district's first tax levy of 42½ cents per $100 of assessed value went smoothly, and only three persons were delinquent by February 1889.[33]

A year and a half after its formation the Turlock Irrigation District had apparently overcome the legal and financial hurdles that came with a new kind of irrigation institution and finally seemed ready to build its canal. So far the only thing that had been accomplished had been a series of surveys. The initial survey had established the general line of the main canal accurately enough to allow the cost estimate to be made. When that job was done, further, more painstaking work was begun to prepare detailed plans for construction. All activity had been suspended and the engineer laid off while the Supreme Court considered the constitutionality of the Wright Act. Once that question was settled surveys were resumed in June 1888, only to be cancelled again three months later because the sale of bonds was still indefinite. When the contract for the bonds was signed the surveyors went back to work. In April 1889 they finished laying out a main canal seventy feet wide and eight feet deep that could carry a flow of 1,500 second-feet from La Grange to the plains near Hickman.[34]

Since its formation the district had been making regular water-rights filings for 4,500 second-feet to be diverted at or near the Wheaton dam, but the on-again, off-again nature of work on the canal had prevented it from proving the diligence needed to hold an appropriative water right. Finally, in early 1889, with financing apparently assured and the surveyors at work, the district began the construction process by purchasing the land where the canal would go, and by February good progress in land aquisition was reported.[35] As usual things did not go smoothly for long. Some landowners along the canal route were cooperative and agreed to either grant the district the land it needed free of charge or they asked only a reasonable sum for their property, but others set prohibitively high prices or demanded an immediate cash payment, which the district could scarcely afford. In case the district and the landowner could not negotiate a satisfactory price for a right-of-way, the district had the power to condemn the land and let a jury determine a fair price, but litigation could be so expensive and time consuming that everything possible was done to reach voluntary agreement. In one case the board's right-of-way committee had to visit an owner at least twenty times to reach a settlement, and in another instance it waited for someone to return from the Hawaiian Islands to conclude a deal.[36] Progress was being made but it was not as rapid as had been hoped or expected.

Progress in other areas was disappointing as well. Bond sales were lagging despite the contract with the San Francisco syndicate. Apparently, statements made by the governor and by enemies of the Wright Act created fresh doubts regarding the ability of irrigation districts to sell bonds and that panicked the would-be purchasers. By April 1889 only $9,000 in cash had been received, but in July a measure of confidence returned and the first substantial sale of $50,000 was completed. In the months that followed it was reported that additional sales amounting to about half the total issue had been made.[37] Survey work had continued and on August 6, 1889, the board formally adopted George Manuel's plans for the canal and ordered him to prepare maps and specifications needed to call for bids on construction.[38] Before calling for bids, however, the board ordered the surveyors to investigate the dam site and canal route outlined by General Bost in 1885 beginning at Thompson's Ferry, about two miles below La Grange.[39] The survey confirmed that the plan already adopted was the best suited to the district's needs, so in December 1889 the board called for bids on Section 1 of the main canal.[40]

Section 1 was the most difficult part of the canal, running through the rugged foothills from about a mile above the town of La Grange to Peaslee Flat. Since the exact location of the diverting dam on the Tuolumne had not been decided, the canal work began at a point away from the river. The bids received in January 1890 all exceeded the amount the directors thought reasonable, prompting a delay in awarding the contract while cost-cutting changes in the canal plan were made. The low bidders agreed to the changes and on January 20, 1890, the first construction contract was awarded to Hamilton W. Gray. The contract covered five of the six subsections of Section 1, while the board rejected all bids on the remaining portion and ordered the work done under district superintendence.[41]

One month later, excavation of the canal began.[42] In April the district awarded the contract for Section 2, extending from Gray's work to near Hickman, to S. S. Watson of Los Angeles who shared the job with the Heitchew, Meadows Company.[43] By the summer of 1890 work was underway along the entire length of the main canal.

The canal work from La Grange to the plains passed through country that made construction difficult. East of Hickman the hills were smooth and rounded but nearer La Grange they became steeper, rockier and deeply cut by creeks and gullies. Hamilton Gray faced the greatest challenges. Deep cuts had to be made through hillsides, dams built where the canal would follow parts of natural creeks and three tunnels bored through rock. Fortunately, some of the heaviest work could be done with water, using techniques borrowed from the hydraulic mining industry, and in fact with water from the La Grange Ditch and Hydraulic Mining Company. Water under pressure was aimed by huge nozzles at hillsides, which were then washed away, leaving a cut for the canal. The method worked well but when an attempt was made to use some of the earth and rocks washed out of one of the cuts to build a dam, the results were disappointing. At Snake Ravine, just above La Grange, plans called for a dam across the ravine to create a small lake through which the canal would flow. A cut was being hydraulicked out just to the west and it was intended to let debris from that work run down into the ravine, where it would form the dam. Unfortunately, the rocks and mud were allowed to simply flow along the ground rather than being carried to the dam site through pipes or flumes. As a result the larger rocks separated from the fine silt and only the silt reached the dam, which made it "constantly shaky and wet, vibrating when jarred."[44] When the fill reached thirty feet in height in November 1890 it collapsed, sliding down into the Tuolumne River.[45] Watson and Heitchew, Meadows had easier terrain to work with, but hardpan near the surface sometimes made blasting necessary before the scrapers could go to work. At other places ten or twelve horses had to be harnessed to a single plow to break the ground.[46]

The contractors finished the work assigned them by May 1891, but the canal was still far from complete. Nothing had been done to replace the ruined Snake Ravine dam and the high wooden trestles and flumes that would carry the canal over some of the deep gulches in its path remained to be built. Wood was subject to decay, so those structures would only be built when the canal was nearly ready to carry water. Before that could happen the district needed a dam.

When work started on the canal in early 1890, plans for a dam were still indefinite. From the first surveys it had been recognized that the best location for a diverting dam was somewhere near Wheaton's dam. A dam was necessary not to store water but to raise the level of the river behind it high enough to permit a gravity flow to all the irrigable lands within the district. Although the Wheaton dam site was excellent, the district had made no progress in negotiations with its owner, M. A. Wheaton. The main obstacle was undoubtedly money. In the 1870s Wheaton had offered to sell the dam for what it had cost him, but now that it appeared that one or both of the local irrigation districts needed the site to complete their works, he placed a much higher value on the dam, the property around it and the water rights he claimed to hold. In February 1890 Director R. M. Williams announced that the Turlock District had found a way out of the deadlock by purchasing a quarter section of land on the river just upstream from Wheaton's holdings that included a satisfactory dam site.[47]

With a dam on its own land, all the district would need from Wheaton was a narrow right-of-way across his property for the canal. Since Wheaton's land straddled the county line, attorney Hazen had to file condemnation suits in both Stanislaus and Tuolumne counties for the right-of-way.[48] A clear picture of the situation on the river can be found in a letter from Turlock District resident J. I. Jones to the *Modesto Daily Evening News* in March 1890. Jones visited the La Grange area and went over the ground with the district's new engineer, E. H. Barton, who had just replaced George Manuel. From an eighty-two-foot-high dam on the district's property, a wooden flume half a mile long would carry water to a tunnel through the steep bluff that stood beside the Wheaton dam. Another flume or canal would connect to the works being built by Hamilton Gray, a little over a mile from the dam site. Jones continued:

> Now as to the unreasonable damages claimed by Mr. Wheaton, which a good many have thought stood in the way of this enterprise. I will say that I have been over the ground – I mean looked over it – (as nothing but a bird could travel over a portion of it) and that it does him absolutely no damage, except the value of the strip of land that it (the canal) occupies, which is fit only for goats and on a portion of which they have no business.[49]

When the condemnation case came up for trial in May the enormity of Wheaton's demands were made clear. He was asking $150,000 for his water rights and an additional $150,000 for a right-of-way over his land. The district had offered him $20,000, presumably for right-of-way only, since his water rights were of questionable validity.[50] The Stanislaus County suit was quickly decided on terms favorable to the district. For the five acres of land in question the jury awarded Wheaton only $75 plus another $20 for damages that might be done to his remaining land by the canal, for a total of $95.[51] The Tuolumne County suit was heard in July, and although the mountain county jury was undoubtedly less sympathetic to irrigation, Wheaton still received only $5,000.[52]

The Turlock District's new dam site and the Stanislaus County jury's decision in the condemnation suit threatened to drastically reduce the value of Wheaton's dam and other property. Under the circumstances he was apparently disposed to make the best deal he could. Less than two weeks after the first condemnation trial, Wheaton sold his water rights, dam and enough land for a new dam and canal to the Modesto Irrigation District for $10,000 in cash and $11,000 in bonds. According to the provisions of the contract, the Modesto District could destroy the existing dam and build a new one up to fifty feet higher than the old one. Wheaton' dispute with the TID was plainly evident in a provision that "the Turlock Irrigation District shall not be permitted to join or share in either the building or use of the proposed dam which the Modesto Irrigation District intends to build" until the district had settled the right-of-way conroversy and Wheaton agreed to the TID's use of the dam.[53]

Although it would seem that with a right-of-way across his land, the Turlock District would have no need for further dealings with Wheaton, they were in fact continuing negotiations with him. The discussions involved the cancellation of suits and judgements and the purchase of Wheaton's entire holdings and on August 1, 1890, Wheâton made a written offer to sell.[54] Wheaton had long been a thorn in the side of the district and it may have seemed worthwhile to simply remove the nuisance completely. The Modesto Irrigation District, having already purchased some of Wheaton's property was also a party to the talks, and on August 5 the Turlock board acknowledged that "certain informal negotiations and propositions" had been exchanged between the two districts. A committee of the whole board was created to negotiate and accept agreements with Modesto.[55] The following day, August 6, the two boards met together and decided to meet again on Saturday, August 9. In the meantime Turlock's engineer had prepared plans for a tunnel from the district's proposed dam to the canal below Wheaton dam. With the board reportedly inclined to call for bids on the tunnel without delay, a final decision on any settlement with Wheaton or the Modesto District had to be reached quickly.[56]

The Turlock and Modesto district boards met together in Modesto on August 9 for a marathon session that lasted from 10:00 a.m. until 9:00 p.m. and produced perhaps the most important agreement in the history of either district. The Memorandum of Agreement signed on August 9, 1890, provided that the two districts would jointly pay $35,000 for all the property and rights of M. A. Wheaton on the Tuolumne River. The Modesto District, having already paid Wheaton $21,000, was to pay an additional $2,500, with Turlock to pay the remaining $32,500. The two districts agreed to share the ownership of the land acquired from Wheaton and the quarter section containing the Turlock District's proposed dam site as equal partners. They further agreed to join in the erection of a single dam at the Wheaton site to serve both districts, also dividing the costs of construction and maintenance equally between themselves. Although the joint dam and the property surrounding it were to be shared equally, it was agreed that all water diverted by the dam would be divided "in proportion to the number of acres in the respective Districts." Since the Turlock District, with 176,210 acres, was just over twice the size of the 80,564 acre Modesto District, Turlock would receive over two-thirds of the water from the joint dam. (The TID's official acreage was given as 176,210 acres. A recalculation in 1983 based on original district boundaries showed actual acreage to have been 179,527 acres.) The two districts also provided that if either of them acquired "any additional water rights, privileges or rights in stored water" above their joint dam, it was obligated to give the other district the right to share the water or property on the same acreage formula that applied to diversions from the dam.[57]

The August 9, 1890 agreement was of tremendous importance. It made the Turlock and Modesto irrigation districts partners in the development of the Tuolumne River. The two districts already had so much in common that the agreement to pool their resources and build a single joint dam seemed entirely logical. After all, one dam, the Wheaton structure, had been at the heart of the first irrigation plans for the plains on both sides of the river, and after the districts were formed there had occasionally been talk of a dam to serve both of them.[58] The 1890 agreement, however, did far more than provide for the construction of a joint dam; it peacefully divided the river between the two districts. That action placed the Tuolumne in sharp contrast to other California or Western streams where rival appropriators battled over the right to water. Irrigation expert Elwood Mead later wrote that, "the irrigator whose water does not furnish grounds for either an inquiry or a grievance is a rare exception."[59] The disputes over water rights regularly wound up in court, but there were times when even a judicial decision was insufficient protection. One Californian explained that to get his share of water "he first got a court decree and then shipped in two men from Arizona who were handy with a gun."[60] In part, the agreement to divide the river was made possible by the very circumstances that had so long restricted irrigation development on the Tuolumne. More easily dammed streams encouraged early and numerous canals, resulting in a multitude of claimants for the water and a dense tangle of conflicting rights. This was the case on the Kings River and on many others, where the total amount of water claimed under various appropriations exceeded what the river carried and the users were locked in a seemingly perpetual struggle to define their respective rights. In sharp contrast, the expense involved in diverting the Tuolumne prevented any development until the Turlock and Modesto irrigation districts were formed, and those two agencies virtually monopolized the irrigable land around the river. That there were only two major appropriators on one of the Central Valley's largest streams simplified matters greatly and made it easier to reach the kind of agreement they did.

Without an agreement to peacefully divide the Tuolumne according to a particular formula the two districts would probably have faced each other in court. Aside from a mining ditch orginating some distance above La Grange and the dubious rights claimed by M. A. Wheaton, the Turlock District had the earliest claim, and according to the appropriation principle the first in time was the first in right. Under those circumstances the TID would have been entitled to take all the water it needed, up to the limit of its rights, before the Modesto District would be allowed to begin its diversions. Arguments over how much water each district was entitled to and over the validity of the various claims would have been virtually inevitable. Instead, the districts agreed that, between themselves, the strict priorities would be disregarded in favor of a division of the river based on the relative size of each district. Not only was the formula applied to water diverted at the joint dam, it was also applied to any future water development anywhere on the watershed above the dam. In that way neither district could acquire any water or water rights on the Tuolumne without allowing the other district to participate and claim its proportionate share. With one document the districts replaced potential conflict with cooperation and laid the foundation for a far-reaching partnership on the river.

The districts intended to proceed immediately on building the dam. The agreement provided that their engineers would each submit a plan for the dam and, in case the boards failed to agree on one of them, a third engineer would be hired to make the choice. E. H. Barton of the TID and Luther Waggoner of the MID quickly drew up similar plans for masonry dams at locations varying only twenty-five feet from one another. Despite the similarities each board stubbornly favored its own engineer's design, so Col. G.H. Mendall of the Army Engineers was chosen to select the final plan. He developed a design based on Luther Waggoner's proposal, but the districts were so eager to begin work on the project that they advertised for bids at the same time they decided to employ Col. Mendall.[61] S.S. Watson was the only bidder, offering to build the dam for $10.45 per cubic yard, with the districts supplying the cement. The bid was quickly dismissed as too high. Although there was some talk of the districts attempting to construct the dam without a contractor, nothing more was done until the middle of 1891.[62]

New bids on the dam and on the cement that the districts were to furnish to the contractor were called for in May 1891. They were opened on June 2, 1891, and district officials must have been disappointed that the low bid, submitted by R.W. Gorrill, was only six cents per cubic yard less than the bid they had summarily rejected six months earlier.[63] The cost was higher than it would have been had the districts been on a firmer financial footing. As it was, Gorrill was forced to agree to payment of $150,000 in cash with the remainder of the estimated $331,480 contract to be payable in bonds of the districts at 90 percent of face value.[64] Unfortunately the bonds were not worth that much on the open market. Investors were still leery of irrigation district securities and the uncertain future of some of the districts organized under the Wright Act added to their caution. The bonds were thought to be a risky investment and as a result they were salable only at a substantial discount. Since Gorrill could not get the ninety cents on the dollar the bonds were supposed to be worth, he, and other contractors who had to accept their pay in bonds rather than cash, had to inflate their charges to compensate for the low value of the bonds. Some years after the dam was completed engineer James D. Schuyler commented that, "Under ordinary conditions of prompt payments in cash the construction should have been done for one-half the actual cost."[65] Before the final contract was signed in June, even the guaranteed cash payment of $150,000 fell victim to the districts' inadequate resources. A new clause promised Gorrill only whatever cash the districts had on hand.[66] The financial shoe-string was stretched tight indeed.

La Grange Dam was built of rubble masonry with huge, irregularly shaped boulders set in concrete and faced with roughly dressed stones. Arched against the current, its upstream face would be nearly perpendicular while its downstream profile would be sloped to a broad, solid base. It was designed to raise the level of the river behind it one hundred feet, and since it was to have no spillways or outlets other than the two irrigation canals, the river would pour over the top the structure and down the lower face. In extreme floods, water up to sixteen feet deep might spill over the crest. To protect the base of the dam from erosion when water went down its face, a small weir about two hundred feet downstream would create a fifteen foot deep "water cushion" to absorb some of the impact of the overflow.

Work on the project began soon after the contract was finally signed. The districts' representatives at the construction site, TID director R.M. Williams and MID director F. A. Cressey, soon located the districts' cement warehouse on a high bluff north of the river. The contractor put a force of sixty men to work building a bridge over the river and opening a rock quarry, and as more machinery was lowered into the deep canyon, the number of men increased to one hundred.[67] As the season of snowmelt passed, pre-parations were made to dry out the dam site so the foundation could be laid. A small cofferdam just upstream diverted the low summer flow into four flumes that carried it past the dam site and over a natural rock barrier where the Wheaton dam had stood. With the riverbed exposed, workers using derricks powered by steam engines removed boulders and softer materials until bed-

Derricks and steam engines at work on the La Grange Dam.
This view probably taken in 1891.

rock was reached. A pothole twenty feet deep was found near the center of the dam and cleaned out so the foundation would rest entirely on solid rock.[68]

The first masonry was placed in September 1891. The narrow canyon was now full of derricks, chutes, hoists, crushers, steam engines and other machinery. Rock for the concrete was quarried from the north side of the river and sand was scooped up from the river nearby. Cement was hauled from the railroad at Waterford to the disrtict's warehouse high above the dam, where the barrels were opened. Loose cement was sent down in a pipe to a mixer where one part of cement was combined with two parts of sand and eight parts of crushed rock, revolved twenty-five or thirty times and then transferred to buckets carried by the derricks. High water from sudden storms was a constant threat, and work on the base of the dam was rushed. By mid-November 180 men were at work and the dam had risen to five feet above the old low water level.[69] Two tunnels, each about four feet wide and six feet high, were built into the base of the dam and one more was installed ten feet higher. These replaced the flumes and allowed the river to flow through the dam at its low stages. The tunnels had gates that would permit them to be closed and cemented when the dam was completed.[70] Since the tunnels could handle only relatively low flows, one side of the dam was always kept lower than the other to let excess water pour over a portion of the structure without halting work on the remainder. In January 1892 high water interupted work, but the river soon fell and the job resumed, only to be halted in February when rains pushed the river over the dam. By February the dam had risen to over twenty-five feet above the old river bed.[71]

With construction of the dam well underway, the TID board made the unhappy announcement on April 5, 1892, that the district's construction fund was exhausted and that another bond issue would be needed to complete the dam and finish the canal system.[72] The board, and the community at large, were undoubtedly reluctant to undertake an even larger debt, and one director, the conservative and cautious Miller McPherson, refused to vote for the new bond issue. However, as attorney P. J. Hazen explained, the partially completed irrigation project was "a thing which you can not let loose of without getting hurt."[73] If additional bonds were not approved, work might stop and no good would come of the $600,000 already spent, even though the principal and interest on the first bond issue would still have to be paid. On the other hand,

La Grange Dam rose 127½ feet above the Tuolumne streambed when it was completed in 1893. It was later raised 2¼ feet.

Hazen pointed out that the Wright Act allowed the directors to complete work in progress by a direct tax levy if necessary, which would have been an intolerable burden on local landowners.[74] The *Modesto News* interviewed prominent citizens in the Turlock District regarding the new bond issue and found they almost uniformly agreed with Hazen's analysis that the voters had little choice but to approve the extra bonds. To allay any suspicions that money had been squandered in the past, the board offered to open its books to anyone who cared to examine them. Only one man accepted the offer and after several hours he went away apparently satisfied that there had been no mismanagement.[75] The election held May 14, 1892, resulted in the approval of another $600,000 in bonds by a vote of 218 to 97.[76] Although the bonds passed by a better than two-to-one margin when only a simple majority was needed, the negative vote was the largest both in number and as a proportion of votes cast that the district had seen. Five years of delays, frustrations and unanticipated expenses had begun to take their toll.

It was late July 1892 before the river went down enough to allow work to resume on the dam.[77] Up to 200 men were on the job by mid-September when the cement warehouse was destroyed by fire. The building itself was worth between $700 and $800 but the 3,200 barrels of cement it held were valued at $14,000. Although the Modesto District's share of the property was covered by $6,000 worth of insurance, the TID's loss was uninsured. Extra wagons and teams were ordered to haul more cement from Waterford so construction could continue.[78] A month after the fire the charred cement was still too hot to handle with bare hands, but insurance adjustors and district workers were sifting through it to see if any of it could be salvaged.[79] The cement was still being tested on November 3 when the floor of the damaged warehouse gave way, dumping 2,000 barrels of loose cement down the hillside.[80] A few weeks later a high wind blew the roof off the new warehouse, sending billowing clouds of cement particles into the air until canvases were brought to cover what was left.[81]

Despite the calamities that befell the cement supply, work on the dam moved rapidly. An overhead cable system was installed in the fall of 1892 to more efficiently carry concrete and other material. Three cable lines spanned the canyon, operated by hoists and a forty horsepower steam engine located on a hill south of the river. Up to ten tons at a time could be hauled in the gondolas suspended from the cables.[82] Newspaper reports indicate that work on the dam, now perhaps fifty feet above the riverbed at its lowest point, continued through the winter of 1892-1893. By June 1893 the Tuolumne was flowing through the tunnels in the base of the dam and over the lower part of the work in a cascade that one exuberent reporter thought would "in sublimity and grandeur . . . rival those in the Yosemite Valley."[83] Unfortunately, the water crossing part of the dam slowed work until August when the entire flow again went through the tunnels.[84] The work force was increased to about 150 men laboring around the clock to complete the job before the beginning of the next rainy season. Two of the three tunnels were closed and cemented while the third was equipped with a gate to close it later.[85] In early December the river went over the top briefly and then subsided long enough for the contractor to officially finish the dam on December 13, 1893.[86]

When it was built La Grange Dam was the highest overflow dam in the country and perhaps in the world. It rose 127½ feet above the streambed and had a crest over 300 feet long. It was ninety feet thick at the base and twelve feet wide at its rounded top. Although the contractor and various consulting engineers were involved in a long-running feud over exactly how much material was placed in the dam (and how much had to be paid for), it consisted of about 39,500 cubic yards of masonry, including over 31,000 barrels of cement. The final cost, including the cement, came to $550,000.[87]

At the time of its completion, the canals it was meant to serve were far from finished but a few local residents looked beyond the day when they would be done and foresaw yet another use for the great dam at La Grange. Believing that some water would always flow over the dam, falling one hundred feet in the process, the editor of the *Modesto Daily Evening News* prophesied that, "Here, then, will be a source of power valuable within itself. This power, no doubt, in time will be utilized in generating electricity and for other purposes." A few days later the paper took note of the large hydroelectric project then underway at Niagara Falls and confidently predicted that what Niagara Falls was for New York, La Grange Dam would be for Stanislaus County.[88] The hydroelectric fervor was shortlived and may have been due to an interest in electricity occasioned by the expansion of Modesto's two year old steam-powered electric system. Whatever the reason, the early interest in hydroelectricity from the Tuolumne is noteworthy, especially in light of the district's eventual entry into the electric business and the construction, many years later, of a power plant just below La Grange Dam.

By the time the dam was finished, it had been seven years since C. C. Wright had prepared his irrigation district bill, but no water had yet reached the farmers of the first irrigation district. Irrigation had so far proven much harder to achieve, and much more expensive, than any but the deepest pessimists would have assumed in 1887. Unfortunately, the legal and financial trials of the district's birth were only a foretaste of what it and all the other Wright Act districts would have to endure in the years just ahead.

CHAPTER 3 – FOOTNOTES

[1] *Modesto Daily Evening News,* Mar. 21, 1887.
[2] *Ibid.*, Mar. 23, 1887. [3] *Ibid.*, Mar. 26, 1887.
[4] *Ibid.*, Apr. 13, Apr. 15, Apr. 16, Apr. 19, Apr. 25, 1887.
[5] Quoted in *Modesto Daily Evening News,* Apr. 30, 1887.
[6] U. S. Dept. of Agric., Office of Experiment Stations, *Report of Irrigation Investigations in California,* Bulletin No. 100 (1901), p. 32.
[7] For an example of anti-irrigationist arguments see the letter from J. W. Davison in *Modesto Daily Evening News,* May 11, 1887.
[8] *Modesto Daily Evening News*, Apr. 26, May 17, 1887. [9] *Ibid.*, May 23, 1887.
[10] *Idem.* [11] *Ibid.,* May 30, 1887.
[12] *Idem.* [13] *Ibid.*, June 7, 1887.
[14] Minutes of the Board of Directors of the Turlock Irrigation District [Hereafter cited as "Board minutes"], vol. 1, pp. 1-6 (June 15, and 27, 1887).
[15] *Modesto Daily Evening News,* July 11, 15, 1887.
[16] *Ibid.,* July 26, 1887. [17] *Ibid.,* July 25, 1887.
[18] Board minutes, vol. 1, pp. 18-19 (Sept. 16, 1887).
[19] *Ibid.*, p. 25 (Oct. 10, 1887). [20] *Ibid.*, p. 28 (Nov. 1, 1887).
[21] *Ibid.*, p. 29 (Nov. 7, 1887).
[22] *Modesto Daily Evening News,* Dec. 13, 1888. [23] *Ibid.*, Nov. 18, 1887.
[24] *Turlock Irrigation District v. Williams,* 76 Cal 360, quoted in *Modesto Daily Evening News,* June 2, 1888.
[25] *Modesto Daily Evening News,* June 1, 1888.
[26] Board minutes, vol. 1, pp. 46-47(June 4, 1888).
[27] *Modesto Daily Evening News,* June 6, 1888.
[28] Board minutes, vol. 1, p. 52 (Aug, 7, 1888). [29] *Ibid.*, p. 68 (Dec. 18, 1888).
[30] *Modesto Daily Evening News,* Dec. 3 and 7, 1888. [31] *Ibid.*, Dec. 11, 17, 1888.
[32] *Ibid.*, Dec. 12, 1888. [33] *Ibid.*, Feb. 27, 1889.
[34] *Ibid.*, Apr. 11 and 26, 1889.
[35] Board minutes, vol. 1 p. 78 (Feb. 5, 1889); *Modesto Daily Evening News,* Feb. 25, 1889.
[36] *Modesto Daily Evening News,* Feb. 6, and 21, 1890.
[37] *Ibid.,* Apr. 4, July 30, Aug. 26, Sept. 16, 27, Nov. 27, 1889.

[38] Board minutes, vol. 1, p. 91 (Aug. 6, 1889).
[39] *Ibid.*, p. 98 (Oct. 1, 1889). [40] *Ibid.*, p. 106 (Dec. 3, 1889).
[41] *Ibid.*, p. 113 (Jan. 8, 1890); p. 115 (Jan. 14, 1890); p. 115 (Jan. 20, 1890).
[42] *Ibid.*, p. 119 (Jan. 28, 1890).
[43] *Ibid.*, p. 133 (Apr. 15, 1890), p. 136 (Apr. 18, 1890); *Modesto Daily Evening News,* Apr. 28, 1890.
[44] J. B. Lippincott, "The Failure of Snake Ravine Dam, Turlock Irrigation District, California," *Engineering News*, XL, no. 16 (1898) , p. 242.
[45] *Modesto Daily Evening News,* Nov. 17, 1890.
[46] *Ibid.*, Dec. 11, 1890. [47] *Ibid.*, Feb. 6, 1890.
[48] *Ibid.*, Feb. 20, 1890. [49] *Ibid.*, Mar. 19, 1890.
[50] *Ibid.*, May 26, 1890. [51] *Ibid.*, June 3, 1890.
[52] *Ibid.*, July 16, 1890.
[53] Contract between M. A. Wheaton and Modesto Irrigation District as printed in *Modesto Daily Evening News,* June 30, 1890.
[54] Board minutes, vol. 1 p. 151 (July 30, 1890) , p. 158 (Aug. 11, 1890).
[55] *Ibid.*, p. 152 (Aug. 5, 1890).
[56] *Modesto Daily Evening News,* Aug 6, 1890.
[57] Memorandum of Agreement, Aug. 9, 1890.
[58] Board minutes, vol. 1, p. 21 (Sep. 16, 1887).
[59] Elwood Mead, *Irrigation Institutions,* (New York, 1910), p. viii.
[60] U.S. Dept. of Agric., Bulletin No. 100, p. 53.
[61] *Modesto Daily Evening News,* Aug. 26, Sep. 3, 1890; C. E. Grunsky to Helen Hohenthal, June 9, 1930 in Hohenthal, Supplementary Material.
[62] *Modesto Daily Evening News,* Sep. 19, 1890; Board minutes, vol. 1, p. 166 (Sept. 18, 1890).
[63] Board minutes, vol. 1, pp. 199-202 (June 2, 1891).
[64] *Stanislaus County Weekly News,* June 12, 1891.
[65] James D. Schuyler, *Reservoirs for Irrigation, Water Power and Domestic Water Supply,* (New York, 1901), p. 178.
[66] *Stanislaus County Weekly News,* June 26, 1891.
[67] *Ibid.*, July 3, 17, 31, 1891. [68] *Ibid.*, Oct. 30, Nov. 13, 1891.
[69] *Idem.* [70] Elias, *Stories of Stanislaus,* p. 65; Schuyler, p. 178.
[71] *Stanislaus County Weekly News,* Jan. 8, Feb. 12, 26, 1892.
[72] Board minutes, vol. 1, pp. 250-252 (Apr. 5, 1892).
[73] *Satnislaus County Weekly News,* Apr. 29, 1892.
[74] *Idem.* [75] *Ibid.*, May 6, 1892.
[76] Board minutes, vol. 1,p. 258 (May 16, 1892).
[77] *Stanislaus County Weekly News,* July 29, 1892.
[78] *Ibid.*, Sep. 16, 1892. [79] *Ibid.*, Oct. 14, 1892.
[80] *Ibid.*, Nov. 11, 1892. [81] *Ibid.*, Dec. 2, 1892. [82] Elias, p. 66.
[83] *Stanislaus County Weekly News,* June 2, 1893.
[84] *Ibid.*, Aug. 11, 1893. [85] *Ibid.*, Nov. 17, 1893; Elias, p. 66.
[86] *Stanislaus County Weekly News,* Dec. 8, 12, 1893.
[87] Elias, pp. 64-66; Schuyler, p. 176.
[88] *Modesto Daily Evening News,* Dec. 5, 13, 1893.

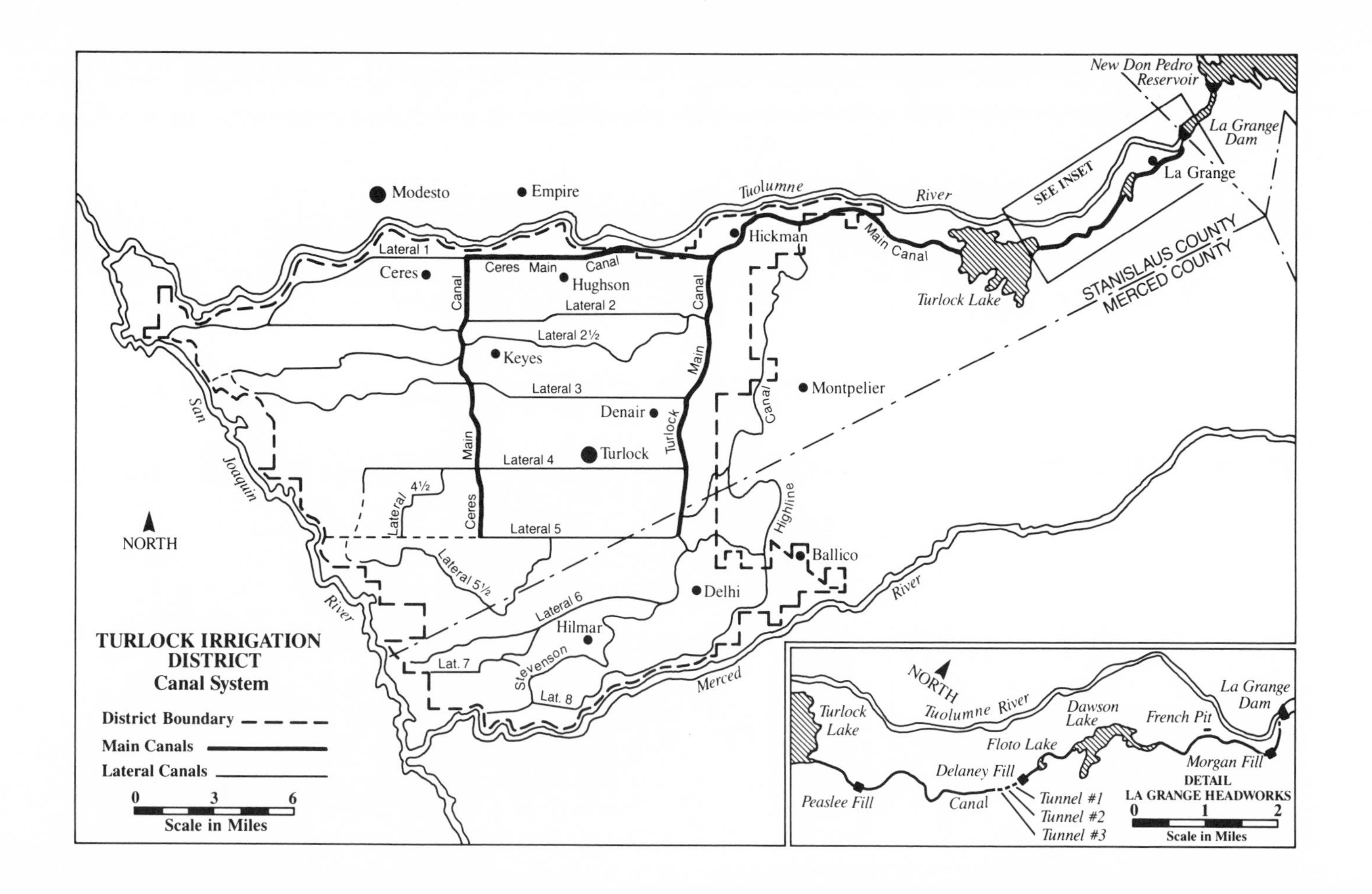
TURLOCK IRRIGATION DISTRICT
Canal System
District Boundary
Main Canals
Lateral Canals
0 3 6
Scale in Miles
NORTH
Modesto
Empire
Ceres
Hughson
Keyes
Denair
Turlock
Hickman
Montpelier
Ballico
Delhi
Hilmar
La Grange
La Grange Dam
New Don Pedro Reservoir
Turlock Lake
SEE INSET
STANISLAUS COUNTY
MERCED COUNTY
Tuolumne River
San Joaquin River
Merced River
Main Canal
Ceres Main Canal
Turlock Main Canal
Ceres Main Canal
Highline Canal
Lateral 1
Lateral 2
Lateral 2½
Lateral 3
Lateral 4
Lateral 4½
Lateral 5
Lateral 5½
Lateral 6
Lat. 7
Lat. 8
Stevenson
NORTH
Turlock Lake
Tuolumne River
Dawson Lake
French Pit
La Grange Dam
Floto Lake
Morgan Fill
Delaney Fill
Peaslee Fill
Canal
Tunnel #1
Tunnel #2
Tunnel #3
DETAIL
LA GRANGE HEADWORKS
0 1 2
Scale in Miles

4

The District in Distress 1894-1900

The completion of La Grange Dam was a significant accomplishment for the irrigation districts, but whatever prestige there was in having a half interest in the highest overflow dam in the country could not conceal the Turlock District's worsening financial situation. When the second issue of bonds was placed on the market in early 1893 the only bid received was from L.M. Hickman who once again agreed to take the bonds only for the purpose of reselling them to legitimate cash buyers.[1] No investors, however, were willing to pay the statutory minimum of 90 percent of face value that Hickman had bid, although a few bonds were peddled to staunch friends of the district or given in lieu of cash payments. For example, Hickman and John Mitchell, who had become a TID director in 1890, bought $9,000 in bonds themselves in April 1893 to enable the district to buy cement for the dam, and later bonds were exchanged directly for cement.[2] Directors and other officials also accepted bonds in payment of their salaries. Prospects for any improvement in the value of the bonds grew bleaker with the onset of the worst depression in the nation's history. What became known as the Panic of 1893 began in May when the stock market collapsed. By the end of the year, some five hundred banks had failed, and sixteen thousand businesses, including some of the leading railroads, had been forced into bankruptcy. Investment capital became virtually nonexistent even for relatively solid enterprises, and simply out of the question for a struggling irrigation district. The district was so short of cash that when the H.S. Crocker Company of San Francisco demanded payment for printing the bonds, the Secretary was ordered to "notify them there was no money in the Treasury but that they would sell them a bond by their paying the difference."[3] There is no indication of whether or not the printer agreed to accept a bond of dubious value and give the district its change in cash.

Despite the fact that their bonds were virtually unsalable, the board boldly planned to finish the canal system. In March 1894 directors Cogswell and Dunn were ordered to ascertain what work was still needed and a month later the board adopted engineer R.H. Goodwin's report on the subject and advertised for bids on a long list of jobs.[4] At the head of the canal, designated as Section 0, virtually nothing had been done. The steep cliff on Turlock's side of the dam made it impractical to connect a canal or flume directly to the face of the dam. The alternative was a six hundred foot tunnel through the bluff to a point where a flume could be built along the hillside. Bids on the tunnel had first been requested in 1892 but when the only one received was too high, the district undertook the job itself. Only forty feet had been finished when work stopped to avoid interferring with the construction of the dam.[5] Besides the tunnel, seventeen hundred feet of flume were needed in Section 0 to connect the dam and tunnel with the rest of the canal. In Sections 1 and 2, excavation of the canal had already been completed but remaining work included a dam at Dawson Lake, lining for the three tunnels, wooden flumes at Morgan Gulch, Delaney Gluch and Peaslee Creek, and a flume or another dam at Snake Ravine. In addition, debris that had accumulated in the three years since any work had been done needed to be cleaned out, bridges had to be built and drops installed. (To maintain a consistent grade and velocity of water in a canal occasional changes in level, resembling a series of stair steps, are needed. Where levels change, structures resembling small dams, known as drops, are used to control the flow.) No work had yet been done on the plains where the main canal would have to be extended southward and a series of lateral canals built west from it.[6]

By May 3, 1894, the time set for opening bids on the completion of the canal system, only one bid had been received. Doe, Hunt and Company of San Francisco offered to undertake the entire job, including the construction of several laterals for an estimated $379,704. The board met again the following day to carefully review the company's detailed proposal. Doe, Hunt made some unspecified concessions and the board then voted three-to-one to award the contract, with McPherson opposed and Roger Williams absent.[7] Although absent at the time of the vote, Williams was publically opposed to what he considered an exhorbitant contract. He claimed that forty-five cents in hard money would do what Doe, Hunt promised to do for a dollar, which if nothing else was a clue that the contractor was not going to be paid in hard money. Even the contract's supporters, like Director E.V. Cogswell, readily admitted that the price was much too high but thought it was the only way to finish the canal.[8] The document was signed on May 9, 1894, and a district survey crew was soon ordered to begin laying out the laterals that would carry water to the land.[9]

By June, subcontractors were at work on the canal. Tunnelling was resumed at La Grange Dam, and J.D. McDougald put 175 horses and 60 men to work with scrapers.[10] Lumber for the flumes was ordered and while delivery was delayed temporarily by a nationwide railroad strike, it began arriving in early August. The job of putting the canal in shape to carry water had hardly begun, though, when a lack of money brought construction to a standstill. On August 7, Doe, Hunt presented the TID board with a bill for work already done. The board approved the claims but made them "subject to payment only after present indebtedness and preferred claims of about $50,000 are paid."[11] Since the district's treasury was as bare as Mother Hubbard's cupboard that provision effectively defered payment to Doe, Hunt indefinitely. The workmen expected to be paid on August 15 and when no money came they laid down their tools. TID engineer R.H. Goodwin had also had his claim for payment denied, prompting him to quit work too and take his maps and plans with him.[12] On September 4, Mr. Hunt came before the board to demand payment and this time, faced with a work stoppage, the board complied, issuing about $19,000 worth of warrants payable to the contractor.[13] Warrants, like checks, are of little value unless they can be redeemed for cash. The real worth of the district's payment can be surmised from the fact that McDougald found money, mostly his own, to pay and discharge everyone but his manager and foreman. He blamed the turn of events on "bond buyers failing to keep their agreement which necessitated Doe, Hunt and Company to be short of the money needed when pay day came around."[14] How the bonds were to have been marketed was never explained. Perhaps arrangements had been made to transfer the bonds to Doe, Hunt for resale, but whether that or some other means of selling them had been attempted the dismal result was just further proof that the district's securities were thought to be a risky and unprofitable investment.

In October, Doe, Hunt wrote the board explaining that they were "compelled to stop work on our contract... because of the inability of the district to pay the warrants issued us for work already done."[15] The letter went on to confirm the company's commitment to resume work again, if and when the district had money. The board, however, responded by cancelling the contract on grounds that Doe, Hunt and Company had abandoned the work for an unreasonable time.[16] Since the district's inability to render payment was to blame for the alleged abandonment, the board's action might seem hypocritical. In fact, it was a coldly calculated move, for by claiming that Doe, Hunt had defaulted on their contract the district was entitled to withhold payments due them, effectively cancelling any debt owed to the company. With bills piling up and no real hope of raising money to pay them, the directors were apparently trying to avoid any obligation that they could.

Trouble brewing within the district added to its difficulties. Like most of the Wright Act districts, the Turlock Irrigation District had a number of anti-irrigationists – men opposed to irrigation in general or irrigation districts in particular – among its residents. Some districts, like the Central Irrigation District in the Sacramento Valley, were eventually destroyed by the hostility of the dominant landowners, and even the Modesto District faced a determined legal challenge and later fell under the control of its enemies.[17] The Turlock District was spared such an active opposition until 1893. In that year, A. F. Underwood, the owner of 676 acres a few miles southwest of Keyes, filed suit against district collector H.A. Osborn to prevent the district from selling property taken for delinquent taxes. He argued that the district's bonds were illegal and that its taxes, used to pay interest on the bonds, were likewise illegal.[18] The local court issued an injunction against the district, but dissolved it a few months later, ordering that the taxes would have to be paid. Undaunted by this setback, a group of thrity-five to forty farmers from the Turlock District gathered in Ceres in November 1893 to form an anti-irrigation organization. J.S. Muncy chaired the meeting, which named Underwood, Hiram Hughson and S.E. Foster a committee to write by-laws. The announced purpose of the group, like that of Underwood's lawsuit, was to test the legality of the bonds and to restrain the collector from collecting taxes.[19] They acted on their intentions in February 1894 when Hiram Hughson and eleven other plaintiffs sued Acting Collector John M. Crane to restrain the sale of delinquent property. Taxes owed by the plaintiffs amounted to about $3,300 out of a total levy of $40,855. Superior Court Judge Minor promptly ruled that the taxes were valid and must be paid, even if under protest, but the decision was appealed to the Supreme Court.[20]

The Panic of 1893 and the prolonged depression that followed it worsened the already declining fortunes of the area's wheat growers, and added to the strength of the anti-irrigation movement. In 1893 and 1894 the price of wheat plunged to levels described as "ruinous" and recovered only moderately before falling again at the end of the decade.[21] In what was becoming an annual event, 1895 brought another lawsuit, this one to enjoin the sale of property delinquent from the 1894 district tax of $1.25 per hundred dollars of assessed valuation. The long list of plaintiffs in the case held some unpleasant surprises. The names of persistent anti-irrigationists like Hiram Hughson, John Fox, I.W. Updike, A.F. Underwood and S.E. Foster naturally appeared on the complaint, but so did the names of TID director N.L. Tomlinson, former director W.L. Fulkerth, former collector A.N. Crow and long time district treasurer C.N. Whitmore. Other plaintiffs were among those who had signed the 1887 petition calling for the formation of the Turlock Irrigation District.[22] Had all of these irrigationists gone over to the opposition? Some, like A.N. Crow did become active anti-irrigationisits, indicating that sentiments could indeed change. Others, including Whitmore, continued to participate in district affairs and were still apparently interested in its survival and success. There is no real evidence concerning Director Tomlinson's sympathies, but the usually reliable Sol Elias reported that none of the Turlock directors were ever part of the anti-irrigation movement.[23] The participation of some of the district's supporters in the suit to effectively stop its tax collection machinery appears to have been a matter of economic necessity. In 1895, after a second year of disasterous wheat prices, many farmers were on the brink of bankruptcy. During the decade of the 1890s half the ranches in the Turlock area were mortgaged and a quarter of them foreclosed.[24] Hard-pressed landowners often tried to do just what the district itself did when it cancelled the Doe, Hunt contract on somewhat specious grounds – escape any obligation they could. Delinquencies became commonplace in the middle and late 1890s, including at one time or another most of the large landowners in the district. Although only a few of the delinquents joined in the annual lawsuits, injunctions preventing the sale of their property and a general uncertainty over the outcome of the legal issues involved, effectively prevented the taking and sale of any delinquent property. Most of the taxes were eventually paid, though in some cases not until after the turn-of-the-century.[25]

In the meantime, the district became unable to meet its most basic obligations. In July 1895 it failed for the first time to pay the interest due on its bonds.[26]

The rising tide of delinquencies and injunction suits and the resulting default on bond interest payments naturally tended to further reduce the value of the district's bonds and make them harder to sell. However, the Turlock District's own misfortunes were not entirely to blame for its inability to market its bonds. Virtually all of the districts organized under the Wright Act were in trouble. Some fell victim to internal opposition, others were plagued by poor and overly optimistic planning, usually in terms of the available water supply, and a few, especially in southern California, were merely the tools of speculators. The districts were entangled in so many lawsuits that one observer suggested that the Wright Irrigation Act should have been called the Wright Litigation Act.[27] Far from settling all the issues that might arise, the case of *Turlock Irrigation District v. Williams* became only the first of many irrigation district cases. The Wright Act's opponents, many of whom were influential wheat barons, land speculators or cattlemen, filed suit after suit challenging every detail of the law and the manner in which individual districts were run. There was, in short, nothing to encourage investment in any irrigation district's bonds and a great deal to frighten away potential buyers.

The situation worsened further when a case challenging the constitutionality of the Wright Act reached the U.S. Supreme Court. It originated in the Modesto Irrigation District, where William Tregea had filed suit in local court concerning the inclusion of the city of Modesto in the district, the later exclusion of certain other lands and the validity of the district's bonds. The case ultimately came before the California Supreme Court, which ruled in favor of the district. In this and other instances, the California courts consistently upheld the propriety and constitutionality of the Wright Act. Since they could not overturn the act in California, its opponents decided to carry the fight to the federal courts. Claiming to find grounds for federal jurisdiction in the *Tregea* case, they brought it to the U.S. Supreme Court. If the highest court in the land found that the irrigation district act violated the United States Constitution, the act would be thrown out, the districts organized under it dissolved and their bonds very likely rendered completely worthless. The Supreme Court heard the case in early 1894, but instead of announcing a decision the Court returned the matter to its docket for reargument. The possibility that the Court was divided in its opinion and might ultimately rule against the Wright Act weighed heavily on the minds of potential bond buyers. In fact, there were reports that some speculators welcomed the delay and uncertainty as a means of forcing bond prices even lower so they could buy cheaply in anticipation of a rise in value when the case was decided, as many thought it would be, in favor of the Wright Act.[28]

The incessant barrage of lawsuits, injunctions and delinquencies that engulfed the TID cast a pall of uncertainty over the Turlock area as a whole, just as similar events growing out of the Wright Act helped paralyze other parts of the Central Valley. Together with low wheat prices and worn-out soil, the travails of the district drove the area deeper into a cycle of debt and depression. Each acre carried a portion of a bonded debt that would, without irrigation, be staggering to repay, and many farmers had lost clear title to their land through delinquency. Early in the twentieth century, Frank Adams recalled the trials of the terrible nineties.

> Land titles were so clouded that sales practically ceased, and thousands of acres of land that had once been considered worth $30 to $40 an acre were carried on the county and district assessment rolls at one-third of these values or less, with no sales at even these figures. Not only could no land be sold at figures approaching its value, but land that had begun to fail before the passage of the district law in 1887 was now even more difficult to cultivate profitably.[29]

From 1893 to 1900 the assessed value of property in the Turlock Irrigation District dropped lower every year until it was not much more than half the value it had been seven years earlier.[30] Not only did property values decline but Stanislaus County even lost population during the decade after 1890 as people fled the deteriorating economy. When fires gutted much of downtown Turlock in 1893 and 1984, large sections were not immediately rebuilt.The charred ruins bore witness to the impact of hard times, and the little town's prospects were so poor that in 1900 the Southern Pacific considered closing its Turlock depot. There is no evidence that the railroad ever carried out the threat, but even the possibility that the depot could have been lost was proof that the grain era had forever lost its vitality.[31] By 1895, the outlook for the Turlock Irrigation District appeared bleak and was growing bleaker.

In a turn of events that may have seemed just short of miraculous, gloom was suddenly replaced by guarded optimism and a hope that the Turlock canal system would soon be completed after all. On May 7, 1895, James A. Waymire, an Oakland attorney, former judge and an investor in irrigation district securities, appeared before the board and "several propositions were advanced" in regard to the Doe, Hunt contract. Although no action was taken, the "propositions" were sufficiently interesting to prompt another meeting on May 11 at the office of the board's attorney, P. J. Hazen.[32] In early June the pattern was repeated, with Waymire joining the board at its regular monthly meeting and again a few days later at Hazen's office.[33] By then it was clear that Waymire was in effect negotiating to take over the Doe, Hunt contract, complete the works and arrange for the sale of bonds and the settlement of the district's debts. Waymire had taken an early interest in irrigation district bonds and had been instrumental in negotiating the first sale of TID bonds in 1888.[34] The extent of his holdings in the district's bonds in 1895 is unknown, but he probably already had some reason to be interested in the district's success. However, negotiating with the Turlock board was no easy task. For the same bonds, the directors wanted more than the Doe, Hunt contract had offered, including one hundred miles of additional laterals and the assurance that the district would have operating money over and above the amount earmarked for construction. Waymire initially balked at some of these new demands, but in the end he gave his consent.[35] On July 2, 1895, the district, Doe, Hunt and Company and Waymire entered into an agreement, with the company assigning its contract to Waymire.

In essence, the agreement called for the completion of the canal system, including laterals, for a sum not to exceed $382,000 in gold coin. The contractor, Waymire, was made responsible for finding a sale for the district's bonds to pay for his work and to provide $60,000 in cash to the district for other purposes. The contract also provided that, "The failure to pay the money due or to become due to the Contractor by reason of inability to obtain such moneys from the proceeds of the sale of said bonds . . . shall not excuse in any way the Contractor from fully completing said works."[36] In other words, Waymire would have to accept whatever the bonds would bring as payment. To earn the maximum $382,000 and provide the $60,000 in cash required by the contract, he would have had to sell all the unsold bonds, $492,000 worth, for about 90

James A. Waymire, an Oakland attorney,
and investor in Turlock Irrigation District bonds.

percent of their face value. The market value of the securities was much lower, at times down to half the face value. The Doe, Hunt contract had been labelled exhorbitant when it was first considered, but that now meant that Waymire could afford to sell the bonds at a greater discount and still have enough money to pay for the work. Any increase in the value of the bonds and more money would flow into the treasury and thus to the contractor; any further drop in value and Waymire would face ruin. It was, on Waymire's part, a fantastic gamble on the future value of the district's bonds. It was also a gamble he repeated at about the same time in the Poso Irrigation District of Kern County. The directors of that district traded their unsold bonds to him for the completion of their dam and canals.[37] Even before his agreement with the Turlock District had been finalized, Waymire was negotiating the prospective sale of the district's bonds in New York and Chicago. Concern over the *Tregea* case pending before the Supreme Court made finding buyers harder, but an opinion by a former federal judge that the outcome would be favorable induced the bankers to take the Turlock and Poso bonds. So confident was Waymire that his arrangements would prove profitable that he made the same offer of construction in exchange for bonds to the Modesto Irrigation District, but that district apparently showed no interest in the scheme.[38]

Waymire's crews began work within weeks of the agreement, beginning at the uncompleted tunnel at the dam.[39] But, like the Perils of Pauline, any good news was followed by a fresh onslaught of misfortune. This time the trouble arose far from Turlock. Maria Bradley, a delinquent landowner in the small Fallbrook Irrigation District in San Diego County, had filed suit in federal court arguing that the district's tax assesments were unconstitutional. Judge Erskine Ross agreed, ruling that irrigation under the Wright Act benefitted only some landowners while taxing all and that the act authorized the taking of private property through taxes without due process of law. In other words, Judge Ross had declared the Wright Act to be in violation of the U.S. Constitution.[40] Waymire quickly professed his belief that Judge Ross had erred and that the decision would eventually be overturned but that did nothing to diminish the impact of the ruling. Bond merchants had been nervous enough over the anticipated *Tregea* decision, now Judge Ross's action added to their unease and put a halt to bond sales until the whole constitutional issue could be disposed of. Waymire used what little money he could raise to continue tunnelling at La Grange. In 1895, he mortgaged his Oakland home for nearly $16,000 to support his irrigation activities and in March 1896 he bought $16,500 worth of TID bonds himself to pay for work done on the tunnel and to fulfill his obligation to make periodic cash payments to the district. At the same time, it was announced that no further work would be done until the Supreme Court rendered its decision in the *Fallbrook* and *Tregea* cases.[41] The usual injunction suits were filed against the Modesto and Turlock collectors in early 1896, but even before an injunction could be issued the collectors voluntarily suspended the sale of delinquent property pending a decision by the Supreme Court.[42] That action by district officials symbolized the impact of the *Fallbrook* case; activity in irrigation districts throughout the state ground to a halt. No one wanted to pay taxes, buy bonds or build canals until the fate of the Wright Act was finally decided in Washington.

The holders of California irrigation district bonds, who stood to lose their investment if the Wright Act was thrown out by the Supreme Court, put together a fund to pay for the defense of the act and entrusted the job of selecting the attorneys to James A. Waymire. He hired two noted jurists, former federal circuit judge John M. Dillon and former California chief justice A. H. Rhodes. Not long after, Waymire chanced to talk with Supreme Court Justice Stephen J. Field, who asked him in a friendly fashion who he had chosen to present the case to the Court. When Waymire told him, Field agreed that Dillon and Rhodes were highly competent but he advised Waymire to get Joseph H. Choate, an eminent lawyer who Field said the justices liked to listen to. Waymire telegraphed Choate, but the reply was disquieting. Choate told Waymire to consult George H. Maxwell, the anti-Wright Act activist who had been involved in the *Tregea* case and was now in charge of the case against the Fallbrook district. Maxwell confirmed that Choate had already been retained for a $10,000 fee by the opponents of the Wright Act.[43] As Waymire recalled:

> It seemed highly important to find a match for the eminent New Yorker. Finally I thought of Honorable Benjamin Harrison, ex-President of the United States. He was the equal of any man as a lawyer, his personality would certainly be as interesting and impressive as that of Mr. Choate and the fact that he had appointed three of the Judges would do no harm. Fortunately I had a personal acquaintance with the General. A letter explaining the nature of the case and offering a retainer met with a favorable response.[44]

The ex-President's fee matched that of his opponent: $10,000.[45] The *Tregea* and *Fallbrook* cases were argued before the Court in early 1896 but to the disappointment of nearly everyone, the Court adjourned for the summer without having announced its decision. At long last, on November 16, 1896, Justice Peckam read the Court's decision in the *Fallbrook* case. Judge Ross's verdict was overturned and the Wright Act was declared constitutional. The *Tregea* case was simply dismissed without a decision.[46]

The reaction in Stanislaus County was described as, "one of relief that the long delayed case had at last been decided, though those in favor or opposed to the law expressed themselves very strongly upon the matter."[47] No cannons boomed, no hats were flung in the air; in fact, the feeling of relief was subdued compared to the boisterous display that greeted news that the law had been upheld by the state Supreme Court in 1888. Good news had failed to insure success before and optimism was in short supply in the depression-racked districts. Only in the little town of La Grange did word of the decision set off a full scale celebration that climaxed in a champagne banquet several days later.[48] The citizens of La Grange had good reason to rejoice, for it was expected the decision would mean the return of a large labor force to the nearby TID canal.

Canal work did resume shortly, but not without difficulties. Sol Elias eloquently summarized Waymire's efforts this way:

> Judge Waymire began the construction of the canals and laterals with the utmost faith in his ability to complete the system in the time specified. Injunctions, internal differences, inability to procure purchasers for the bonds, strikes on the works due to failure to meet obligations, lack of general cooperation, inability to secure supplies, insufficient financing and a host of other handicaps delayed the completion. The residents of the Turlock District will never know the difficulties that Waymire encountered. The inside story of Waymire's operations, the expedients to which he resorted, the indomitable efforts he made, and the almost superhuman faith in himself and the final outcome – these would constitute a story that would fill many chapters and would read stranger than fiction.[49]

Elias hardly overstated the problems Waymire faced. Certainly the bonds did not sell well or for as much as they should have. The unabated opposition of the anti-irrigationists, organized in 1897 into the Tax Payers Defense Association of Modesto and Turlock Irrigation Districts, kept the district enmeshed in injunction suits and other actions. Even with the law declared constitutional, local harassment and the poor reputation of irrigation districts in general kept bond buyers wary and prices low. The manner in which any of the bonds were finally sold for cash remains one of the mysteries Elias alluded to. The bonds had originally been sold, at least technically, to L.M. Hickman, but following a California Supreme Court decision that ruled the alleged sale illegal, the remaining bonds were readvertised and sold, for 90 percent of face value, to R. A. Friedrich of San Francisco.[50] Like Hickman, Friedrich appears to have only held the bonds until Waymire arranged for their sale to bona fide investors. Presumably, the bonds were sold for the going price, which was well below ninety cents on the dollar, even though Friedrich was supposedly obligated to pay the district his full bid price for the bonds. There is nothing to indicate how this sleight of hand was accomplished although Waymire himself may have been the "purchaser," taking the bonds to pay for his work and then selling them for whatever they would bring.

Sol Elias said that Waymire's story "would fill many chapters and would read stranger than fiction." Unfortunately, the "chapters" went largely unrecorded save for a few anecdotes. In one celebrated incident, the ex-judge had managed to raise $2,000 in San Francisco to pay the workers on the canal, but the sum was in two large checks which he did not have an opportunity to cash before leaving the city. Arriving in Modesto on a Saturday evening, he found the bank vaults closed until Monday by time locks. Since his men expected their pay sooner than that, Waymire asked Judge A. Hewel for help. He turned the checks over to Hewel and the two men went all over town cashing checks on the judge's personal account until they had the coin needed. In another instance, C. N. Whitmore, the district's treasurer and one of its largest landowners, was in such financial distress that his property was in the hands of a bank and he owed $3,000 in irrigation taxes. Waymire meanwhile was without money to pay his men and a strike threatened to stop work completely. He appealed to Whitmore, and the two of them visited the bank, which agreed to advance Whitmore enough to pay his taxes. Taking possession of the money immediately, Waymire bought a pistol and headed for the construction camp.[51] Variations on the same theme may have occurred over and over again as Waymire sought ways to finance the project. However it may have been accomplished, it was never much more than a hand-to-mouth proposition that taxed Waymire's ingenuity and his purse.

Waymire resumed construction in January 1897 with a small crew at the tunnel by La Grange Dam.[52] It was no more than a token effort, however, intended merely to maintain the contract while he arranged additional financing. In June, Waymire, who had been in the East in an unsuccessful attempt to get a position in the new McKinley administration, announced that he had sold some bonds in Chicago.[53] Soon after, subcontractors and their crews began arriving, along with the wagon loads of lumber needed for the flumes. At the end of 1897, 160 men were at work, building flumes, lining tunnels and putting the earthworks in shape to carry water. The tunnel at the dam was virtually finished and the control gate was installed.[54] As sections of the canal were finished they were puddled with small amounts of water to settle the fills and loose dirt and check for squirrel holes or fissures in the rock cuts. Some of the water used for puddling came from Dawson Lake, where a creek had been dammed to form part of the canal, and when that was exhausted the head-gate on the tunnel was opened to admit the first Tuolumne River water to the canal.

By mid-June 1898 all the high flumes were in place and the canal was thought to be ready to carry the water down to Hickman, on the eastern edge of the plains. The process was expected to take some time since only a small stream would be used and large areas behind dams at Snake Ravine and Dawson Lake would have to be filled before water could be released further down the canal.[55] At Snake Ravine, where the dam had washed out in 1890, another dam had been built using essentially the same methods but with greater care in construction. When water first reached the ravine in July 1898 it eroded a hillside next to the dam, causing a minor leak. For two days, the canal carried water eighteen inches deep and the dam held, but when three inches more were added on July 13, the dam failed again. The whole mass of earth and rock slid down the ravine at a rate of six to ten feet per second for a thousand feet into the Tuolumne River. The subcontractor responsible for the dam and two dogs were on top of it when it collapsed and they had a wild ride down to the river. All escaped unhurt.[56] The second dam failure at the site meant that a fifteen hundred foot flume would have to be built around the ravine before any more work could be done on the canal.

The Snake Ravine flume was finished in September, and water was once again turned into the canal. Within a few days another washout occurred, this time on the Turpin place about five miles

The Turlock Irrigation District's tallest wooden flume
carried the main canal over Morgan Gulch near La Grange.

east of Hickman, where another dam, much smaller than the one at Snake Ravine, had collapsed.[57] While repairs were underway, a small ditch was dug around the break so puddling could continue below. By late October 1898 the break had been repaired and water was once again turned into the canal from the dam. The water reached Hickman for the first time at dawn on November 12, 1898. Water from the Tuolumne was now flowing through the entire length of the main canal; through tunnels, cuts and embankments, along creek beds near La Grange and over the tall, trestle-like flumes. At the end of the canal about a mile southwest of Hickman, it was allowed to flood out across the open ground.[58]

With the completion of the twenty-three mile main canal from La Grange Dam to Hickman the most difficult part of the canal system was completed. At division gates about a mile below Hickman, the main canal would split into branches. One, known as the Turlock Main Canal, would run south, bending a little this way or that to follow the contours of the land, toward the Merced River. The route along the edge of the foothills kept it higher than the plains that sloped away toward the San Joaquin River. Smaller lateral canals would leave it at intervals to carry water westward into the district. When they were completed, irrigators would connect their ditches to these laterals to complete the distribution system. The other branch of the main canal, the Ceres Main, would run along the district's northern edge not far from the Tuolumne River to a point northeast of Ceres, where it would turn south to replenish the laterals it crossed. The main canals on the plains were to be seventy feet wide at the bottom and one hundred feet wide on top. As soon as the main canal above Hickman was tested, work began in earnest on the main canal branches and laterals. Subcontractor R.W. Morgan put four graders to work south of Hickman in December, 1898. The graders shaped the sides of the canal and left the dirt in the center to be removed later by scrapers. L.M. Hickman put his extra stock, numbering about a hundred head, to work cleaning out the remaining earth with two plows and twenty scrapers. The graders on the Turlock Main reached Elmdale (Denair) at the beginning of 1899, and by February, 4 graders, 25 scrapers, 230 horses and 85 men were laboring on the canal, moving over 5,000 cubic yards of earth per day. At the same time, thirty-five men were patching weak spots in the main canal in the foothills.[59]

The anti-irrigationist Defense Association, still trying to impede what it had had no success in stopping, urged its members to refuse the district a right-of-way for any canals that would cross their land. In only two places, however, did anti-irrigationists actually delay construction; along Lateral 3 northwest of Denair, and northeast of Hughson where Hiram Hughson and John Fox tried to prevent the excavation of the Ceres Main. In each case, the district filed condemnation actions to get possession of the property needed for the canal. The continuing difficulty of raising money to pay for construction was a more pressing problem than the anti-irrigationists. To avoid spending hard-to-come-by cash, Waymire offered to use teams belonging to local landowners, applying the money they earned to their owners' delinquent taxes.[60] Soon after, the force of hired labor was replaced almost completely by district residents under the direction of T. J. Harp, and the press reported "a scramble for work in excess of demand."[61] By May, work was going on all over the system. Water up to four feet deep was running from La Grange Dam to Hickman, where a wasteway dumped it back into the river. The Turlock Main Canal had reached the Merced River and work on the Ceres Main was proceeding except across the Fox and Hughson lands, where condemnation suits were still pending. Lateral 2, which ran south of Ceres, had been finished, and so had Lateral 3, except for the section being condemned.[62] By the fall of 1899 the way had been cleared for the completion of the Ceres Main. Another contractor, E. M. Roberts, was hired by Waymire at the end of the year and quickly finished the work near Ceres and then moved the construction camp of two hundred horses and sixty men to the Hickman property to finish all that remained to be done east of the railroad.[63]

The year 1900 began on an ominous note when Roberts' men, on the job only a month, quit because they had not been paid.[64] Waymire found enough money to get them back to work, but in early March the carpenters walked off after two months without pay. As usual, the situation was blamed on the low value of TID bonds and the difficulty of finding buyers even at reduced prices. The work stoppage lasted almost two weeks until money finally materialized to pay the men. Meanwhile, work on the main canals and lateral continued. South and west of Turlock, the laterals were built by H. S. Crane and George Bloss, who were among those controlling the vast Mitchell estate that covered much of the area.[65]

By February, a large head of water (The term "head" refers to a flow of water) was turned into the Ceres Main to test the canal. When the water in the lateral near Ceres reached the vicinity of the Southern Pacific tracks, it attracted throngs of curious observers. The holiday mood that marked the arrival of water in the canal was captured by a little parody of Robert Southey's poem "How Does the Water Come Down to Ladore?"

> 'Twas the greatest event of the season,
> The ladies were out in their gala attire,
> The men they had out every rig they could hire,
> To outdo each other they all did aspire.
> Every young fellow had his best Sunday girl,
> Every dear lass had her hair done in a curl,
> Every thing went with a bustle and a whirl —
> When the water came down the canal.[66]

Finally, in the middle of March 1900 water from the Tuolumne River was at last diverted from the canals onto the land. The *News* reported:

> The water of the Turlock canal is now being used on a piece of land one mile north of Ceres, Henry Stirring having taken it out of lateral 0 to irrigate a piece of land for corn. He had quite a large head running on Tuesday, Wednesday and Thursday of last week. Many Modestoans have driven out to see the sight.
>
> Mr. Stirring is entitled to what honor there is in being the first to use the Turlock water.[67]

Others were reportedly getting their land ready to irrigate and even planting alfalfa, a forage crop that would not grow without irrigation. In June, the water came down the lateral to Turlock, and by the end of the year the entire system of laterals had been finished.[68] No records were kept of who irrigated or how much land was watered in 1900. Certainly, the totals were not large. Those who did use the water found it was subject to interuption. In June, for example, the main canal washed out near Peaslee Creek, and no sooner was the damage repaired than another break occurred at the Hickman drop.[69] Despite such occasional inconveniences, though, the canal system worked and irrigation was finally a reality on the plains south of the Tuolumne.

Elwood Mead, one of the country's best known irrigation experts, once called the Wright Act "a disgrace to any self-governing people."[70] Mead was far from alone in his scornful appraisal of the West's first widely used irrigation district law, and with good reason. Only eight of the forty-nine districts formed under the act survived into the twentieth century. Some of the failed districts had been speculative in nature, meant to give an illusion of irrigation while their promoters reaped quick profits from land sales. Other districts were set up to operate or improve existing ditches, and these too offered an opportunity for a form of abuse as the owners of canals or water rights encouraged the formation of districts to buy their properties at a handsome profit. Twenty-seven of the districts were intended to do what the law envisioned – enable farmers to finance and build irrigation systems. Of these only three succeeded – the Turlock and Modesto districts and the small Browns Valley Irrigation District on the Yuba River. Some of the other districts discovered that their proposed water supply was inadequate or unaffordable, while others drowned in a flood of lawsuits brought by landowners who wanted nothing to do with an irrigation district or its taxes. The law itself was in some ways to blame for these misfortunes. The districts were strictly autonomous, local institutions, entirely without engineering or financial supervision by the state, and as a result many were allowed to form and issue bonds in the face of circumstances that made their success doubtful. Indeed, after William Hammond Hall's resignation in 1889, California had no state engineer to whom the districts could have turned for advice had they been so inclined. As it was, the law provided too few safeguards against poor planning and over-optimism by inexperienced directors, or against opportunists trying to derive private profits from a public institution. In 1897, the legislature repealed the Wright Act and put in its place another, more restrictive law.[71]

The revised irrigation district law of 1897, known variously as the Bridgeford Act or the Wright-Bridgeford Act, did not tamper with the central concepts of the Wright Act, but tried to implement them in a way that would avoid its acrimonious results. For one thing, it appeared that the Wright Act had made it too easy to organize an irrigation district. Its requirement of fifty signatures on the freeholders petition had resulted in the formation of districts opposed by the landowners who owned most of the land and would have to pay most of the taxes. Such districts produced little more than litigation, uncertainty and ruin. The 1897 law stipulated that the organization petition had to carry the names of a majority of property owners, representing a majority in value of real property in the proposed district. That made it harder to set up new districts, but in so doing attempted to insure that once established they would have sufficient support from their taxpayers. Even that provision, however, would not prevent troublesome opposition from a minority of landowners. The new law therefore provided a mechanism for legal challenges both before final action by the Board of Supervisors and after the organization election. Such procedures would hopefully put to rest questions of the district's legitimacy before it was ready to go into operation. Finally, instead of twenty year, 6 percent bonds, the revised statute authorized thirty year, 5 percent bonds. No new irrigation districts were established for over ten years after the passage of the 1897 law, and it was not until 1915, when financing had been made easier, that California saw a rebirth of enthusiasm for irrigation districts.[72]

The generally dismal results of the Wright Act, which nonetheless served at least as a conceptual model for irrigation district statutes across the West, make it all the more important to understand how the Turlock District survived its stormy infancy. the district had substantial advantages from the start. One somewhat intangible one was the fact that C. C. Wright probably had his own locality foremost in his mind when he wrote his legislation, so it suited no place better. The Tuolumne River itself was a valuable asset. There was no question of its sufficiency as a water supply, and the near absence of competing water right holders made possible the simple and far-reaching division of the river. The annual injunction suits and other forms of guerilla warfare waged against the Turlock District had an undeniable effect, but compared to other districts, the antagonists were rather weak. In some districts, the large landowners were so united in opposition that the district could not survive, but in the Turlock District the three largest owners – Mitchell, Hickman and Whitmore – were active supporters, and, except for Hiram Hughson, the most ardent anti-irrigationists were not major property owners. To a large extent, then, the Turlock Irrigation District suffered more from external than internal problems. An atmosphere of opposition and uncertainty engulfed the Wright Act and all the districts founded under it. Constant legal challenges plagued the districts and at times threatened to overturn the law, which made it difficult for any of the districts to function or finance their activities. The national Panic of 1893 and the moribund California wheat industry added to the problem by drying up credit and forcing even some of the friends of irrigation to let their taxes become delinquent. That these circumstances did not destroy the Turlock District was due in part to the advantages of local support and water supply cited above, to conservative management by the board, and to a measure of luck. The luck was personified by James A. Waymire, without whom the canal would almost certainly not have been finished when it was. Waymire is something of an enigma; widely revered as the district's saviour, the ex-judge was also a speculator who could have reaped a large profit had events gone more favorably. Whatever his motives, Waymire did persevere, and in 1900 the Turlock Irrigation District stood virtually alone as an example of what the district system could achieve. But although the water had arrived, it remained to be seen if irrigation would bring the benefits its advocates had long promised.

CHAPTER 4 – FOOTNOTES

[1] Board minutes, vol. 1. pp. 289-290 (Feb. 7, 1893).
[2] *Stanislaus County Weekly News,* Apr. 7, 1893; Board minutes, vol. 1, p. 315 (Dec. 5, 1893).
[3] Board minutes, vol. 1, p. 324 (Apr. 3, 1894).
[4] *Ibid.,* p. 322 (Mar. 6, 1894), p. 325 (Apr. 4, 1894), pp. 327-329 (May 3, 1894).
[5] *Ibid.,* p. 264 (July 5, 1892), p. 266 (July 14, 1892), p. 270 (Sept. 6, 1892); *Stanislaus County Weekly News,* Aug. 26, 1892.
[6] *Stanislaus County Weekly News,* Apr. 13, 1894.
[7] Board minutes, vol. 1, p. 334 (May 4, 1894); *Stanislaus County Weekly News,* May 11, 1894.
[8] *Stanislaus County Weekly News,* May 11, 1894.
[9] *Ibid.*, June 8, 1894.
[10] *Ibid.*, June 29, July 23, 1894.
[11] Board minutes, vol. 1, p. 341 (Aug. 7, 1894).
[12] *Stanislaus County Weekly News,* Aug. 24,1894.
[13] Board minutes, vol. 1, p. 343 (Sept. 4, 1894).
[14] *Stanislaus County Weekly News,* Sept. 7, 1894.
[15] Board minutes, vol. 1, p. 350 (Oct. 17, 1894).
[16] *Idem.*
[17] See Calif. Dept. of Engineering, Bulletin No. 2, *Irrigation Districts in California, 1887-1915,* by Frank Adams (1916), p. 20 [Hereafter cited as "Adams, 1916"] for general information; and Rhodes, "Thirsty Land,", pp. 85-101 for details on the Modesto District.
[18] *Stanislaus County Weekly News,* Mar. 3, 1893.
[19] *Ibid.,* Nov. 10, 1893.
[20] *Ibid.,* Feb. 23, May 4, 1894.
[21] U. S. Dept. of Agric., Office of Experiment Stations, *Annual Report of Irrigation and Drainage Investigations, 1904,* "The Distribution and Use of Water in Modesto and Turlock Irrigation Districts, California," by Frank Adams, Separate No. 3 (1905), p. 98. [Hereafter cited as "Adams, 1904"]
[22] *Stanislaus County Weekly News,* Feb. 27, 1895.
[23] Elias, p. 145.

[24] R. L. Adams and W. W. Bedford, *The Marvel of Irrigation: A Record of a Quarter Century in the Turlock and Modesto Districts, California,* (San Francisco, 1921), p. 48.
[25] Turlock Irrigation District, Delinquent Tax Lists.
[26] *Stanislaus County Weekly News,* July 5, 1895.
[27] Lippincott, p. 242.
[28] *Stanislaus County Weekly News,* Oct. 19, 1894, Apr. 5, July 12, 1895.
[29] Adams, 1904, p. 98.
[30] Computed from Board minutes showing annual tax rate and the amount to be raised by the tax.
[31] G. E. Gaylord to Howard Whipple, Sept. 13, 1921 in Turlock Public Library.
[32] Board minutes, vol. 1, p. 367 (May 7, 1895).
[33] *Ibid.*, p. 368 (June 4, 1895).
[34] *Modesto Daily Evening News,* Dec. 7, 10, 1888.
[35] *Stanislaus County Weekly News,* June 7, 1895.
[36] Board minutes, vol. 1, p. 372 (July 2, 1895).
[37] Adams, 1916, p. 20; *Stanislaus County Weekly News,* July 12, Nov. 8, 1895.
[38] *Stanislaus County Weekly News,* July 12, 1895.
[39] *Ibid.*, July 18, 1895.
[40] Adams, 1916, p. 106; *Stanislaus County Weekly News,* July 26, 1895.
[41] *Stanislaus County Weekly News,* Sept. 13, Nov. 22, 1895, Mar. 13, 1896, Mar. 15, 1907.
[42] *Stanislaus County Weekly News,* Feb. 21, 28, 1896.
[43] *Modesto Morning Herald,* Supplement, Apr. 22, 1904.
[44] *Ibid.*
[45] *Stanislaus County Weekly News,* Oct. 16,1896.
[46] *Ibid.*, Nov. 20, 1896
[47] *Modesto Daily Evening News,* Nov. 16, 1896.
[48] *Stanislaus County Weekly News,* Nov. 20, 27, 1896.
[49] Elias, p. 147.
[50] The case was that of *Hughson v. Crane,* begun in 1894. See *Stanislaus County Weekly News,* Dec. 25, 1896; Board minutes, vol. 2, p. 3 (Mar. 2, 1897).
[51] *Modesto Morning Herald,* Supplement, Apr. 22, 1904.
[52] *Stanislaus County Weekly News,* Feb. 19, Apr. 23, 1897.
[53] *Ibid.*, June 4, 1897.
[54] *Ibid.*, Aug. 30, Oct. 17, Dec. 10, 1897.
[55] *Modesto Daily Evening News,* June 14, 1898.
[56] Lippincott, pp. 242-243; *Modesto Daily Evening News,* July 14, 1898.
[57] *Modesto Daily Evening News,* Sept. 21, 1898.
[58] *Ibid.*, Nov. 12, 1898; *Stanislaus County Weekly News,* Nov. 18, 1898.
[59] *Stanislaus County Weekly News,* Dec. 9, 1898, Feb. 3, 1899.
[60] *Ibid.*, Mar. 3, 1899.
[61] *Ibid.*, Mar. 17, 1899.
[62] *Ibid.*, May12, 1899.
[63] *Ibid.*, Dec. 8, 1899.
[64] *Ibid.*, Jan. 26, 1900.
[65] Horace Crane interview in Hohenthal, Supplementary Material.
[66] *Stanislaus County Weekly News,* Feb. 23, 1900.
[67] *Ibid.*, Mar. 16, 1900.
[68] *Stanislaus County Weekly News,* June 8, 1900; California, Irrigation District Bond Commission, *Report on Turlock Irrigation District,* (1914), p. 17.
[69] *Stanislaus County Weekly News,* June 22, 29, 1900.
[70] Quoted in William Lilley and Lewis L. Gould, "The Western Irrigation Movement, 1878-1902: A Reappraisal," in Gene M. Gressley, editor, *The American West: A Reorientation,* (Laramie, 1966), p. 64.

[71] Adams, 1916, pp. 8-47; Donald J. Pisani, *From The Family Farm to Agribusiness: The Irrigation Crusade in California and the West, 1850-1931* (Berkeley, 1984), pp. 267-270.

[72] *Cal. Stats.* 1897, 254; for general information on post-Wright Act California irrigation districts see Calif. Dept. of Public Works, Div. of Engineering and Irrigation, Bulletin No. 21, *Irrigation Districts of California,* by Frank Adams (1930), pp. 17, 32-50. [Hereafter cited as "Adams, 1930"]

5

The Irrigation Age

When water arrived in the Turlock Irrigation District in 1900, the area was not much different from what it had been in 1887, except poorer. The total number of taxpayers had increased from 236 to 313, and there were now 57 parcels of less than 80 acres compared to only a few thirteen years earlier. The minor increase in small parcels was of no real consequence since all fifty-seven combined added up to less than two thousand acres, or just over one percent of the land in the district.[1] It was still a big grain farming country, although the newspapers reported that some, like Henry Stirring, had recently bought farms in anticipation of water.[2] If the hopes of those who had promoted irrigation as the answer to the area's economic doldrums were to be fulfilled, the large ranches would have to be broken up.

Small farms were the heart and soul of the irrigation dogma; in fact, the irrigation crusade, locally and nationally was dedicated as much to their creation as to the simple extension of irrigated agriculture. They implied family farms and men working their own land rather than the rootless hired hands who planted and harvested the wheat. The process implied new crops as well, for the orchards, vineyards and row crops that could be grown under irrigation were seldom cultivated on the same grand scale as wheat. Such crops were better suited to farmers able to devote a few acres and considerable effort to them. And small farms meant more people, more towns and more trade. This vision of irrigation had propelled the local crusade for the Wright Act, and sustained the friends of irrigation through the dark days of the 1890's. The same vision was part of the national reclamation movement working, at the turn-of-the-century, toward a federal irrigation program. It was an expansive concept, but hardly an original one. The Jeffersonian ideal of the yeoman farmer, who drew a host of democratic virtues from the tilling of his own soil, had long been a fixture in the national value system. The irrigation movement built on that heritage, and in fact promised to regenerate it.

Irrigation and the small farm were thought to be capable of improving the quality of rural life. Kevin Starr, a recent historian of the California dream, described some aspects of the small farm mythology in these terms:

> Intensive farming had made a new way of life possible, one possessing the benefits of country life and at the same time preserving values of diversity, leisure, and family living. With the subdividsion of its large holdings underway, California offered the middle class a way out of the increasingly burdensome work loads of business and the professions. They could return to the land as scientific farmers. Photographs of snug ranches played up the new style of rural life. Flower and vegetable gardens surrounded well-designed bungalows. Walnut and almond trees or prize dairy herds filled out the middle distance, while on the horizon rolled row upon row of vines and blossoming fruit trees. It was both garden and industry, a way of making a living and a way of life in total contrast to what had characterized the previous agricultural frontier.[3]

Irrigation was an integral part of this lifestyle. The trees and vines and dairies, the symbols of a higher type of rural life, grew in most parts of California only where water was brought to them. The finest example of what irrigation could do was in the citrus-growing colonies of southern California, where the Mediterrean imagry of sunshine and oranges blended with the irrigationist dream of small, family farms and a thriving, close-knit suburban community life.

Although frequently overstated by its advocates, the irrigation ideal established a goal, a standard against which the reality of all that irrigation did could be judged. Life "under the ditch" was seldom an idyllic existence and the result only rarely looked as elegant as those near-mythic southern California homes surrounded by roses and blossoming orange groves. Stripped of its hyperbole, however, the ideal centered on the creation of a new generation of family farms in the arid West, and the success of the irrigation movement could in large part be judged by the extent to which that happened, and whether or not it produced the economic and social benefits prophesied.

At times it seemed that the irrigation crusaders somehow believed that small farms – and the definition of what was "small" varied considerably – were an almost automatic consequence of irrigation. That was not true. There was no good reason why large, even immense farms could not be irrigated by hired hands and their fields tended not by family farmers but by tenants and laborers. It is true that farms in the premier irrigated regions, like Fresno and southern California, were small, having been subdivided by developers who found that a profitable way to market their land and water. And it was also true that without water small farms were impossible in the San Joaquin Valley. However, irrigation by itself would not dissolve large landholdings. Many of the state's irrigated farms were relatively large and even giant enterprises like Kern County Land Company or Miller & Lux, the antithesis of the family farm ideal, irrigated their holdings when it suited them.[4]

By the turn-of-the-century, if not well before, irrigation reformers realized that specific steps would have to be taken to insure that irrigation was indeed equated with small or family farms. The 1902 federal reclamation law contained one such mechanism; the requirement that individual ownership in reclamation projects be limited to 160 acres. The provision hardly guaranteed small farms, but it was designed to implement the irrigation ideal of family farms. The 160 acre limitation has been one of the most controversial elements of reclamation policy; a fact that by itself should confirm the lack of any reliable connection between irrigation and small farms. The Wright Act, too, had a mechanism for the encouragement of small farms, though it was indirect. The law required that all land in an irrigation district be subject to district taxes, whether it used water or not. That placed an added burden on anyone trying to continue dry farming on worn-out land, and provided at the very least an incentive to irrigate, and usually a good reason to subdivide larger holdings. In the Turlock District, the irrigation tax was not the only thing that encouraged the selling off of the great wheat ranches. The collapse of the grain economy dictated a change in crop pattern, but farmers already in debt could scarcely afford to prepare vast acreages for irrigation. They could, however, profit from irrigation and higher land values by selling, and many of the major landowners, including the heirs of the huge Mitchell estate and C.N. Whitmore, were ready to do just that. There is no evidence that any of the larger ranchers entertained hopes of keeping all of their land and irrigating it. Instead, in 1900 farmers, businessmen and real estate agents alike anticipated a land boom now that water had at last arrived.

The boom did not occur overnight. Advertising and promotion were needed to acquaint people with the opportunities available in the Turlock District and, of course, it was not the only irrigated land in California or the West trying to attract settlers. The first advertisements appeared even before all the laterals were completed. The July 1900 issue of *Land of Sunshine*, a magazine promoting the Southwest, carried realtor J. W. Bell's message that the Turlock area had "the cheapest good land and the best cheap land" in California, and C.N. Whitmore offered ten acre farms with the advice that, "If you're looking for a home, don't overlook CERES."[5] The following year it was announced that the giant Hickman ranch, half of it in the TID, would be subdivided and sold by a San Francisco real estate firm.[6] By the end of 1901, however, the impact of irrigation was still mostly prospective. Less than four thousand acres had been irrigated, land sales, though promising, were still slow and the towns had hardly changed in appearance. It would take more, and more ingenious, promotion to fulfill the promise of irrigation.

The Turlock region's most successful single promoter arrived in Turlock in January 1902. Nels O. Hultberg had been a missionary for the Swedish Mission Covenant Church in Alaska, and had played a central, if not uncontroversial, role in the discovery of gold near Nome.[7] He came to Turlock, which he said "looked like an old forsaken mining town," seeking a place to establish a settlers' colony, and he later recalled that the first man he met, a storekeeper, asked if he, Hultberg, would like to buy him out.[8] At that time there was a surfeit of potential sellers waiting to be bought out and a shortage of buyers. The biggest would-be seller was the Fin de Siecle Investment Corporation, which controlled the estate of John Mitchell. When Mitchell died in 1893, his property passed to three nieces and was run by their husbands – Stephen Crane, H. F. Geer and George Bloss. They tried and failed to sell off parts of Mitchell's empire during the depression of the 1890s, and in 1899 formed the Fin de Siecle company to manage and dispose of their holdings.[9] J.W. Bell, the real estate man, put Hultberg in touch with Horace Crane, the company's secretary-treasurer, and a deal was soon struck; if Hultberg could sell a thousand acres of the company's land for $25 an acre by the end of May 1902 Fin de Siecle would lay out a seventeen thousand acre tract for him. Hultberg aimed his advertising compaign exclusively at the Swedish-American community. His ads appeared in Swedish language papers, including the church-related *Mission Friends*, beginning in February, and by the end of March twenty-two families had arrived, and in April people came by the carload.[10] Having proven his

The Homeseekers' Opportunity

Irrigated Land, Water Right Included

$35.00 to $60.00 an Acre . . . . .

Ceres, Stanislaus County, California

Eight thousand acres in the famed Whitmore Tract, 135 miles of canal and irrigating ditches in this district, and the Whitmore Tract is conceded the best of all. This tract of 8,000 acres lies on either side of the Southern Pacific, and the Santa Fe runs through it on the east. Modesto is the county seat, a flourishing and prosperous city of over three thousand inhabitants. Every acre of the Whitmore Tract is tillable; deed to the land includes a perpetual water right.

TITLE—The title has been vested in the Whitmores for more than thirty years, and comes directly from the United States Government.

PRODUCTS—Sweet potatoes, corn, watermelons, alfalfa, and all the fruits possible in the Golden State of California.

ORANGES—Many homes have their orange trees, and though not raised as yet here in commercial quantities, we have the perfect conditions for the orange; and fruit matures four weeks earlier than in Southern California.

OLIVES—Apricots, nectarines, figs and walnuts also attain perfection; table and wine grapes do well; a great cattle and stock country.

CLIMATE—The average temperature for June, July and August at 7 A. M. is 62 degrees, at 2 P. M. 90 degrees, and at 9 P. M. 70 degrees.

Send for illustrated matter which will be mailed free and for further detail information correspond with

C. N. WHITMORE, Ceres, California

A 1904 real estate advertisement.

prowess as a promoter, the Fin de Siecle company subdivided the colony site Hultberg named "Hilmar," after his son.

Although he later claimed to have secured thirty-five thousand acres of land around Turlock by purchase, lease or option, Hultberg apparently owned little land himself, acting instead as an agent and promoter.[11] During his first year in Turlock, he formed a real estate partnership with Walter Soderberg, an acquaintence from Alaska, and brought in the Rev. Andrew Hallner, a former editor of the *Mission Friends,* to help manage and advertise the Hilmar colony. Until early in 1904, Hultberg held the price at $25 per acre and advertised only in Swedish publications. Rev. Hallner reported at the beginning of 1904 that in just 15 months, 975 colonists had arrived in Turlock.[12] Some of the newcomers were from farming communities in the East, who came to Hilmar because it represented a chance to own their own land. During the hard years of the 1890s, many farms on the Great Plains had been foreclosed, turning farmers into tenants.[13] Other colonists came from urban backgrounds, like those who followed Rev J. O. Boden from the steel mills of Youngstown, Ohio. The minister was a friend of Andrew Hallner and visited Turlock first in early 1903. He ultimately bought 1,200 acres for himself and members of his congregation who came out to form the Youngstown colony.[14]

The first immigrants arriving on the new irrigation frontier found that the work of pioneering could be discouraging. Blowing sand buried crops and sandblasted young trees, filled in ditches and filtered through cracks in the crude houses and barns where the settlers made their first homes. The soil in Hilmar was some of the sandiest in the Turlock District, and to hold it in place windbreaks of bamboo or cottonwood and eucalyptus trees had to be planted, and farmers learned to seed rye and other grains between the rows of their crops to keep the soil from blowing around. Jack rabbits were numerous and inflicted heavy damage on young orchards by chewing the bark off the trees, and in 1903 grasshoppers devastated what the sand and the rabbits missed. Rabbit drives were held to corral and kill that pest, fences were built and farmers learned that for some reason hogs liver rubbed on the trees acted as a rabbit repellent. There was little that could be done about the grasshopper plague, but it soon abated.[15] Undoubtedly, those who came expected that it would take hard work and some discomfort to create new farms from the worn-out grain ranches, but many were discouraged by the unaccustomed soil conditions and the ravages of pests. Some of the colonists might have left if they had been able to afford it, but they had usually put all they had into their land and had no choice but to remain. Many years later, Esther Hall Crowell, the Hultberg and Soderberg secretary who dealt daily with both land sales and the problems of the Hilmar colony, gave talks to local groups about those difficult early days. Among her notes were these comments:

> At one time the settlers had a mass meeting at the Hilmar Church and were going to sue Mr. Hultberg for misrepresentation, but nothing came of it. Both men and women worked awfully hard, the women helped with the chores and raised chickens; sold butter and eggs for groceries. If it hadn't been for the local merchants that gave them credit they would have starved.[16]

Hultberg understood the seriousness of the situation and said that between what he called "crank" ministers who stirred up the dissidents, and the grasshoppers his colony was almost destroyed in 1903.[17]

Perhaps one source of the trouble was the gap between the claims of the promoters and the reality settlers faced, at least in the early years. Hultberg sent pictures of prosperous farms in the Swedish community of Kingsburg in the San Joaquin Valley or San Jose to show what Hilmar would someday be. In her notes, Mrs. Crowell wrote, and they weakly lined out, the following sentence: "I worked at the land office at that time and was sending out literature telling people how grand it was, and many times I asked God to forgive me for helping to fool those poor people."[18] Of course, many came to look over the colony before buying land, but even so they probably would not have experienced the effects of a sandstorm or any of the other day-to-day difficulties that would face them once they moved onto the land. In their eagerness to sell real estate, promoters commonly sold the future rather than the present. Modesto newspaper publisher T.C. Hocking later recalled his own efforts along those lines:

> ... later when C. N. Whitmore was ready to breakup all his holdings I wrote and printed for him, a pamphlet describing those lands. When the old gentleman read it he said, "Do you believe that?" and I said I did or I wouldn't write it. It was, however, purely a visionary article. The land was then just played out grain land, but I pictured for prospective buyers, irrigation, orchards, dairies, vineyards and fine homes.[19]

It was that kind of "visionary" promotion that attracted settlers to the Turlock District in the first years of irrigation.

At the end of 1903, the Modesto Irrigation District finished the construction of its canal system and prepared to begin irrigation in the spring of 1904. Preparations were also begun for an Irrigation Jubilee to celebrate the completion of the two Tuolumne River projects, and not incidentally, advertise the lands of the Modesto and Turlock districts. Although water had been running in Turlock's canals for three years, only about 12,000 out of over 176,000 acres in the district had used water by 1903.[20] Most of the district was still being farmed in grain and except for the settlement of Fin de Siecle lands south and west of Turlock and the sale of some farms near Ceres, subdivision and settlement were just beginning to change life in the Turlock Irrigation District. The towns were starting to show some definite signs of growth. At Ceres, a "new town site with fine broad streets and a park" had been laid out, and in Turlock the rapidly increasing rural population, principally in the Hilmar colony, had finally begun to erase the charred and desolate appearance of the 1890s.[21] Like their brethern in Modesto, Turlock businessmen and real estate agents welcomed the publicity a jubilee would bring.

The logo chosen for the jubilee was a clue to what the area's promoters believed was in store for it. The drawing of a navel orange formed the background with a smaller peach superimposed on top of it, and on the peach was a fig, and at the center of the fig was a picture of the La Grange Dam.[22] These represented some of the fruits it was hoped would soon be grown with water from the dam, for fruit growing had long been regarded as the most advanced type of agriculture. Badges to be worn during the celebration did not carry the 'fruit salad' logo, but instead showed a water and manufacturing scene with a woman in the foreground and the words "Water Is Wealth," all attached to ribbons of green, orange and purple representing alfalfa, oranges and figs.

The grand event was scheduled for April 22-23, 1904. The state Board of Trade and the California Promotion Committee planned special excursions to the jubilee. The Southern Pacific Railroad offered special rates to Modesto for the occasion, and ran trains from Modesto to the Turlock area to give visitors a chance to see more of the land and the canals. In Modesto, no effort was spared to put on a spectacular show, from thousands of electric lights strung through the town to the American flag that would be fired into the air to float down on a parachute. A working model of La Grange Dam was built, and the green, orange and purple jubilee colors were hung everywhere. On Friday, the 22nd, Governor George Pardee and irrigation expert Elwood Mead spoke, and the following day, after tours of the districts, James A. Waymire, University of California president Benjamin Ide Wheeler and others exhausted their eloquence. The whole affair was a glorious success aside from a few complaints about price-gouging by Modesto merchants and the presence of pickpockets on the excursion trains travelling through the districts.[23] There is no evidence that any substantial and sales were made as a direct result of the jubilee, but it was a symbol that optimism and a sense of change had replaced the depression that had been so much a part of the area only five years earlier.

The pace of subdivision and settlement began to quicken after 1904. The experience of one real estate agent, Edward J. Cadwallader of Turlock, may have been somewhat typical of the period of irrigation settlement. Cadwallder recalled forty years later that,

> My first impression of Turlock was not very favorable. In 1904 I was working in Ceres... and as the Modesto papers were playing up Turlock pretty strong, about settlers coming in from the middlewest, I decided to go there and have a look myself....
>
> Turlock sure didn't look very good... The streets were sand a foot deep, a few board walks, a horse trough, a lot of hitching racks and scattered frame buildings was the picture of Turlock, except for two brick buildings...
>
> The Vignolias were running a Hotel under great difficulties,as there were no public utilities. Each familiy had to depend on windmills for their water supply and some times the wind failed to blow and there was the question of the disposal of the sewage. There was no regular septic tanks, only vaults and outside toilets – the town was in bad shape.
>
> Since this was a farming community, nearly every family had a pig in a pen in the backyard to use up the swill. The Vignolias were no exception and they had the Geer lot fenced where they had a bunch of hogs and all the chore boy had to do was carry the swill out the back door and across the alley and dump it in the hog's trough right on East Main Street.[24]

In the spring of 1905, the C. N. Whitmore company, having sold much of their land around Ceres, bought the Big Ed Davis ranch, covering four sections northeast of Turlock. The property included the "town" of Elmwood, a station located on the railroad built in the 1890s to challenge the Southern Pacific's monopoly and then sold to the Santa Fe. Cadwallader landed the job as the Whitmore company's sales representative in Elmwood, and moved into the first house in the community. Sales were "pretty low" the first year because the company wanted $75 an acre for their lands. Elsewhere, land in Hilmar could be had for from $20 to $50 per acre, the Modesto Bank was selling a section north of Turlock for $40 per acre and Charles Geer put a section west of the Elmwood tract on the market for $50 an acre. Cadwallader took a job at the warehouse in Elmwood and did other work until sales picked up.[25]

One of his first customers was the remarkable Emma Abott Simon. In 1905 she came out from her home in Iowa to visit her sister-in-law in Ceres and soon fell in love with the land on the Hughson and Tully ranches, both located north of Turlock. When she returned home she did so as a missionary, exhorting her husband, a hardware store owner, and friends to move to California. George Simon agreed, and his wife, who wanted people she already knew to be her neighbors in California, convinced nine other persons to join her colonization plan. She was to come to Turlock as their agent, buy land, improve it and care for it for up to five years until they came to take it, although they had the right to reject what she had done. She bought over two hundred acres of the Elmwood tract for $57.50 per acre, which meant the Whitmore company had lowered its asking price somewhat. The land was overgrown with weeds big enough to defeat the discs and mowers sent to clear them. Finally, a team of horses was hitched to each end of a length of railroad rail and it was dragged over the property to smash down the rank growth. The Simon's took sixty acres for themselves, and the other settlers arrived a few at a time until the Iowa colony Mrs. Simon had started numbered sixty residents by 1910.[26]

According to Cadwallader, about 1906,

> things commenced to move in Turlock. A considerable acreage of melons were grown on Colorado Avenue. They were top quality and the market was good and the results were that the growers that paid $60.00 per acre for the land had cleared $100 per acre, and that wasn't hard to explain to a prospect.[27]

Other things, however, were harder to explain to a potential buyer. For example, in the spring of 1906, Cadwallader received a telegram from the Whitmore company's Los Angeles office telling him to meet a client at the Turlock depot the following day. He needed a commission rather badly just then, but when the time came to go to Turlock to meet the train, a strong north wind blowing over fallowed fields made the air so black with dust the team could scarcely find its way across the open country from Elmwood to Turlock. Convincing anyone that they were in the heart of a growing irrigation empire would obviously be no easy task under the circumstances, but much to Cadwallader's relief the prospect was from further south in the valley and knew sandstroms were only temporary. The following day the man bought twenty acres.[28]

Cadwallader opened his own real estate office in Turlock in 1906, just as Elmwood began to grow. That year, representatives of the Friends (Quaker) Church from the southern California town of Whittier took an option on the townsite. Though some of their members did come to the area, the group failed to exercise its option, and in 1906 or 1907, John Denair bought the town and gave it his name. Denair was a division superintendent for the Santa Fe Railway at Needles, California, who in 1906 formed the Denair Land and Development Company to purchase and subdivide some three thousand acres west of Elmwood. The company later acquired and sold other tracts in the Turlock area as well as the Denair townsite. The era of the small farm was also one of the small town, and Denair grew into a typical rural village where farmers did their shopping, sent their children to school and shipped their crops.[29]

North of Denair along the Santa Fe tracks and due east of Ceres another town, Hughson, was established. Hiram Hughson had deeded a right-of-way and depot site when the railroad was built, and a school was in operation in the 1890s, but until 1907 nothing further was done.[30] Hughson, an intransigent anti-irrigationist, sold over two thousand acres, including the townsite, to the Hughson Town Company of Charles Flack and C. W. Minniear. They laid out the town and planned to subdivide the surrounding acreage, while TID engineer Burton Smith surveyed a community ditch from the Ceres Main Canal to serve the proposed settlement. Development began as soon as the last crop of grain was harvested in the summer of 1907.[31] To enhance the town's status and prospects, especially in relation to Ceres, its more established rival to the west, Hughson's promoters purchased the two-story Gilette Hotel in Ceres, removed it from its foundation and started to haul to Hughson. It moved so poorly that it had to be cut in two, and the halves were then pulled at a rate of two or three miles a day by mules and steam tractors to Hughson. There it was reassembled atop a new lower story.[32]

Land near the town sold for $100 per acre, and apparently sold well. One of the leading salesmen, William Coward, was also one of the area's most unusual characters. He was over six feet in height, and Cadwallader recalled that he "had a powerful big hand and when he shook hands with one he would put quite a little pressure on and almost raise one from the ground and you would get a funny feeling."[33] David Lane, also a real estate man, said Coward always carried an ice pick, which he used with some dexterity, and was the only man Lane could remember who could eat apples and dry crackers at once. As Lane wrote, Coward "talked socialism, fought railroads, used his ice pick, ate his apples and crackers and sold real estate."[34] He was also a little demented and highly excitable, and on one occasion knocked one of the officers sent to restrain him in the head with the ice pick. A rest cure allowed him to return to work, and in 1908 or 1909 he accompanied other local real estate agents on a trip to Los Angeles. Coward and Minniear suggested inviting their hosts from the Los Angeles Chamber of Commerce on a return excursion to the irrigation district with an eye to selling them some real estate. Cadwallader and the others on the trip rebuffed the idea, pointing to the lack of suitable hotel accomodations in Turlock. Coward was not easily dissuaded, and he and Minniear went ahead with their plan alone, chartering a sleeper and club car to solve the hotel problem. The scheme resulted in the sale of "more than a hundred thousand dollars worth of the Hughson Colony."[35]

Through the efforts of men like N. O. Hultberg, David Lane, Edward Cadwallader and William Coward, the old wheat ranches were carved into irrigated farms of ten, twenty or forty acres in size. The first decade of irrigation was the most active phase of colonization but the process continued for another twenty years as outlying or hold-out tracts were sold. The impact of irrigation and the creation of small farms was plainly evident in the increasing number of taxpayers in the district. In 1900, there were only 313 listed on TID assessment rolls, but by 1910 there were over 2,000.[36] In the same period, the population of Stanislaus County more than doubled, rising from 9,550 in 1900 to 22,522 in 1910.[37] In 1912, an analysis of farm sizes in the Turlock Irrigation District showed that of 2,385 farms, only 225 had 80 acres or more and all but 17 of those were under 220 acres. There were 117 five acre farms, 363 with ten acres, 164 of fifteen acres and 644 with twenty acres, which meant half the farms in the district covered twenty acres or less. There were also 250 farms of thirty acres and 410 with forty acres.[38] These figures contain some gaps and approximations but nonetheless their meaning is clear; irrigation, in the Turlock District at least, did mean small farms, just as its advocates had always said it would.

The visible evidence of irrigation included not only the new houses, barns, orchards and towns dotting the landscape, but the network of ditches running from the river to the fields. The first job that greeted the irrigation settlers was getting their land ready to receive the water that had drawn them there. Among other things that usually meant building the ditches that linked their farms to the district's works. The Turlock Irrigation District built and maintained only the main canals and widely-spaced laterals, which

Fresno scraper completing a farmer's ditch.
Source: U.S. Department of Agriculture.

constituted the central skeleton of a water distribution system. Except for a fortunate few whose property adjoined a canal and could be served directly from a sidegate in it, farmers depended on the so-called community ditch system to connect their farms to the water supply. The community ditches – hundreds of miles of them – were built and maintained by the irrigators using them, usually without any formal organization. Some of the ditches, like the Hughson ditch, served hundreds of acres and scores of individual farms, while others reached only a handful of parcels. Not all irrigation agencies used this kind of dual system that left the final link in the canal network up to the irrigators themselves, but it was certainly a common approach, even where corporate developers owned the water rights and the main canals.[39] It was also an inexpensive arrangement, because the farmers could do much of their own construction and maintenance, and one that carried to each neighborhood the cardinal irrigation principles of cooperation and local responsibility.

There were a number of ways in which water could be handled once it reached the farm. Taking the turn-of-the-century West as a whole, the most common irrigation method was known as wild flooding. Supply ditches running along the high ground were temporarily dammed to divert small streams into field ditches dug down the slopes. These smaller ditches were in turn plugged at intervals to force water out onto the field, letting it flood without restraint down the hill. Used with skill, wild flooding could be a reasonably effective technique, though the distribution of water tended to be uneven. It was widely used in alfalfa, grain and pasture crops in the Rocky Mountain states, but rarely used in California. Wild flooding required only a small stream, or head, of water, perhaps two or three cubic feet per second (about fifteen to twenty gallons per second), and considerable hard work to open and close all the little dams and cuts needed in the field ditches. On the other hand, it required little capital investment since no land levelling and only the simplest kind of ditches, often dug with one pass with a specially built lateral plow, were needed.

Another method using a small head of water was furrow irrigation. It was suited to certain row crops that could be grown between the furrows and to orchards. It was frequently used in the citrus groves of southern California, where elaborate pipe systems were installed to control the flow into the furrows.[40]

The final irrigation system was the check method of flooding and

Building levees between irrigated checks, 1904.
Source: U.S. Department of Agriculture.

its variants. Under that method, the land was divided into a series of basins called checks, each one level or nearly so and surrounded by levees. A large flow of water was turned into each check until it was just covered. The technique was an ancient one that in North America had long been practiced by Mexican irrigators along the Rio Grande River and elsewhere in the Southwest. There the checks were very small with from ten to fifty of them per acre. When the method was first adopted by California irrigators the penchant for large scale farming seemed to dictate equally large checks. Instead of ten checks or more to the acre, some Californians tried to put ten acres into each check with disasterous results. By the time water reached the Turlock region, the mania for oversized checks had abated and the standard practice was to create checks of up to an acre in size.[41] The basin method of irrigation found in some orchards, in which each tree was surrounded by levees, was just a way of using very small checks. Similarly, the so-called border method used levees to confine water to a narrow strip running down a gentle slope, and was like the basic check system in practice.

At the end of the nineteenth century, when the Turlock canal system was being built and placed in operation, the check system was used primarily in the San Joaquin Valley, where it was the predominant irrigation method, and, under the name of the border method, in the Imperial Valley and Arizona. A Department of Agriculture bulletin clearly explained the advantages the check system offered in the San Joaquin Valley.

> The soil in many parts is porous, containing a high percentage of fine sand. In such districts it is doubtful if any other method of applying water would be so successful. As a rule, the slope is slight, which enables the farmer to form check after check with only a few inches of difference in elevation. It is due, however to the character of the streams which furnish the water supply for the valley that the check system is so generally used. These streams head in the Sierra Nevada Mountains, where the precipitation, particularly in the form of snow, is heavy and they are all subject to floods in the spring. . . Irrigation works have accordingly to be planned to take care of a large volume of water during the spring months. The Tuolumne River, to cite a somewhat extreme case, frequently discharges enough water to cover 20,000 acres a foot deep in a single day in May, while the total discharge for the month of August may be little more than this. In great fluctuations of this nature not only must the canal engineer and superintendent adapt their structures to carry large volumes, but the irrigator is under the same necessity to form his checks, sluice boxes, and lateral ditches in such a way as to accomodate large volumes for short periods of time. There is no other system practiced in the West which enables one man to handle from 10 to 20 cubic feet per second without assistance and with little waste.[42]

All of these conditions – porous soil, level terrain and a seasonally abundant water supply – were present in the Turlock District. The canal system was apparently designed with the check system in mind by engineers, notably George Manuel from Fresno, who were familiar with that approach, and no other method appears to have been considered by the district or by its irrigators.

Laying out a farm irrigation system began with the location of one or more ditches along the high ground or perhaps through the center of a level parcel. After the ditches were dug, check boxes made of redwood lumber with two-by-four framing were set into them. Flashboards fitted into these boxes could be removed to let water flow from the ditch into the check. The checks themselves could be either rectangular or irregularly shaped to follow the contours of the land. On steeper slopes, contour checks were simpler to build, but in the generally level Turlock region, rectangular checks, which were also neater looking and better suited to orchards and vineyards, predominated. The amount of dirt that it was practical to move with horse-drawn scrapers was limited, so where even a gentle slope was encountered it was often easier to build several small checks than to level a larger area. The result, especially in some orchards, could be a staggering number of checks. North of Denair, Carl Muller had one piece of land with only four trees in each check, and on the George Paterson place near Hughson, an apricot orchard was originally laid out with cross-levees every two rows down the slope and a dozen trees per check.[43] In these and other cases, checks were created that did not adjoin the farm ditch. These were irrigated by running water from one check to another, often through small redwood gates dug into the cross-levees that separated the checks. In later years, tractors and heavy earthmoving equipment permitted extensive relevelling, greatly simplifying the check system on many farms.

Maintaining the dirt farm and community ditches was a never ending job. Ditchbreaks were an all too common occurance, often the result of burrowing gophers, muskrats or ground squirrels that weakened the banks. Less dramatic was the threat posed by weeds growing in the ditches. If left alone, they could slow and finally choke off a flow of water down the ditch. The only remedy was periodically plowing the ditch to destroy the vegetation. Practice varied from place to place, but all ditches seem to have been cleaned at least once a year, usually in the spring.[44] The job was done with a two-horse team pulling a single plow. The first furrow was run along the top of the ditch and succeeding trips worked down the sides. The wide banks of a community ditch would require as many as fourteen trips on each side to finish the job. Handling a plow on a slope was exacting work that required a good team, "considerable patience and usually a liberal amount of profanity at times."[45] On some community ditches the landowners took turns cleaning the ditch, on others the work was informally divided between some or all of the irrigators, while in still other instances an expert plowman might be hired to do the job.[46] Perhaps because ditch maintenance was such a difficult and time consuming chore, it was not always done as often as it should have been. In early 1914, the TID board took note of the fact that many ditches were in deplorable condition, and the district's rules in effect shortly thereafter required ditchtenders to refuse water to farmers whose ditches were not adequately cleaned.[47]

The Turlock Irrigation District delivered water to farmers on a rotation system, with each irrigator getting his turn to use the water according to a fixed pattern. When water was turned into a ditch, the irrigator at the end of the ditch got it first, and when he was done it went to the next farm up the ditch, and so on until it reached the head of the ditch. The "up the ditch" pattern was used so that if the ditch broke, the water could be transferred to an irrigator above the washout until repairs could be made. According to the rules, adjoining irrigators could trade places in the schedule with the ditchtender's permission, but if for any reason a farmer refused the water when it was his turn, he would not have another opportunity until the next rotation. The rules established by the district were not always strictly adhered to in practice. A great deal depended on the ditchtender and on the circumstances on the individual ditches. Some farmers were apparently able to get water more or less when they needed it, though not oftener than they were entitled to by the rotation schedule, while others found it nearly impossible to break out of the prescribed rotation pattern. Even when the rotation rule was enforced, some ditchtenders would drop back at the end of a run to pick up farmers who had had to refuse water in their regular turn for good reason, such as having hay down in their fields.[48] The rotation system, like the check system, was predicated on the use of a large head of water for a short time on each farm. In the early years, the TID was not always able to deliver uniform heads, but by 1916 the standard alfalfa head was twenty second-feet, and district rules permitted no more than half an hour per acre with that much water.[49]

Irrigation was as much an art as it was a science. Though neighbors and even the local ditchtender might be on hand to offer advice during the first irrigation, the most important lessons were taught by experience. The new irrigators learned how to guide water over their land, how to adjust the boards in the check boxes to divide a head into several checks at once, and how to determine when each check had enough water. They learned too that on an irrigated farm, the shovel was an important implement. It was used to fight minor breaks, shore up check boxes, cut or reinforce levees, and even to lean on while watching the water finish filling a check. Since water ran night and day, a kerosene lantern was nearly as necessary as the ever-present shovel.

Almost as common a sight as the shovel-toting irrigator was that of people swimming in the canals. The water in Lateral 4 had no sooner reached Turlock in the spring of 1900 than some of the local boys were out swimming in it.[50] The cold water fresh from the melting snowpack was a welcome antidote to the heat of the valley's late spring and summer, especially at a time when there were few other ways to keep cool. The dirt ditches were ideal for swimming. Their gently sloped sides and relatively shallow water made it easy to get in and out. Where water poured over drops, it scooped out deep holes that allowed diving and made for the best swimming. These became the most familiar gathering spots, each known for its own peculiarities. At some drops, the swimmers made improvements, in the form of simple diving boards, and in one case, sand was even collected from the canal during the winter to make a beach.[51] Although youngsters made the most frequent use of the canals, almost anyone might go for an occasional swim, and seeing neighbors at the local drop on a warm evening made it a social as well as a recreational pleasure.

What crops did the water grow? In the first few years of irrigation, the answer was primarily alfalfa. It is a deep-rooted perennial forage plant that when cut will quickly send out new growth. In its first year, it could usually be cut at least twice, and in suceeding years would yield approximately four cuttings annually, which could produce a total of five or six tons of good quality hay per acre. Alfalfa needed regular irrigation to prosper, and when the Tuolumne dropped too low to keep water in the canals, usually in August in the earliest years, the plants sent out only a weak growth that was best grazed off by livestock rather than cut.[52] Along with alfalfa came dairy herds fed on alfalfa hay. By the turn-of-the-century, the cream separator and modern butter making equipment, along with refrigerated railway cars, had opened an expanding market for butter, and for the cream needed to manufacture it. Local creameries bought the cream produced by the herds of Holstein and Jersey cows, while the skimmed milk was fed to hogs that became yet another source of income. Taking into consideration the value of the cream, sales of milk-fed hogs and the value of calves, a Modesto newspaper estimated in 1904 that "an acre of alfalfa will produce six tons of hay per year. This means that the net profits of an acre of alfalfa run through the cow is $50 per year."[53] Needless to say that, in its eagerness to attract settlers, the paper probably made the most optimistic assumptions in reaching its conclusion, but there did seem to be considerable merit in raising alfalfa and cows. Although irrigation advocates, like the designers of the Irrigation Jubilee's orange, peach and fig emblem, had long assumed that irrigation would mean orchards and vineyards, trees and vines took a large initial investment and several years to mature before they began to show a profit. New settlers had often invested everything they owned in their land, homes and ditches, and needed a quick cash income. Alfalfa offered just that in a crop that was easy to grow in the Turlock District. It was a perfect pioneer crop, and in 1904 the TID crop report showed it growing on 11,600 of the district's 12,000 irrigated acres.[54]

In the first couple of years that water was available, irrigation meant alfalfa almost as much as dry farming meant grain, but a variety of other crops soon appeared. Alfalfa acreage continued to increase until 1914, when it reached a peak of 67,682 acres, or about 72 percent of the irrigated cropland in the Turlock District that year. Thereafter, it lost ground rapidly, and by 1920 it accounted for less than 31,000 acres. The decline in alfalfa acreage was matched by an increase in the irrigation of grain, which in 1920 covered nearly as many acres as alfalfa.[55] Experts like Frank Adams of the U.S. Department of Agriculture, who conducted a detailed investigation of the Turlock and Modesto Districts in 1904, had been amazed at the initial reluctance to irrigate grains, even though they were irrigated successfully in other parts of the West.[56] By 1909, when five thousand acres of grain were irrigated, farmers had learned that feed grains, like barley, could be a useful crop, and they could be double-cropped with beans, corn or Egyptian("gyp") corn. Alfalfa, grain, beans and corn were the backbone of district farming in the first two decades of irrigation, making up about 70 percent of the irrigated acreage in 1920, and an even greater proportion in earlier years. Orchards and vineyards accounted for 20 percent of the acreage receiving water in 1909, although the rapid expansion of other crops soon reduced that proportion.[57] Peaches and apricots were grown in small orchards in several parts of the district, and Ceres quickly became a center for Smyrna figs. Oranges, which had so often been mentioned as one of the crops irrigation would bring, were never grown on a commercial basis, although orange trees graced the gardens of many settlers' homes. The most dramatic success stories, however, came from two specialty crops: sweet potatoes and melons.

Alfalfa fields and hay wagons were common sights in the early days of irrigation.
Source: McHenry Museum.

For some reason, the early crop reports prepared by the TID failed to list some of the crops that were important in the Turlock region, making it more difficult to trace their history. One such crop was sweet potatoes, which preferred loose, sandy soils like those found from around Turlock south to Delhi.[58] They were planted at least as early as 1901.[59] Among the earliest growers were the Goulate brothers of Turlock. In 1902, their crop was lost to a sandstorm that buried most of their plants. They were already in debt on their sixty acres as well as for horses and equipment, but the Turlock mercantile firm of John Osborn and Son advanced them the credit needed for another year. They tried again, this time with rows of oats to hold the soil in place and serve as windbreaks for the sweet potatoes, and succeeded in growing twenty-five acres of them. The crop proved so valuable that "they realized enough to pay the balance due on the land, some $1600 with interest; to pay for the horses, wagons, plows and cultivators, and living expenses; to pay for a barn and for a potato cellar, and the cost of checking and planting 32 acres to alfalfa. And they had $300 cash on hand after paying for everything."[60] Other sweet potato growers, primarily Portugese, had equally good luck with the 1903 crop, often paying for their land in a single season. The following year 1,100 acres were planted to sweet potatoes, and by 1912 the acreage had grown to over 2,500 acres and peaked in 1922 at 5,144 acres.[61] Compared to alfalfa or some of the other, more easily grown field crops, the extent of sweet potato cultivation was not large, but its economic impact made it an important crop for a number of early farmers.

Like sweet potatoes, melons, including watermelons, cantaloupes, casabas and honey-dews, do not appear on TID crop statistics until 1912 when they had already become one of the most important crops in the district. Frank Adams reported that in 1904 melons as well as other crops were being planted on a more or less experimental basis.[62] Edward Cadwallader's comments on the melon boom that began about 1906 have already been quoted, and they illustrated the impact the crop had on land sales and the local economy. Cantaloupes were probably predominant from the start in terms of melon acreage, and in 1907 most of Stanislaus County's 764 acres of that crop were grown near Turlock, and there were reports of watermelons being grown between the rows of young orchards.[63] Melon growing took advantage of the light soils and of a generally unanticipated side effect of irrigation, a high groundwater table. As irrigation water was turned onto the soil, often in large and wasteful quantities by inexperienced irrigators, or when it seeped from sandy, unlined canals, it collected underground and soon rose to within a few feet of the surface in many places. While these conditions could be disasterous for deep-rooted tree crops or alfalfa, the high water table irrigated melons from below, without the use of any additional surface irrigation, and many growers came to believe that subirrigation produced the best melons.

Most of the melons grown in the early years appear to have been cantaloupes and watermelons, although from 1912 to 1916 the TID lumped all melons into a single category in its crop reports. In 1912, there were reported to be 1,250 acres of melons in the district, rising to nearly 3,000 in 1914, 3,840 in 1915 and 5,143 in 1916. In 1917, the district listed 5,866 acres of cantaloupes, 111 acres of casabas and 1,937 acres of watermelons for a total of 7,914 acres of melons. The acreage dropped by nearly half in the following year, but in 1919 it soared to its highest level, 13,645 acres, or about 11 percent of the irrigated acreage in the district. Thereafter, the total melon acreage varied between 8,000 and 11,000 acres until 1933.[64]

Even more than sweet potatoes, the acreage statistics understate the economic importance of the melon crop. Good yields and high quality quickly made Turlock famous as a melon center. So important were melons to the Turlock region that beginning in 1911 the town held an annual melon festival, and about 1915 the Turlock Board of Trade published a promotional song entitled "Way Down in Turlock Where the Watermelons Grow." The little song's verses are worthy of notice not because of any great artistic merit, but as an indication of the significance of the melon boom and an expression of the civic boosterism that was so prevalent following the introduction of irrigation.

> Way down in Turlock in the fertile San Joaquin,
> Where cantaloupes are cash, and the melon vines are green.
> We carry off the banner in a most decided manner,
> Right here in Turlock, where the watermelons grow.
> Come make your home in Turlock and figure up your gains,
> Where every day is sunshine, excepting when it rains
> Everybody's coming, business simply humming,
> Down here in Turlock where the watermelons grow.[65]

Melons and sweet potatoes, along with peaches, figs, grapes and a wider array of field crops were a sign of a gradually maturing irrigated economy. The initial dependence on alfalfa and dairying was reduced by the success of other crops that could produce a large return per acre and were as well suited to the district's small farms as they were to its sandy soil.

Some idea of what the district's farms looked like after a decade or so of irrigation can be gleaned from contemporary accounts of selected farms. The earliest descriptions were frankly promotional in purpose. Writing in the February 1910 edition of *The Earth*, a Santa Fe Railway paper designed to encourage immigration to areas served by the line, F. L. Vandergrift told the story of Mrs. Simon's successful colonizing effort and gave brief descriptions of the Simon property and other farms as they appeared at that time. In 1910, Mr. and Mrs. Simon owned sixty acres, one-third of which was devoted to the growing of nursery stock. Twelve acres were in grapes and an equal number were devoted to orchards, with half in apricots and the remainder in alternate rows of peaches and figs. There were six acres of alfalfa on the property, and two in eucalyptus trees which at the time was widely promoted as a valuable wood for a wide variety of uses to which it later proved unsuitable. The rest of the land was in gardens, the barn yard and a poultry yard, and landscaping of citrus and ornamental trees around the house, which still stands at the corner of Berkeley and Hawkeye roads in Turlock. Others in the area included J. C. Evans, who had arrived in 1907 and had sixty-two acres, with twenty-one and a half in alfalfa, fifteen acres in peaches and sixteen acres in grapes with the rest in "one thing and another." He also kept cows, pigs and poultry. Nearby, O. J. Abbott had twenty acres divided between a one and a half acre eucalyptus grove, five acres of plums, six of Tokay grapes, three in alfalfa and the rest, about four acres, in melons. G. R. Price had four and a half acres each of alfalfa, Tokay grapes and peaches, and five acres in wine grapes. In 1909, he had raised three acres of sweet potatoes and three acres of cantaloupes between the rows of grapes. South of Turlock, James E. Arthur was less diversified, having only alfalfa and cows on his forty acres. He had twenty-one cows in 1910, milking fifteen of them, and he planned to double the size of the milking herd. North of town, L.C. Cooper told Vandegrift that he rented out his 151 acres for $30 per acre to tenants who grew melons, squash, alfalfa, corn, pumpkins, watermelons and sweet potatoes. His land was watered mostly by subirrigation, like many others, and the high water table made the soil so moist that posts made of willow sticks sprouted and began to grow.[66] Several of the farmers interviewed said they were "holding" their land for prices perhaps five times their original investment, illustrating not only the rapid rise in land prices but perhaps even a willingness on the part of some settlers to profit from the real estate boom themselves.

The story in *The Earth* was unabashedly optimistic, painting as bright a picture as possible of profitable farms, rising real estate values and comfortable homes. The 1915 investigation of small farms in the Modesto and Turlock irrigation districts sponsored by the U.S. Department of Agriculture's Office of Experiment Stations was presumably more objective. Performed by Wells Hutchins, who later became a leading expert on irrigation institutions and Western water law, the investigation involved ten farms in each of the districts, ranging from five to thirty acres in size with an emphasis on those of about twenty acres. Farmers surveyed were apparently selected at random, but in the Turlock District they came primarily from the Denair area with a couple from Ceres and one from Hughson. Among those interviewed was Henry Thake, who had come from Iowa in 1907 to settle on twenty acres a mile northeast of Turlock in Mrs. Simon's colony. He had paid $100 per acre for his land with 50 percent down and the rest due in one year at 8 percent interest. In 1915, the land was worth $400 an acre, and he was growing ten acres of alfalfa, four acres of grain, three acres of watermelons and two acres of cantaloupes. He also had livestock fed on his own hay. As a sign of his prosperity, Hutchins noted that he drove a Ford car. On the other hand, A.W. Johnson's fortunes presented a less satisfactory picture. He farmed twenty acres two miles north of Denair, growing eight and a half acres each of alfalfa and grain or corn, with two acres in garden and orchard and an acre in buildings. He was deeply in debt and was planning to leave the area. When he arrived from Oregon, he had first raised melons in Turlock, but did poorly at that. His job as a TID ditchtender kept him from devoting full time to his farm, so he sold his hay and grain for cash rather than feeding it to stock himself, which Hutchins believed made his farm less profitable. Hutchins considered him a "natural kicker," complaining about everything. Johnson thought twenty acres was not enough land to support a family without outside income. Another farmer north of Denair, J.G. Strong, farmed twenty-two and a half acres, with fifteen of them in alfalfa,two and a half in melons and one and a half in orchard. He also fed thirty Jersey and Guernsey cows. From his survey, Hutchins found that farmers in the Turlock District were more likely to rotate alfalfa with "gyp" corn and oats than their counerparts in the Modesto District. Although cash crops like melons were widely grown, most farmers still depended for much of their income on dairy cows and the sale of surplus hay and corn, and the most successful small farmers were those like Thake and Strong who maintained their own dairy herds. The specialty crops could be rewarding, but risky. In 1915, Hutchins noted that watermelons had been worth a healthy $18.00 a ton at the start of the season, but at the end were selling for only $2.00 a ton. Peaches, the most popular orchard crop, were hardly paying expenses.[67]

The eight farms described above were not necessarily typical or an accurate cross-section of the Turlock Irrigation District. Most of them were located in the Denair area, and what was true there might not have been true in Hilmar or Ceres or elsewhere. None of the farmers seem to have been the Swedish or Portugese settlers so important to the area. Even with such omissions, however, these brief case studies can help form an impression of how the irrigation age was progressing. Clearly, the pioneer period for this second generation of agricultural settlement south of the Tuolumne had been brief, lasting no more than ten or fifteen years. The years of devastating sandstorms, rabbit drives and other hardships passed quickly. So did the initial dependence on alfalfa, and while a livestock and forage crop economy still predominated in terms of acreage, the economic value of melons, sweet potatoes, orchards and vineyards confirmed the belief in diversification held by the early advocates of irrigation. The summer expanse of stubble and fallow was being replaced by the cool green of alfalfa or the shade of young orchards. The far separated houses of the grain era were succeeded by closely spaced small farms, their houses often surrounded by the palms and oranges that bespoke a luxuriant image of California. The changes spread to the towns, too. In twenty years, Turlock had grown from the sand-blown village of two hundred inhabitants that greeted the first irrigation immigrants to a prosperous community of about three thousand people.[68]

Ever since irrigation was little more than a distant vision, its advocates had proclaimed that it would, as if by magic, bring forth a new era of growth and prosperity. At times, the claims must have seemed extravagant but the reality, for the Turlock area, turned

A hotel, a hardware store and a newspaper office
marked Hughson's beginnings in 1907.
Source: Margaret Sturtevant.

out to be just what had been promised; the goals established by the irrigation ideal were realized. Elsewhere, the results would not always be so satisfactory, as corporate farms and large landholdings sometimes became the beneficiaries of federal and state irrigation projects. For the Turlock area, however, irrigation was the catalyst for the region's second great transformation. The first, thirty years earlier, had turned the plains from a grazing country to an agricultural one, and established a population and a way of life that remained stable until it was no longer economically viable. Not only did irrigation bring with it a new population and a different way of life, but it altered the relationship between man and his environment on the plains south of the Tuolumne. The canals and ditches and levelled checks changed the face of the land and, with the water they distributed, modified one of the outstanding features of the San Joaquin Valley climate; the summer drought. The impact of the irrigation revolution on every level – economic, social, environmental – was so profound that irrigation easily became the most important single fact in the history of the Turlock region.

CHAPTER 5 – FOOTNOTES

[1] TID Assesment Roll, 1900.

[2] *Stanislaus County Weekly News,* July 13, 1900. However, TID records do not show Henry Stirring as a property owner.

[3] Kevin Starr, *Americans and the California Dream, 1850-1915,* (New York, 1973), p. 202.

[4] U. S. Dept. of the Inter., Census Office, *Report on Agriculture by Irrigation in the Western Part of the United States at the Eleventh Census, 1890* by F. H. Newell, (1894), p. 33.

[5] *Land of Sunshine,* July 1900.

[6] *Stanislaus County Weekly News,* Nov. 1, 1901.

[7] On Hultberg's role in the Alaska Gold Rush see L. H. Carlson, "The Discovery of Gold at Nome, Alaska," in Morgan B. Sherwood, editor, *Alaska and Its History,* (Seattle, 1967), pp. 353- 380.

[8] N. O. Hultberg to Harry B. Ansted, Mar. 29, 1925, in Hohenthal, Supplementary Material.

[9] Robert C. Doherty, "The Fin de Siecle Investment Corporation," (student paper, Stanislaus State College, 1972), pp. 3-8.

[10] Hultberg to Anstead, Mar. 29 1925.

[11] *Idem.*

[12] *Stanislaus County Weekly News,* Jan. 1, 1904.

[13] Author's interview with Arvid E. Lundell, Mar. 4, 1983.

[14] Tinkham, p. 596.

[15] Lundell interview; author's interview with William Tell, May 10, 1983; *Streams in a Thirsty Land,* p. 77.

[16] Esther Hall Crowell notes in Turlock Public Library, p. 7.
[17] Hultberg to Ansted, March 29, 1925. [18] Crowell notes, pp. 6-7.
[19] T. C. Hocking interviewed by Helen Hohenthal, June 12, 1930, in Hohenthal, Supplementary Material. [20] Adams, 1904, p. 113.
[21] *Modesto Morning Herald,* Supplement, Apr. 22, 1904.
[22] *Stanislaus County Weekly News,* Jan. 22, 1904.
[23] *Ibid.,* Apr. 22, 29, 1904; *Modesto Morning Herald,* Supplement, Apr. 22, 1904.
[24] E. J. Cadwallader, "History of Turlock," (typescript, 1945), p. 24.
[25] Caldwallader, pp. 4-5.
[26] F. L. Vandergrift, " How a Woman Led a Colony to Turlock-Denair," *The Earth,* (Feb. 1910), pp.2-5, 12; *Turlock Journal,* Feb. 11, 1910; Helen Hohenthal, "New Settlers in the Turlock District," (unpublished chapter prepared for *Streams in a Thirsty Land*).
[27] Cadwallader, p. 6. [28] *Ibid.,* p. 8.
[29] *Streams in a Thirsty Land,* pp. 165-170. [30] *Ibid.,* pp. 162-163.
[31] *Stanislaus County Weekly News,* June 21, 1907.
[32] Lloyd Gross interviewed by Margaret Sturtevant, 1974.
[33] Cadwallader, pp. 9-10.
[34] David F. Lane to Helen Hohenthal, June 18, 1930, in Hohenthal, Supplementary Material.
[35] Cadwallader, pp. 9-10.
[36] TID Assessment Rolls, 1900, 1910. [37] Elias, p. 161.
[38] James J. Rhea, *The Turlock District, Stanislaus County, California,* (San Francisco, 1912), p. 7.
[39] Maass and Anderson, pp. 162, 191, 287,289.
[40] B. A. Etcheverry, *Irrigation Practice and Engineering,* vol. 1, (New York, 1915), pp. 104-110; U. S. Dept. of Agric., Office of Experiment Stations, Irrigation Investigations, Bulletin No. 145, pp. 36, 80-82, in Elwood Mead, *Textbook for Reading Courses in Irrigation Practice,* (Berkeley, 1906).
[41] Bulletin No. 145, pp. 28-29, in Mead, *Textbook for Reading Course.*
[42] *Ibid.,* p. 29.
[43] Author's interview with Carl Muller, Feb. 24, 1983; author's interview with Grant Paterson, Jan. 26, 1983. Adams, 1904, gives additional information on laying out and constructing farm irrigation systems.
[44] Author's interview with Grant and Mildred Lucas, Feb. 17, 1983.
[45] Muller interview.
[46] Lundell interview; Muller interview; author's interview with Arthur Starn, Feb. 25, 1983; author's interview with Henry Schendel, Mar. 17, 1983.
[47] Board minutes, vol. 5, p. 62 (Nov. 12, 1914); S.T. Harding, *Operation and Maintenance of Irrigation Systems,* (New York , 1917),pp. 258-259.
[48] Author's interview with Norman Moore, Jan. 19, 1983.
[49] Adams, 1916, p. 83.
[50] *Stanislaus County Weekly News,* June 8, 1900. [51] Lucas interview.
[52] *Modesto Morning Herald,* Supplement, Apr. 22, 1904; Adams, 1904, p. 116.
[53] *Modesto Morning Herald,* Supplement, Apr. 22, 1904.
[54] "Irrigated Crops: Turlock Irrigation District," Meikle files, vol. 4, Exhibit 524. Figures prior to 1914 may contain inaccuracies.
[55] *Idem.*
[56] Adams, 1904, p. 118. [57] "Irrigated Crops, " Exhibit 524.

[58] *Soil Survey of the Modesto-Turlock Area* (1909), p. 53.
[59] *Stanislaus County Weekly News,* Aug. 23, 1901.
[60] *Modesto Morning Herald,* Supplement, Apr. 22, 1904.
[61] Adams, 1904, p. 117; "Irrigated Crops," Exhibit 524.
[62] Adams, 1904, p. 117.
[63] *Soil Survey of the Modesto-Turlock* (1909), pp. 52-53.
[64] "Irrigated Crops," Exhibit 524.
[65] Songsheet in Turlock Public Library. [66] Vandergrift,pp. 2-5, 12.
[67] U. S. Dept. of Agric., Office of Experiment Stations, Irrigation Investigations, *The Settlement of Small Farms in Modesto and Turlock Irrigation Districts,* by Wells Hutchins, (Berkeley, Aug. 5, 1915), pp. 1-34.
[68] Bramhall, p. 39.

6

The Turlock Irrigation District 1900-1913

The completion of the TID's canals and laterals was a notable achievement and one that greatly benefitted the Turlock region, but it did not automatically bring an end to the problems that had beset the district itself. In 1900 the TID's finances were still in a shambles and the anti-irrigationists, who were in part responsible for that situation, showed little inclination to abandon the battle they had already lost. In addition to clearing away the debris left by more than a decade of struggle and controversy, district officials had to learn how to run their new irrigation system. The district's survival was no longer in question but it would still take over a decade to end the era of divisiveness.

Money was the most pressing concern. Injunction suits to prevent the district from enforcing the collection of its taxes had been filed annually since the middle of the 1890s and the injunction continued in effect. The inability to collect taxes in turn meant that interest on the bonds went unpaid. More serious still was the fact that, in accordance with the Wright Act, the district had issued twenty year bonds bearing 6 percent interest that were to be repaid on the schedule provided in the act. It required the payment of interest only for the first ten years and then payments covering interest and principal for the remaining ten years. The purpose of the provision was to allow districts to construct their works at the most minimal initial cost, repaying the expense later, when irrigation had had time to raise an area's population, land values and prosperity, which would make the burden easier to bear. Unfortunately, C.C.Wright had underestimated the difficulties that irrigation districts would face. Even though the canal system was largely finished in 1900, it would be years before the Turlock region was fully subdivided and irrigated and its settlers well enough established to easily afford the repayment of the bonds. Legally, however, repayment would have to be made at an increasing rate as more and more bonds passed the ten year mark. Unless something was done, the prospects seemed good for a continuation of the financial crisis of the nineties.

The district's first move to unsnarl its tax collections came in February 1900 when the board began a legal action to remove the blanket injunction on the collection of 1895 taxes. In a circular addressed to district landowners, Collector W. H. Cameron announced that the action was being taken and offered delinquents a chance to pay that year's taxes plus a 5 percent penalty and fifty cents for costs or face the possibility of higher penalties and costs once the injunction was lifted.[1] The following month the court dissolved the injunction on 1895 taxes, effective April 7, 1900 and by April 20 some $13,000 had been paid into the district treasury.[2] The 1895 injunction had been a blanket one, preventing the taking of any property delinquent on that year's taxes, but in other years only the sale of land belonging to the plaintiffs named in the suit was enjoined. The existence of the blanket injunction, even though it covered only one year, had been one of the things that prevented the district from acting promptly to seize and sell the property delinquent in other years that was not specifically protected by the courts. With the 1895 injunction disposed of, the district could move more forcefully against the remaining delinquents. In April 1900 it was announced that all delinquent property for the years 1893 through 1899 had been sold to the district except that which was still shielded by the remaining injunctions. Almost a year later, in March 1901, the district made the first sale to a private party of land taken for nonpayment of irrigation taxes, in this case to heirs of the delinquent taxpayer.[3] It was not until April 1901 that the board ordered its attorneys to begin proceedings to find out if the remaining injunctions were going to be "perpetual and eternal" or if they could be removed and the taxes collected on the remaining delinquent lands.[4] By the time the board issued those orders, however, the previous injunction cases were about to be over-shadowed by yet another suit.

Every year with clockwork-like regularity the anti-irrigationists sued the TID collector to enjoin the sale of their delinquent property on the grounds that the district was illegal or its bonds were invalid. The year 1900 was no exception, and in April 1901 the case of *Baldwin et. al. v. Turlock Irrigation District* came to trial in superior court. Usually the proceedings attracted little notice. Injunctions were routinely issued but the cases were never pressed to a conclusion. The *Baldwin* case was different, for this time the bondholders intervened.[5] At stake was the question of the validity of the bonds and, of course, the ability of the district to collect the taxes needed to pay the interest due on them. James A. Waymire, who had become one of the largest owners of TID bonds, was among the attorneys representing the bondholders. The case was argued in April, but visiting Judge Lorigan of Santa Clara did not issue his decision until late June. In the meantime, the precarious nature of the Turlock District's finances became all too apparent. For at least three months in early 1901, Modesto banker Oramil McHenry had been cashing TID warrants even though the district had no money in its treasury to cover them. In June, however, he announced that he could extend no more credit to the TID. McHenry had hoped for a prompt and favorable decision in the *Baldwin* case since the delinquents would not pay up until they were forced to, and he was confident that if the decision had come sooner the public might never have known that a crisis was at hand. The crisis he was referring to was a threatened strike by the unpaid ditchtenders.[6] Shortly after news of the impending walkout was made public, Judge Lorigan ruled that the district's bonds were valid and the 1900 irrigation tax was legal. Although the outcome was entirely favorable to the district, the judge ordered that no sales of delinquent property were to be made until he made his final decree. Even though the district had won the case, the functioning of its tax collection apparatus was temporarily halted as surely as if another injunction had been handed down.[7] As a result, the district remained nearly destitute, and in August the ditchtenders made good their threat and walked off the job. The following day, the board found enough money to pay them one month's salary and they went back to work. Water was out of the system only two days as a result of the work stoppage.[8] The final decision in the *Baldwin* case did not come until September, when Judge Lorigan lifted all restraints against the collection of the 1900 tax. Even then, the district was still short of cash, forcing it to temporarily discharge the canal superintendent and his men, leaving only one man at the dam in November 1901.[9]

According to irrigation expert Frank Adams, the *Baldwin* case was an especially significant one because it "finally demonstrated to the people of the Turlock irrigation district in clear terms their obligation to the bondholders."[10] The dwindling number of die-hard anti-irrigationists were apparently still not convinced, for at the end of 1901 they filed the usual lawsuit to enjoin the collection of district taxes. This time, however, their efforts had little effect and 90 percent of the 1901 levy was paid on time and no injunction was issued.[11] While the district was working to dissolve the injunctions that still prevented the collection of some taxes from previous years, the board extended its earlier offer to accept the payment of the delinquent tax plus the small 5 percent penalty and costs to those still covered by injunctions.[12] How many of the hold-out delinquents accepted the offer is unknown but most of the delinquencies were paid by 1902, and thereafter opposition to district taxes ceased.

At the same time as it was settling its tax collection difficulties the district was working on a plan to refinance its bonded indebtedness. A mechanism to do just that was approved by the legislature in 1897. It allowed an irrigation district, with the approval of its bondholders, to exchange its old bonds for new ones with longer maturities and lower interest rates. Refunding, as the practice was known, was first proposed in the Turlock District in 1899, when petitions were circulated calling for the substitution of $1,200,000 worth of new 5 percent bonds maturing in from twenty to forty years for the same value in the old 6 percent bonds that were about to come due. Organizers of the petition drive noted that in 1900 the tax levy would reflect an obligation to pay $72,000 in interest and $66,000 in principal, which would require a relatively high tax rate. Understandably, the petition was signed by some of the largest landowners in the district.[13] The TID board took up the matter of refunding on September 21, 1899, but deferred action until October 3 when, after a long discussion, the petition was withdrawn.[14] Little more was done until 1901, after amendments to the law made refunding easier. In June 1901 the board resolved that the outstanding bonds should be refunded as soon as possible.[15] The bondholders were obviously receptive, especially after Judge Lorigan confirmed the validity of the district's bonds. They met soon after that decision in San Francisco and agreed to refinance the debt owed them.[16] In a letter to the TID board, Daniel Meyer, representing the bondholders, wrote that their meeting had considered a proposal by the district's attorney, P.J. Hazen, as well as one from James A. Waymire. A subsequent meeting had resulted in the appointment of a bondholder's committee consisting of Meyer, Waymire and Horace Crane, Secretary of the Fin de Siecle company, which owned a quarter of the land in the district and held $52,000 in its bonds. In response, the board appointed directors

Miller McPherson and S. H. Crane and attorney Hazen to negotiate a settlement.[17]

There appears to have been substantial agreement from the start on terms that would be satisfactory to both the bondholders and the district. Those terms were included in the petition that had to be signed by two-thirds of the property owners representing two-thirds of the assessed value of the district before an election could be called on the refunding bonds. The bondholders would be asked to accept $1,200,000 worth of new 5 percent bonds in exchange for all the outstanding old bonds, amounting to $1,170,000, as well as interest that would have been due in 1902 amounting to $63,720, another $75,749.26 in interest unpaid as of July 1901 and certain unpaid warrants issued by the district in the sum of $34,951.58. In other words, the district's major creditors would be paid $1,200,000 in new bonds to cancel $1,344,420.84 in debts.[18] At its November 1901 meeting, the board accepted the petition, and on December 10, 1901, the bonds won easily on a vote of 171 to 22.[19] The approval of the bonds meant that only a final agreement with the bondholders was still needed. By April 1902 that agreement was largely completed, lacking only the ratification of enough of the bondholders to put it into effect. By that time, American bondholders with $771,000 worth of TID bonds had already consented and European bondholders, mostly in London and Zurich, holding $260,000 were expected to agree soon, despite an initial reluctance to accept some parts of the pact.[20] In its final form, the contract signed on May 7, 1902 differed only in detail from the proposal contained in the bond petition. New bonds with a par value of $1,109,000 were to be exchanged for $1,170,000 worth of old bonds, the interest due on the new bonds in July 1902 would be cancelled and an additional $30,000 in interest coupons on the old bonds would be handed over to the district. In addition, the bondholders consented to pay a cash premium of $30,000 to the district at the time of the exchange.[21] In all, the district paid off its debts for about eighty cents on the dollar and retained enough cash to cover its operating expenses.[22] Although as a result of the settlement the bondholders received less than they had originally expected to get, the compromise was hardly unfair since many of them had paid much less than 80 percent of face value in the first place and by refunding they guaranteed the value of their investment.

Refunding settled affairs with the purchasers of the district's first two bond issues, but it had little effect on the overall marketability of the district's securities. Bonds totalling $300,000 issued in 1905 and 1910 for work on the canal system were often sold to contractors to pay for their work, and a large bloc of the 1905 issue was sold to W.R. High only "so that the district may dispose of them to bona fide purchasers without the necessity for readvertising."[23] As bonds from a $1,206,000 issue approved in late 1910 were offered for sale they were purchased by Turlock banker and Mitchell heir Horace Crane, who was usually the only bidder.[24]

Irrigation district bonds in general, not just those issued by the TID, remained difficult to sell at anything like their par value. Through the initial efforts of the Oakdale and South San Joaquin irrigation districts, legislation authorizing the state certification of irrigation district bonds was enacted in 1911. A commission consisting of the State Engineer, the Attorney General and the Superintendent of Banks was established to examine the status and feasibility of the districts and of the projects their bonds were intended to pay for. If the projects were physically and financially prudent, the bonds were certified by the commission and could be purchased by banks, trusts, insurance companies and state school funds. Not only did certification provide a wider market for irrigation district bonds, it suggested, though it did not guarantee, that they were a safe investment. The bond certification law went through several revisions before reaching a form acceptable to most districts in 1913. The Turlock District's bonds were certified the following year, and thereafter they were more readily and profitably sold.[25]

Bonds paid for the district's major capital expenditures, but the day-to-day operating money came from taxes. The irrigation district law allowed the ordinary tax levy to be used only for the payment of the interest and principal due on the bonds. Taxes for salaries, maintenance, and operation expenses had to be submitted to the voters annually in special elections. The operating assessments usually passed handily, and only in 1901 and 1909 did a special assessment fail to win the needed two-thirds majority. In 1901, refunding the bonds apparently provided the funds the tax would have raised, and in 1909 the directors simply called another election and that time the voters approved the levy.[26] In the early years of irrigation, the directors, according to Frank Adams, "hesitated to spend even pennies where dollars were required for efficient administration."[27] Their conservatism extended to assessing enough money to operate the system, and in 1905 two directors had to be appointed "to see about getting money to run the District until taxes are collected and paid into the Treasury," and a similar money famine occurred in 1910 and probably in other years as well.[28] It was not until sometime after 1910 that the directors learned to assess enough to operate the district and avoid an annual financial crisis.

The last piece of financial business remaining from the calamitous nineties was James A. Waymire. The ex-judge was widely credited with the salvation of the TID but disputes over his contract and compensation dragged on for years. In the final stages of construction, Waymire ran out of money and some of his creditors took their claims against him to the district. In January 1901 the board, tiring of the whole business, referred the problem to its attorney. The following month, the directors voted that the district had "been compelled to expend money and incur liabilities" to finish the canals because Waymire had failed to complete his contract. They ordered a review of the records to see how much the district had spent, and declared that Waymire's right to control any part of the work was at an end.[29] Waymire, on the other hand, insisted that the district still owed him a substantial sum for the work he had done. The claims lay before the board for some time until the directors finally rejected them in October 1904. It was reported at the time that although they appreciated Waymire's steadfast dedication to his job, the board members insisted they could not legally pay him any more than his contract required.[30] Waymire assigned his claims to San Francisco attorney W.T. Baggett, who then sued the district for $126,700, including $74,700 for additional work done on the canal, and various sums for damages and to repay Waymire for his assistance in refunding the bonds.[31] The trial had not yet ended in September 1906 when it was learned Waymire might soon be evicted from his Oakland home. He had mortgaged the property in 1895 to finance his work on the Turlock canal, and lost it to foreclosure in 1900, remaining only as a tenant.[32] In early 1907, Waymire and Baggett won a $20,000 judgement against the district, and although the board voted to appeal the award, a compromise was soon reached calling for the payment of the whole sum in bonds.[33] The victory came too late for Waymire. In December 1907 neighbors watched as he and his wife were evicted from their home and their furniture piled on the lawn.[34] Despite his losses, Waymire continued to dream great dreams. By 1908 he was actively promoting a scheme to bring water from Lake Tahoe to San Francisco. He was still optimistic about the chances of his latest plan when he died in April 1910.[35]

While the district was struggling to put its financial affairs on a business-like basis it was also struggling with the operation of its canal system. Water had apparently been run in 1900 without any sort of formal organization whatsoever; the few irrigators may have simply taken what they needed from the newly completed canals. As the district's first full irrigation season began in 1901, ditch-tenders were hired and B.W. Child was named superintendent of the canal to oversee the distribution of water.[36] Child was both an attorney and a civil engineer but there is no indication of what, if any, experience he had in operating a canal system. Even if he was qualified, he lacked the equipment needed to do the job. The TID had no devices for measuring the water flowing into various parts of the canal system or seeping out of the dirt ditches. Without instruments, Child and his men had to rely on nothing more than their own judgement to apportion water coming down the main canal between the laterals and then between the irrigators on the various ditches. It must have been a frustrating job, for in August, Child tendered his resignation but the board refused to accept it.[37] When operating money ran out late in the year, Child and his crew were laid off, only to be rehired in January 1902, but, for Child at least, at a lower salary.[38] There is no evidence to indicate that the district was any better equipped to run water in the 1902 season than it had been in 1901, but the increasing number of irrigators meant that distribution by guesswork was likely to be even less satisfactory. The animosities that were bound to result from the situation were apparent in a petition presented to the board in early 1903 demanding Child's discharge without citing any specific reason why he should be fired.[39] Child remained superintendant until September 1904 when he was replaced by L.F. Hastings.[40]

In response to a request from the Modesto Irrigation District for assistance in organizing its water distribution system, the Department of Agriculture dispatched Frank Adams from its Irrigation and Drainage Investigations staff in Berkeley to both of the Tuolumne River districts in 1904. Adams was appalled at what he found in the Turlock District. After three years in the irrigation business, the district's distribution methods were still as crude as they had been in the beginning. The district had no office; Child furnished his own "harness-shop desk room."[41] No one was keeping any written records of who the water users were or how much water they received and, of course, without any gauging equipment there was no way to record the flow of water through the canal system. Adams concluded that, "It is believed that the irrigation and drainage investigation can do Modesto and Turlock districts no greater service than to urge upon them the necessity of giving greater attention than they have in the past to the details of water distribution."[42] He recommended that the district procure the necessary equipment to measure the water being carried by the canals and delivered to each farmer and he urged the keeping of accurate records on the operation of the canal system and the use of water by irrigators. Finally, Adams argued that ditchtenders, who would use the measuring devices and keep the records, should be paid more than the $50 per month rate then prevailing in the districts and given every other encouragement to do their job well.[43]

Adams' advice was scarcely revolutionary. Irrigation may have been new to the Tuolumne River region, but it was far from a novelty in California or other parts of the West. The Turlock District certainly could have drawn on the long experience of the Fresno area canals or on other parts of the San Joaquin Valley, to say nothing of southern California. Why it did not do so may be explained either by the inexperience and insular thinking of the board and superintendent, or by the reluctance of the board to spend money on anything but the barest necessities. Adams' prodding and the obvious need for better management as larger irrigated acreages began to tax the canal system finally resulted in some overdue improvements. In 1906 the board authorized its third superintendent, Burton Smith, to purchase gauging equipment and in succeeding years efforts were begun to keep the kind of records Adams had recommended.[44] Despite its obvious shortcomings, the Turlock Irrigation District was not completely backward. Soon after irrigation began, the district started installing its own telephone system to coordinate operations, and in early 1908 the board authorized the purchase of the district's first automobile. By 1910 the first car had been replaced and its successor was being driven 15,000 miles a year over the sandy roads and canal banks on district business.[45]

Frank Adams noted that in 1904 the district did not have an office. Steps were soon taken to acquire one, but the process was complicated by a rivalry between Turlock and Ceres. As early as

1889, C.N. Whitmore had offered to build the district a building in Ceres if it would make that town its headquarters, but the board continued to hold its meetings in the public rooms of one of Turlock's hotels.[46] In May 1903 it voted to move its office to Modesto, where it could get two rooms in the same building as its attorney. The decision brought a howl of protest from Turlock District residents and produced offers to provide the board with an office. The real estate firm of Hultberg and Soderberg offered to rent the district space in their building in Turlock and C.N. Whitmore again proposed to build the board a brick building in Ceres.[47] Nothing was done at the time, although the board abandoned its plan to move to Modesto. No more was heard of the matter until May 1905 when Ora McHenry offered the district an office in the new bank building in Turlock.[48] At the same time C.N. Whitmore was trying to get Ceres area residents to subscribe the $1,200 needed to erect a twenty-foot by thirty -six-foot building in that town for the use of the district for as long as its office remained in Ceres.[49]

At its November 1905 meeting the board voted to accept the Turlock bank's offer but the following month the newspapers reported that the board had voted three-to-two to go to Ceres instead.[50] According to the local press, Turlock's boosters were lulled into a false sense of security by the November decision to accept the bank's proposal and in December nothing was said about the office until the 2:13 p.m. train whistle was heard. Some of the board members were to take that train, so waiting to the last minute, they quickly voted to accept the Whitmore offer and then just as quickly left town.[51] Despite entreaties from Turlock residents and a great deal of indignant muttering, the board refused to change its new position and it soon moved into a new building with room for its records and its secretary, Anna Sorensen. The office remained in the brick building on Fourth Steet in Ceres only from 1906 until 1910, when Turlock businessmen offered to pay $1,000 in cash and buy $9,000 worth of the district's bonds if the office were returned to Turlock. Although the Ceres Board of Trade contended that since the bonds would have to be repaid with interest the Turlock offer was less generous than it seemed, the board promptly voted to move back to Turlock. It met in the Commercial Bank building on Main Street and when the bank itself moved to a new building, the board decided to purchase the old one as a permanent home.[52] District offices were to remain in that location for nearly half a century.

The Turlock Irrigation District was wracked by controversies during the early years of irrigation, and most were not as simple, nor as trivial, as the squabble over which town would lay claim to the district's office. Ineffective management of the canals and the persistent shortage of cash undoubtedly contributed to dissension between irrigators and district officials. In 1910, for example, a Turlock Water Users Association was formed that seemed to thoroughly distrust district officials, especially Superintendent Burton Smith. Smith spoke at one meeting of the group in an effort to put to rest rumors circulating about a recent canal break but the gathering became so heated that the press reported a hundred people left because they were denied the right to speak. At another meeting, the propriety of land purchases involving a proposed reservoir site was discussed and charges of graft were heard. In September of that year, a petition with 150 names on it demanding Smith's removal on grounds of incompetence and extravagence was presented to the board and an angry discussion ensued but Smith retained his position. A short time later, the Water Users Association continued its efforts to root out what it perceived as waste or mismanagement on the part of the district by requesting and being granted permission to audit the district's financial records. The group did discuss plans for canal enlargement and for reservoirs, but it seems such discussions all too easily deteriorated into a tirade against Burton Smith.[53]

Further stirring the waters was the presence on the board of the outspoken Andrew Hallner, who had come to Turlock to assist N. O. Hultberg in establishing the Hilmar colony. Hallner was elected to the board in 1909, but by early 1911 he was at odds with the other members and the superintendent and even suggested he was ready to quit.[54] At a tense board meeting following publication of a so-called resignation statement, Director T. A. Owen demanded an explanation of certain of Hallner's comments. The *Turlock Journal* reported that in response, "Director Hallner talked all the time, but evaded answering the questions. Director Owen . . . finally told Mr. Hallner that he was a deliberate liar – and it wasn't the first time either."[55] The *Journal* obviously had little sympathy for Hallner but it nevertheless devoted considerable space to a "resignation meeting" he held in Hilmar to air a long list of complaints touching virtually every aspect of district operations. In the end, however, he announced that his constitutents had persuaded him not to resign. Ironically, Hallner did quit the board three months later in a dispute over the payment of a small sum he claimed the district owed him.[56]

The details of these disputes, for which there are no really unprejudiced sources of information, are less significant than their broader causes. Frank Adams was a keen observer of conditions in the Turlock region, and he concluded that much of the problem lay with the attitudes of the irrigators themselves. He wrote in retrospect that,

> Few in the district had had experience in irrigation, and in their ignorance of their own difficulties and of the difficulties of the district officers many expected far more from the district than it was possible for the district to give. The first few years were therefore naturally stormy ones and meetings of dissenters were frequent and often acrimonious . . . In spite of continual agitation there was only slow improvement in the earlier crude conditions for seven or eight years, and neither the first nor the second nor the third superintendent was able to come up to the expectations of the water users.[57]

Along the same lines he had written in 1904 that,

> These men have not only to learn by experience and careful study what distributive systems are best suited to their needs and conditions, but perhaps more difficult, they have to learn to stand by each other and their officers in carrying into effect the systems they choose. Each individual irrigator is conscious of his right to help manage the affairs of the district, but is not always careful to be sure of his facts before complaining, or, in the face of complaints from others, to stand by the officers he helps to elect. Very naturally the result is not always happy.[58]

Indeed, the results were not happy in the Turlock Irrigation District, but fortunately they were also temporary.

The turmoil, and the reasons for it given by Frank Adams, suggest that attention needed to be given to the relationship between irrigation agencies and the people they serve. It has often been said, and rightly so, that irrigation is a cooperative enterprise, founded on the common use of a water supply and the ditches that carry it. As essential as cooperation is to irrigation, however, the TID's early history shows that it does not always come easily. The Turlock District may have had a particularly difficult time because it suffered from poor management of its canals and because the community of irrigation farmers that it served was still in the process of settlement and many of its members had not yet developed a sense of common interest with other irrigators and with the district. Like the general period of irrigation pioneering in the Turlock area, the era of turmoil surrounding the TID proved mercifully brief. The district's ability to manage its canals improved and the irrigators gained the kind of experience and forebearance alluded to by Frank Adams. In 1914 a series of dramatic events, related in a later chapter, proved that the district's years of internal strife were over.

The district faced more problems than just running the canals it already had; it also had to improve and expand its canal system and water supply. The reliability of the canals, especially the main canal through the foothills, was hardly satisfactory. Its weakest links were the 2,800 feet of wooden flumes along steep hillsides and on tall trestles across the ravines. As if to underscore the vulnerability of those structures, the tallest flume, at Morgan Gulch about half a mile below La Grange Dam, collapsed in August 1901. There were reports that splintered timbers found in the rubble suggested the use of dynamite, and the board eventually posted a reward of $500 for the apprehension of those "who were instrumental in the blowing up of the Morgan Gulch flume."[59] Even though they voted the reward, there was apparently considerable feeling on the board that the structure was a victim of leakage and rot rather than explosives, and no one ever claimed the reward.[60] Morgan Gulch flume was rebuilt, but since wood was subject to decay, it was only a matter of time before another such collapse would occur. Not only was the canal's reliability questionable but its capacity needed to be improved. Although it was supposedly designed to carry 1,500 second-feet, only about 1,000 second-feet could be run safely.[61] As more and more settlers moved into the district it would be necessary to carry more water to the plains to serve more irrigators at once and thus keep the interval between irrigations as short as possible.

Before improvements in the original canal system could be considered, there was an immediate need to expand the lateral system on the plains. At the south end of the district, Lateral 6, the southernmost canal, passed to the north of the Hilmar Colony. Before irrigation could begin in much of the colony, additional laterals were needed. In November 1902 the board gave Horace Crane the contract to build a lateral south of Lateral 6, in effect paying for the work by giving him bonds.[62] Lateral 7 was completed in 1903 and the next year another short lateral, numbered 8, was built off of Lateral 7.[63]

No substantial work on the main canal was done until June 1904 when the Delaney Gulch flume broke. Some 525 second-feet had been flowing through it when it sprang a leak, which washed out the base and caused the tall flume to collapse. It was said that the flume had long been considered a weak structure, in part because it had been built during the troubled times of the 1890s and had warped before use.[64] The break prompted the board to hire a consulting engineer, Samuel Fortier of Berkeley, to suggest ways of strengthening the main canal and bringing it up to its planned capacity. Fortier reported in August 1904 that all wooden structures should be replaced with steel flumes over the ravines and concrete walls where the flumes were built along hillsides, as they were at Snake Ravine.[65] Even before Fortier had made his formal recommendations, the board had arranged to have something done about Snake Ravine. The La Grange Ditch and Hydraulic Mining Company had agreed to fill in a portion of the ravine in exchange for district bonds and whatever gold they unearthed in the process.[66] Within weeks of Fortier's report, petitions had been signed and an election called to vote $200,000 in bonds for canal improvement. The bonds passed by an overwhelming 229-to-21 margin and by October 1904 bids were being opened for several projects, including the rebuilding of Peaslee and Morgan flumes and scraper work to increase capacity.[67] Unfortunately, the district was unable to take Fortier's advice concerning steel flumes, opting instead for less expensive wooden flumes that were nonetheless sturdier than the ones they replaced.

The construction of Laterals 7 and 8 had solved, at least partially, the problems of supplying water to the southern reaches of the district, but there was still a substantial area around Delhi and east of the Turlock Main Canal south of Hickman that paid district taxes yet had no access to water. Construction of a canal to irrigate these areas was first considered in early 1904, but the break at Delaney flume and the urgent need to make repairs on the main canal prevented any action at that time.[68] By the beginning of 1907 demands for a new canal to the south and east sections of the district were renewed. At first, the board responded only that there was no money available for such work, but continued requests resulted in an order to Superintendent Burton Smith to begin a survey.[69] By August 1907 the board announced that a new canal costing approximately $60,000 would be built to serve high ground near the Merced River, and in April 1908 plans for a $100,000 bond issue were drawn up.[70] At the August 1908 bond election, three of the district's five divisions voted heavily against the measure, the vote in Division 3 (west of Ceres and Turlock) running 40-to-0 against the bonds. Vigorous support from Division 4, which included Turlock, and from Division 5, where the lands to be benefitted were located, was enough to offset opposition and carry the day.[71] Surveys for what became known as the Highline Canal were soon completed but the route chosen stirred so many protests that adjustments were ordered.[72] It was not until April 1909 that the board could finally call for bids on construction of the canal. Unfortunately, all the bids were rejected and the work was readvertised but repeatedly drew no bidders.[73] The reason for the lack of interest was a flaw found in the 1908 bond election by a Modesto attorney that threatened to invalidate the bonds. As a result, the process of circulating petitions and calling an election had to be repeated, though this time with more support from throughout the district.[74] Contracts for two of the five sections of the canal had been awarded to T. K. Beard even before the bonds were reapproved.[75] In August 1910 both the $100,000 in bonds and the remainder of the work on the Highline project, coincidentally valued at $100,000, were awarded to Albert Chatom, and the canal was finished in 1911.[76] It ran from the main canal several miles above Hickman, south through the low foothills to the east of the Turlock Main Canal to a point east of Delhi where it turned to follow the Merced River and eventually spill into it. In 1912, two cross ditches were built to link the Highline to the Turlock Main southeast of Turlock, and in 1913 and 1914 Lateral 5½ and most of Lateral 2½ were built and the Ceres Main Canal was extended further south.[77] These final extensions and additions substantially completed the district's network of canals.

Considerable work, most of it difficult and expensive, remained to be done on the main canal. The repairs and improvements made in 1904 were not much more than a band-aid meant to hold things together until more permanent changes could be made. The trestle and flume at Peaslee Creek, 60 feet high and 360 feet long, was the first major structure to receive further attention. As early as April 1909 consideration had been given to building a hydraulic fill there in place of the trestle. Under that plan, earth and rocks washed out of adjacent hills would be deposited around the trestle until the embankment reached the height of the canal. The wooden flume imbedded in the man-made hill would then be removed and replaced by a concrete canal. A contract for the job was awarded in late 1909 and the fill was completed the following year.[78] At the end of 1910 district voters approved a $1,206,000 bond issue designed to pay for a variety of improvements, including a reservoir and major work on the main canal that would replace the remaining wooden flumes at Morgan and Delaney gulches with hydraulic fills and make other changes necessary to raise the capacity of the canal to 2,000 second-feet. The bond issue won by a nearly three-to-one margin but although some work was done in 1912, most of it was not undertaken until 1913.[79] Besides the substitution of fills for the old wooden flumes, plans included the enlargement of the intake tunnel, widening of the canal from a width of eighteen feet to a new minimum of thirty feet, and concrete lining wherever excessive seepage seemed likely. The contract for most of the work went to T. K. Beard, who finished the job in 1914.[80]

Although canal improvements received first priority the district was soon considering ways to increase the water supply that fed the canals. Hopeful promoters of the Turlock and Modesto districts had once believed that the Tuolumne River was an almost inexhuastible resource, but once the canals were in place its limits became increasingly obvious. In the first couple of years, when only a few thousand acres needed irrigation, there was enough water in the river even in August to make operation of the canal worthwhile. As more settlers arrived to claim their share of the river's bounty, the irrigation season began to shorten. There was not less water available but when the river began to fall in early summer the amount entering the canal dropped to the point that only a few farmers a day could hope to get the water. For practical purposes, the more farmers were using the water, the earlier regular irrigation would end. By 1910 the irrigation season was often over by early July. Enlarging the canals ensured that a reasonable rotation interval could be maintained when there was ample water reaching La Grange Dam, but larger canals were no help when there was only a trickle to put in them. The only way to stretch the irrigation season was to store in a reservoir some of the water that poured over the dam every spring, for use when the river's natural flow was no longer adequate.

The question was where to build a reservoir. In 1889, surveyors from the Irrigation Branch of the U.S. Geological Survey collected information on reservoir sites in the upper Tuolumne watershed, including Lake Eleanor and Tuolumne Meadows. Ten years later,

The reliability of the main canal was improved by replacing flumes with earthen fills, like this one at Morgan Gulch.

government surveyors led by J.B. Lippincott and F.H. Newell proposed a reservoir on the Tuolumne at Hetch Hetchy Valley and examined other sites as well.[81] San Francisco soon began a campaign, discussed at length in the following chapter, to secure Hetch Hetchy for its municipal water supply, although the Turlock and Modesto irrigation districts insisted that they should have the site if San Francisco's efforts failed. Another potential dam site on the main stem of the Tuolumne was located in the river canyon a few miles upstream above La Grange Dam. There is no way to tell when it was first determined that a reservoir could be built there, but by 1909 a private power company had acquired a portion of the site and was planning to build a dam to store water for power generation. The company's proposal in time became the irrigation districts' Don Pedro reservoir. Although the Turlock District especially soon took an active interest in the Don Pedro site, it was not until a few years later that the districts could afford to develop it. In the meantime, two other possibilities, one high in the mountains and the other along the main canal, were considered for storage of the water that the Turlock Irrigation District needed.

The first formal action toward a reservoir came in 1908. The year was a dry one, and that as much as anything else turned attention to storage. On August 4, 1908, the Turlock board had just dispensed with the routine business before it and had begun to discuss water storage when a telephone call came from the Modesto District's directors requesting a conference on the investigation of reservoir sites above La Grange Dam. The Turlock directors took the train to Modesto that very afternoon and soon agreed to join in a preliminary survey of potential sites to be conducted by the engineers for the two districts.[82] Later that month the engineers were in the mountains. They visited at least a dozen places where dams could be built, including Hetch Hetchy, Poopenaut Valley, located on the river directly below Hetch Hetchy, and three sites – Hardin ranch, Ackerson Meadow and Stone Meadow – on the south and middle forks of the Tuolumne. Except for Tuolumne Meadows, all the remaining sites were on creeks north of Hetch Hetchy and river's grand canyon. The following summer, the directors of the two districts met again and authorized additional surveys of Poopenaut Valley, Hardin ranch, Tuolumne Meadows and Benson Lake, with the Turlock District paying two-thirds of the cost in general accordance with the provisions of the 1890 agreement covering the division of future projects above La Grange Dam. Contour maps of the four sites were completed, but since the districts contemplated an application to the government for one or more reservoirs on federal park or forest land, the survey maps had to be coordinated with U.S. land survey lines. Unfortunately, the work at Benson Lake, located on Piute Creek amid 10,000 foot peaks, could not be completed so late in the brief high elevation summer season. It was not until September 1910 that the Benson Lake survey maps were finally completed and a report made to the districts' boards.[83]

Although located in an isolated section of Yosemite National Park, Benson Lake was deemed the best suited for development by the engineers. An inexpensive timber crib or masonry dam there could impound 75,000 to 100,000 acre-feet of water. The difficulty of getting permission to build a dam in a national park was offset by the fact that no costly private lands would have to be purchased. The engineers admitted Benson Lake did have other disadvantages. They reported that,

> The only drawback to the site is its difficult accessibility, but this is more than compensated for (by) the excellence of the reservoir and dam site and the further fact that nearly all of the material lies on the ground and that only tools and supplies will have to be transported. It is not feasible to construct a road to the site but all supplies and materials can readily be taken in on pack animals.[84]

As soon as they received the engineers' report, the boards ordered that the documents needed to file a request for the use of the Benson Lake site be prepared.[85] The final reference to Benson Lake came in a joint meeting of the Turlock and Modesto boards on May 11, 1911, when the attorneys for the districts were directed to file on the site, but there is no record of their ever having done so.[86]

Less than a month after sending its superintendent to the high country in search of reservoir sites, the TID board ordered him to survey two areas along the district's main canal where reservoirs might be built.[87] These sites, known as Morley and Hughson lakes, were near one another about mid-way between La Grange Dam and Hickman in the rolling foothills just south of the Tuolumne River. They were natural basins that, with dams to fill the gaps between surrounding hills, could be turned into relatively inexpensive reservoirs. The survey of the two sites was completed by December 1908. According to Burton Smith, a reservoir at Morley Lake would innundate 3,000 acres and hold over 48,000 acre-feet of water. The Hughson, or Dickinson, Lake site to the west had a capacity of about 35,800 acre feet. Estimated costs for earthen levees around the lakes totalled about $24,000 for Morley Lake and over $18,000 for the lower reservoir. When the board heard Smith's report on Dickinson Lake on December 1, 1908, it also heard considerable argument on whether or not the district should proceed with construction. David Lane, C.N. Whitmore, T.A. Owen and T. K. Caswell, all prominent businessmen or farmers, appeared as representatives of a recent citizens meeting to urge that action be taken to begin construction of a reservoir. On the other hand, Andrew Hallner presented a petition from the Hilmar Colony Association opposing the construction of a reservoir at that time.[88]

Nothing was done in regard to the foothill reservoirs until early 1910, when Superintendent Smith was ordered to make more studies and cost estimates.[89] North of the river, the Modesto District had already begun construction of a similar reservoir along its main canal.[90] By the end of May 1910 Smith and consulting engineer Edwin S. Duryea, Jr. of San Francisco had completed their work and reported in favor of the development of both of the prospective sites. At Morley Lake, the largest landowner was Alfred Davis and the purchase of his and other properties at the site was expected to cost $150,000, which was twice as much as the dams, gates and other work would cost. The development of Dickinson Lake, which could be expanded to store 60,000 acre feet, was more expensive, with an estimated $160,000 for works and over $138,000 to buy the lands of Hudelson, Hughson and others. A mass meeting in Turlock on May 31, 1910, to consider the reservoir question was attended by three hundred people and produced a letter to the board asking the district to circulate petitions for a bond election to cover the cost of reservoir construction, canal enlargement and other improvements.[91] Opposition to the bonds came from the recently formed Water Users Association, which, true to its basic distrust of anything done by the board or Burton Smith, argued that the foothill sites were unsuitable because they would lose too much water to evaporation and seepage.[92] Most irrigators, however, favored the plans and on December 6,1910, the bonds passed with over half of the $1,026,000 total earmarked for reservoir construction.

Considerable time passed before anything further was done toward actually building a reservoir. There was some discussion of purchasing land in 1911 but for some reason it was deferred to 1912, at which time the board was petitioned by a majority of the district's property holders, representing a majority of land value, to begin buying land.[93] By that time only the development of Morley Lake was still being cisidered, deferring indefinitely the smaller, more expensive Dickinson Lake reservoir. Land was soon acquired and at the end of 1912 a contract was awarded to T.K. Beard for the costruction of job of building the seventeen dams and levees, the highest of which was twenty-eight feet high, that would enclose the reservoir.[94] The remainder of 1913 was spent constructing the dams and the outlet gate, and as the year ended the project was nearly finished.

Increasing the water supply and canal capacities to bring more water to the land was an obvious and predictable necessity, but getting rid of excess water was equally crucial to the TID's success. Drainage is always an important aspect of irrigation, but its significance is not always fully appreciated. Surface drainage, which removes the runoff from ends of irrigated fields, is a normal procedure with some irrigation methods; the problem is the control of subsurface water. Even the most efficient irrigation methods require the application of at least a little more water to the soil than the crops can use. The excess percolates into the subsoil, where in time it accumulates. The rate of accumulation and its effects vary with the geologic structure of the subsoil, but a rising groundwater table is the likely result. As the water table rises into the plant root zone, roots can be suffocated and the plant stunted or even killed. Rising groundwater can also dissolve the salts found in some soils and carry them upward, concentrating them at the upper limits of its advance. When that advance approaches the root zone, plants can be poisoned by the salts as well as injured by the waterlogged soil. Without drainage, either natural or artificial, irrigation is eventually a self-destructive practice.

The Turlock Irrigation District did not, at least in its early years, have a surface drainage problem because the check systems of flooding did not produce field runoff. However, it was soon apparent that it had a serious groundwater problem. The check system was not inherently wasteful, but inexperience and poorly levelled checks that took extra water to cover the high spots, contributed to excessive water use. When water was plentiful, irrigators were simply not cautious about its management. In fact, it was common to apply excess water in the spring in hopes that the soil would store it to compensate for the early close of the irrigation season. In heavy soils, water might stand on the surface for days, warning the farmer against overirrigation, but the sandy, porous soils of the Turlock region readily absorbed what was given them, perhaps even encouraging excessive irrigation. Seepage from unlined canals and laterals added even more water to the subsoil. Groundwater under the Turlock District did move, draining slowly toward the valley trough and some of the excess eventually seeped into the San Joaquin, Tuolumne and Merced rivers, but the sudden addition of so much water to the soil resulted primarily in a rapid rise in the water table. Before irrigation began, groundwater was found at a depth of twenty to twenty-five feet. By the end of 1903 that level had risen by sixteen to eighteen feet so that water was often only five or ten feet below the surface.[95] In Miller McPherson's well west of Ceres, the water level stood at eighteen feet when it was dug in 1882. In 1906 it was up to within eight feet of the surface, and within six feet by 1908. At the same time, the water level near Turlock was only five feet below ground.[96]

The rapidly rising water table had several effects. The most obvious was the appearance of lakes in low-lying areas, especially in the region southwest of Ceres and Keyes and around Turlock. The oldest and largest of these was Gilstrap Lake, located about a mile southwest of Ceres. The lake first appeared in June 1904. Its water level fluctuated, but at a minimum the lake covered forty acres and could grow to seventy acres in extent. Owner W. A. H. Gilstrap sounded the lake and found its maximum depth to be nine and a half feet. A short distance west of Gilstrap Lake, Hendy Lake covered seven acres about eight inches deep. On R.R. Moore's land three and a quarter miles southwest of Ceres on Crows Landing Road, a ten to fifteen acre lake appeared in 1906.[97] These were only a few of the lakes that formed by 1907 or 1908 in the Turlock Irrigation District. Some only filled small pot-holes, but others covered substantial areas and were more or less permanent, with a lush growth of tules and cattails around them.

Even where no water appeared on the surface, sections of some irrigated fields "became so soft that they could not be crossed with teams and machinery."[98] The only more or less useful thing to come out of this accumulation of excess water was subirrigation, which relied on underground water for irrigation rather than the usual application of surface water. Some contended that melons grown on subirrigated land were higher in quality than those irrigated directly from the canal. The supposed virtues of subirrigation may in some cases have reflected a hope that the rising water table posed no real threat to crops or to future land sales and settlement. The value of subirrigation was easily exaggerated, and in reference to similar conditions in the Fresno area, U.S. Department of Agriculture investigators warned that "in too many cases subirrigation but marked a stage in the rise of ground water from a deep subsoil to within a few inches of the surface. The subirrigated lands of one season usually become the waterlogged lands of succeeding seasons."[99] Subirrigation was successful only if the water table remained just below the root zone. In the Turlock District, the water table varied as much as several feet a year in response to irrigation and rainfall.[100] Deep-rooted crops like alfalfa and orchards were especially susceptible to a high or fluctuating groundwater table and these were the most important crops grown in the Turlock area.

The need for drainage to control the water table did not attract much attention until 1906. At the beginning of the year, George Manuel returned briefly to discuss drainage, and both Turlock and Modesto district officials visited Fresno to see what was being done there.[101] In June 1906 Director Edward Kiernan and Superintendent Smith were directed to install pumping plants for drainage in the hope that pulling water out of the underground strata and releasing it into a nearby canal would lower the water table in the vicinity of the pump. One pump was installed near Turlock during the summer of 1906, but Kiernan and Smith were forced to report that it was too small to do much good. The board ordered them to get a larger pump and keep trying.[102] At the time the first experiment was made, there was some doubt that an irrigation district could legally engage in drainage work since it was not specifically authorized by the irrigation district act. TID attorney P. J. Hazen drafted legislation that became law in March 1907, permitting districts to add drainage to their other responsibilities.[103] With full legal authority and the experience of the previous year behind it, the district drilled a drainage well about three miles west of Turlock in May 1907, and hooked up a 26 horsepower gasoline engine capable of pumping 1,000 gallons per minute from the shallow well. The experiment lasted a month, but the engine broke down so often that the pump ran only intermittantly. The water table was not affected, and pumping as a means of drainage was temporarily abandoned.[104]

As early as February 1907 some months before pumping was

discarded, the board authorized construction of a drainage ditch from R. R. Moore's farm to Lateral 3, with the county paying half the cost because the lake had made Crows Landing Road impassable.[105] As more and more landowners appeared before the board to ask that something be done to drain their lands, the directors decided to purchase a clamshell dredge to dig drainage ditches. The $5,000 machine was manufactured in Stockton and had a 2,800-pound, one-cubic-yard bucket. The dredger boom, engine and turntable were mounted on an eighteen-foot by thirty-foot sled, which was winched along on movable rollers and planks laid on the ground.[106]

The dredger went to work in the summer of 1907 on the south side of Moore Lake, enlarging a shallow ditch cut with scrapers earlier that year. The drainage channel created by the clamshell was eight feet wide on the bottom and up to fourteen feet deep. It ran south about a mile to Lateral 3, which was also deepened for several miles by the dredger. Work on the Moore drain and Lateral 3 was completed in early 1908 and the machine was moved to the north side of Moore Lake to begin digging a ditch from there to Gilstrap Lake. It took a year to dredge the three-and-half-mile-long Gilstrap drain and a quarter-mile branch to nearby Swain Lake. At the same time that work began on the Moore drain in 1907, horse-drawn scrapers were used to dig a shallow ditch from McCabe Lake, located further west than the others, to a nearby slough of the San Joaquin River.[107] By the spring of 1909, the district had over eleven miles of drainage ditches but their effect on the water table was for the most part minimal. McCabe Lake was completely emptied but Moore Lake and Swain Lake were reduced in size rather than eliminated and Gilstrap Lake was scarcely affected at all. The Hatch Lakes, a quarter to a half a mile away from the drain, were unaffected and Ephraim Hatch even claimed another small lake appeared after the construction of the drain.[108] The results were so discouraging that in 1910 the district sold its dredger.[109] The drains that had been built were not maintained and the ditches were soon choked by cave-ins and tules. Though their efforts had largely failed, district officials recognized that a solution to the high water table was still needed and the bond issue voted in December 1910 included over $145,000 for drainage.[110] Just how the money would be spent was not specified, but Superintendent Smith had advocated the use of electric pumps if a means could be found to supply the power at reasonable cost. For the time being, however, nothing further was done to drain the waterlogged lands of the Turlock District.

Gilstrap Lake, southwest of Ceres, was the largest of the lakes created by a rising water table under the newly irrigated TID.

A TID dredger at work on a drainage canal, 1907.
Source: U.S. Department of Agriculture.

The Turlock Irrigation District had made substantial progress by 1913. Its finances were stablized, the size and reliability of its canals had been improved and construction was underway on the district's first reservior. More importantly, the district was becoming a more stable and harmonious place. The atmosphere during the first decade of the century was so full of distrust and animosity that the functioning of the district was almost certainly impaired. For a public irrigation agency, just as for farmer-owned mutual water companies, a basic consensus was needed. In a community of newcomers and inexperienced irrigators that consensus proved difficult to achieve. By 1913, however, the Turlock District had improved its operations and thereby removed some of the immediate causes for argument, while its irrigators had grown more accustomed to their role as members of an irrigation community. One sign of the emerging maturity of its constituents was the voting pattern on bonds for the Highline Canal. At the first election, only those divisions which would directly benefit from the canal voted in favor of it, but when a second election proved necessary two years later, the vote was overwhelmingly favorable. Bitter dissent, however, would flare one more time as the district faced a serious challenge to its Tuolumne River water supply.

CHAPTER 6 – FOOTNOTES

[1] *Stanislaus County Weekly News,* Mar. 2, 1900
[2] *Ibid.,* Mar. 23, Apr. 20, 1900. [3] *Ibid.,* Apr. 13, 1900, Mar. 5, 1901.
[4] Board minutes, vol. 2, p. 152 (Apr. 2, 1901).
[5] *Stanislaus County Weekly News,* Apr. 26, 1901. [6] *Ibid.,* June 14, 1901.
[7] *Ibid.,* June 28, July 19, 1901. [8] *Ibid.,* Aug. 9, 1901.
[9] Board minutes, vol. 2, p. 193 (Nov. 5, 1901). [10] Adams, 1904, p. 100.
[11] *Stanislaus County Weekly News,* Nov. 22, 1901, Jan. 3, 1902.
[12] Board minutes, vol. 2, pp. 219-220 (Mar. 4, 1902).
[13] *Stanislaus County Weekly News,* Sept. 1, 1899.
[14] Board minutes, vol. 2, p. 67 (Sept. 21, 1899), vol. 2, p. 75 (Oct. 3, 1899).
[15] Board minutes, vol. 2, pp. 160-161 (June 4, 1901).
[16] *Stanislaus County Weekly News,* July 5, 1901.
[17] Board minutes, vol. 2, pp. 164-165 (Aug. 6, 1901).
[18] *Stanislaus County Weekly News,* Oct. 11, 1901; Board minutes, vol. 2, p. 191 (Oct. 15, 1901), vol. 2, pp. 194-201 (Nov. 5, 1901).
[19] Board minutes, vol. 2, p. 204 (Dec. 16, 1901).
[20] *Stanislaus County Weekly News,* Apr. 25, 1902.
[21] Board minutes, vol. 2, pp. 227-230 (June 3, 1902).
[22] Adams, 1916, p. 19.

[23] Board minutes, vol. 2, p. 466 (Sept. 16, 1905), p. 476 (Dec. 5, 1905), vol. 3, pp. 338-339 (Aug. 29, 1910). The quotation is from *Stanislaus County Weekly News,* June 28, 1907.
[24] For example, see Board minutes, vol. 3, p. 592 (Apr. 24, 1911), vol. 4, p. 218 (Oct. 21, 1912).
[25] Adams, 1930, pp. 32, 40; Adams, 1916, pp. 48-49.
[26] *Stanislaus County Weekly News,* Sept. 13, 1901; Board minutes, vol. 3, pp. 105-106 (Aug. 16, 1909), p. 124 (Sept. 20, 1909).
[27] Adams, 1916, p. 79.
[28] Board minutes, vol. 2, p. 468 (Oct. 3, 1905); *Turlock Journal,* Oct. 14, 1910.
[29] Board minutes, vol. 2, p. 135 (Jan. 7, 1901), p. 137 (Feb. 5,1901).
[30] *Stanislaus County Weekly News,* Oct. 7, 1904.
[31] *Ibid.,* Nov. 4, 1904.
[32] *Ibid.,* Sept. 21, 1906, Mar. 15, 1907.
[33] *Ibid.,* Apr. 24, 1907; Board minutes, vol. 2, p. 610 (June 4, 1907).
[34] *Turlock Journal,* December 27, 1907.
[35] W. Turrentine Jackson and Donald J. Pisani, *A Case Study in Interstate Resource Management: The California-Nevada Water Controversy, 1865-1955,* (Davis, Calif., 1973), pp. 5-6.
[36] Board minutes, vol. 2, p. 144 (Mar. 5, 1901).
[37] *Ibid.,* p. 167 (Aug. 6, 1901).
[38] *Ibid.,* p. 193 (Nov. 5, 1901), p. 211 (Jan. 7, 1902).
[39] *Ibid.,* p. 286 (Mar. 3, 1903).
[40] *Ibid.,* p. 380 (Sep. 6, 1904).
[41] Adams, 1916, p. 80.
[42] Adams, 1904, pp. 129-130.
[43] Adams, 1904, pp. 130-134.
[44] Board minutes, vol. 2, p. 505 (June 5, 1906).
[45] *Ibid.,* p. 668 (Jan. 7, 1908); *Turlock Journal,* July 22, 1910.
[46] *Modesto Daily Evening News,* Jan. 25 1889.
[47] Board minutes, vol. 2, p. 294 (May 5, 1903), p. 298 (June 2, 1903), p. 332 (Nov. 3, 1903).
[48] *Ibid.,* p. 435 (May 2, 1905).
[49] Mildred D. Lucas, *From Amber Grain . . . to Fruited Plain: A History of Ceres, California and Its Surroundings, 1776-1976,* (Ceres, Calif., 1976), pp. 96-97.
[50] Board minutes, vol. 2, p. 473 (Nov. 7, 1905); *Stanislaus County Weekly News,* Dec. 15, 1905.
[51] *Stanislaus County Weekly News,* Jan. 26, 1906.
[52] Board minutes, vol. 3, p. 248 (May 9, 1910), p. 432 (Dec. 13, 1910); *Turlock Journal,* May 13, 1910.
[53] *Turlock Journal,* July 1, Sept. 16, Oct. 14, Dec. 2, 1910.
[54] *Ibid.,* Mar. 3, 1911.
[55] *Ibid.,* Apr. 14, 1911.
[56] *Ibid.,* Apr. 28, July 28, 1911.
[57] Adams, 1916, p. 79.
[58] Adams, 1904, p. 129.
[59] Board minutes, vol. 2, p. 176 (Sept. 3, 1901); *Stanislaus County Weekly News,* August 23, 30, 1901.
[60] *Stanislaus County Weekly News,* Sept. 6, 1901.
[61] Irrigation District Bond Commission, *Report on Turlock Irrigation District,* p. 16.
[62] Board minutes, vol. 2, p. 262 (Nov. 5, 1902).
[63] Irrigation District Bond Commission, p. 19.

[64] *Stanislaus County Weekly News,* June 17, 1904.
[65] Board minutes, vol. 2, p. 361 (July 5, 1904); *Stanislaus County Weekly News,* Aug. 5, 1904.
[66] *Stanislaus County Weekly News,* July 15, 1904.
[67] Board minutes, vol. 2, p. 395 (Oct. 11, 1904), p. 397 (Oct. 18, 1904); *Stanislaus County Weekly News,* Dec. 9, 1904.
[68] Board minutes, vol. 2, pp. 354-355 (Apr. 5, 1904).
[69] *Ibid.,* p. 563 (Jan. 2, 1907), p. 572 (Feb. 5, 1907).
[70] *Stanislaus County Weekly News,* Aug. 9, 1907; Board minutes, vol. 2, p. 692 (Apr. 7, 1908).
[71] Board minutes, vol. 2, p. 737 (Aug. 31, 1908).
[72] *Ibid.,* vol. 3, p. 6 (Jan. 5, 1909).
[73] *Ibid.,* pp. 79-80 (July 6, 1909), p. 115 (Sept. 7, 1909).
[74] *Turlock Journal,* Nov. 19, 1909; Board minutes, vol. 3, p. 236 (Apr. 18, 1910).
[75] Board minutes, vol. 3, p. 186 (Jan. 13, 1910).
[76] *Ibid.,* pp. 338-339 (Aug. 29, 1910).
[77] Irrigation District Bond Commission, p. 16.
[78] Board minutes, vol. 3, p. 47 (Apr. 6, 1909), p. 153(Nov. 10,1909).
[79] *Ibid.,* p. 425 (Dec. 12, 1910).
[80] Irrigation District Bond Commission, p. 16.
[81] *Modesto Daily Evening News,* Oct. 21, 1889; *Stanislaus County Weekly News,* Sept. 8, 1899; Schuyler, *Reservoirs,* p. 386.
[82] Board minutes, vol. 2 p. 727 (Aug. 4, 1908); *Stanislaus County Weekly News,* Aug. 7, 1908.
[83] *Turlock Journal,* Oct. 11, 1910; Board minutes, vol. 3, p. 99 (Aug. 3, 1909).
[84] *Turlock Journal,* Oct. 11, 1910.
[85] Board minutes, vol. 3, pp. 371-373 (Oct. 10, 1910).
[86] *Ibid.,* p. 625 (May 11, 1911).
[87] *Ibid.,* vol. 2, p. 738 (Sept. 1, 1908).
[88] *Ibid.,* pp. 758-759 (Nov. 4, 1908), p. 771 (Dec. 1, 1908).
[89] *Ibid.,* vol. 3, p. 192 (Feb. 14, 1910).
[90] Rhodes, "Thirsty Land," 140.
[91] Board minutes, vol. 3, pp. 244-245 (May 9, 1910), pp. 259-260 (May 31, 1910), p. 262 (June 7, 1910); *Turlock Journal,* June 3, 1910.
[92] *Turlock Journal,* Nov. 25, 1910, Dec. 2, 1910.
[93] *Ibid.,* June 16, 1911; Board minutes, vol. 4, pp. 126-131 (Aug. 5, 1912).
[94] Board minutes, vol. 4, p. 253 (Dec. 2, 1912), p. 286 (Jan. 13, 1913).
[95] Adams, 1904, p. 128.
[96] U. S. Dept. of Agric., Office of Experiment Stations, *Drainage of Irrigated Lands in the San Joaquin Valley, California,* by Samuel Fortier and Victor M. Cone, Bulletin No. 217 (1909), p. 47.
[97] J. H. Dockweiler, *Water Needs of the Turlock and Modesto Irrigation Districts and Quantity of Water Remaining Available for Storage and Diversion to the Cities around San Francisco Bay,* (San Francisco, 1912), pp. 138-143.
[98] Sweet, Warner and Holmes, *Soil Survey of the Modesto-Turlock Area,* p. 44.
[99] Fortier and Cone, p. 14. [100] Dockweiler, p. 135.
[101] Board minutes, vol. 2, p. 483 (Jan. 8, 1906), p. 497 (Apr. 3, 1906).

[102] *Ibid.,* p. 507 (June 5, 1906), p. 529 (Sept. 4, 1906).
[103] *Stanislaus County Weekly News,* Mar. 22, 1907.
[104] Fortier and Cone, pp. 50-51.
[105] Board minutes, vol. 2, p. 573 (Feb. 5, 1907); *Stanislaus County Weekly News,* Feb. 8, 1907.
[106] Fortier and Cone, pp. 53-54.
[107] Dockweiler, p. 145; Fortier and Cone, p. 54.
[108] Dockweiler, pp. 145, 155.
[109] Board minutes, vol. 3, p. 211 (Mar. 14, 1910).
[110] *Ibid.,* pp. 304-305 (July 28, 1910).

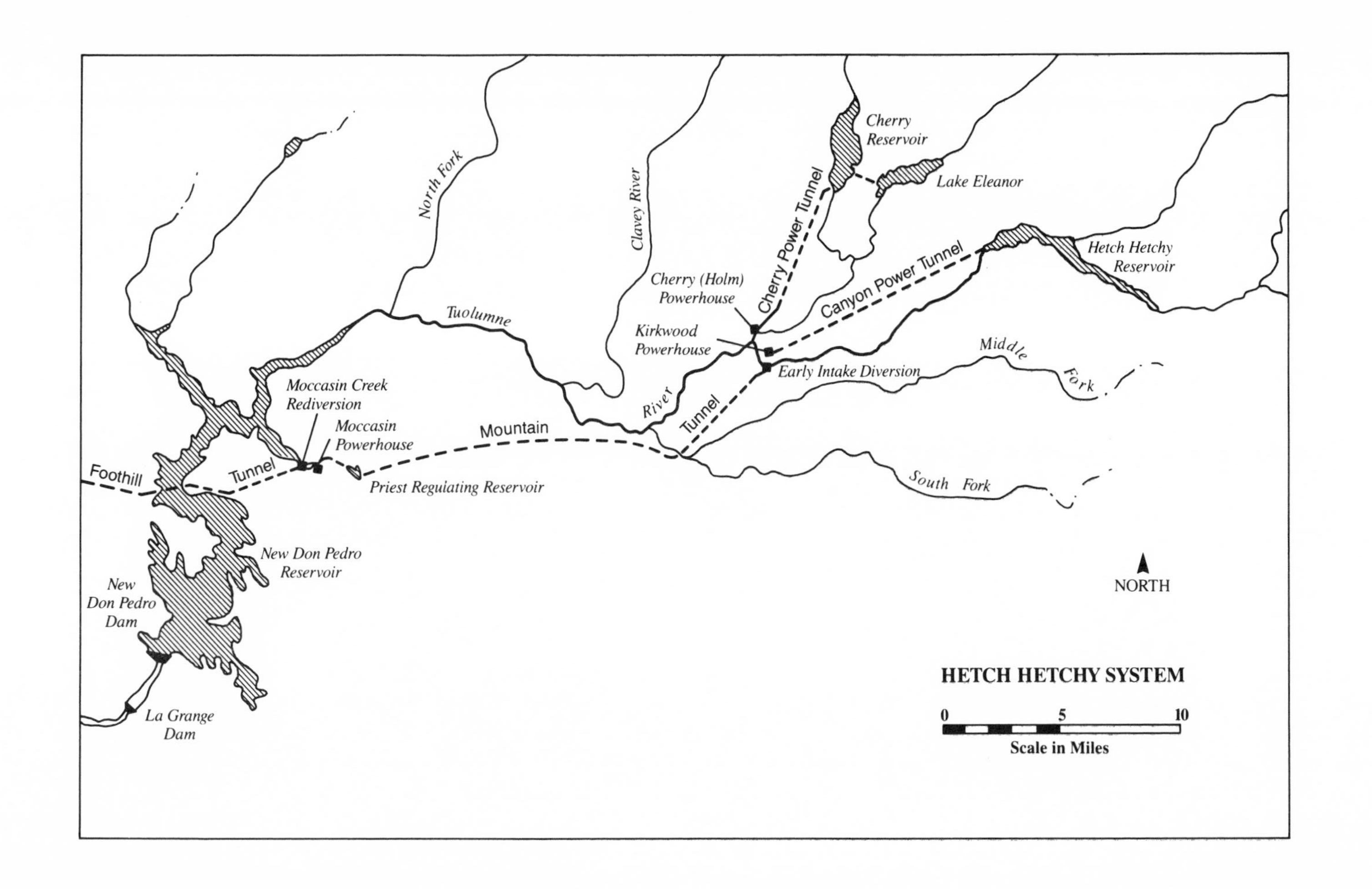
HETCH HETCHY SYSTEM
0
5
10
Scale in Miles
NORTH
Cherry Reservoir
Lake Eleanor
Hetch Hetchy Reservoir
Cherry Power Tunnel
Canyon Power Tunnel
Cherry (Holm) Powerhouse
Kirkwood Powerhouse
Early Intake Diversion
Clavey River
North Fork
Tuolumne
River
Middle Fork
South Fork
Tunnel
Mountain
Moccasin Creek Rediversion
Moccasin Powerhouse
Priest Regulating Reservoir
Foothill
Tunnel
New Don Pedro Reservoir
New Don Pedro Dam
La Grange Dam

7

Hetch Hetchy

In 1900 the Turlock and Modesto irrigation districts had the only major water rights on the Tuolumne River. They enjoyed a situation rare in the arid West; the prospect of expanding their diversions without any serious interference from rival appropriators. All that changed in 1901, when notices were posted in the name of James D. Phelan, the mayor of San Francisco, for water rights on the Tuolumne River and its Eleanor Creek tributary. Phelan was acting on behalf of his city, laying the groundwork for a project that would bring drinking water from the upper Tuolumne to the shores of San Francisco Bay. Phelan's claims, and the request for rights-of-way and reservoir sites on federal lands that followed, called for a dam at Hetch Hetchy Valley, a place naturalist John Muir called the Tuolumne Yosemite. Though not as breathtaking as its famous counterpart on the Merced River, Hetch Hetchy, like Yosemite, had steep granite walls and rock domes rising over 2,000 feet above the river. It had its waterfalls, too, including the misty Tueeulala and the roaring Wapama, both 1,700 feet or more in height.[1] And it had, at the lower end of the three and half mile long valley floor, a narrow outlet perfectly suited to dam building.

Despite the fact that Hetch Hetchy and the watershed above it had become part of Yosemite National Park when it was created in 1890, surveyors had still scrambled over the high country looking for places to put reservoirs to store some of the river's abundant water. Hetch Hetchy had been officially identified as one such site as early as 1891 by the U.S. Geological Survey, while the first faint suggestion that water from the Tuolumne could be used to supply San Francisco had come even earlier. A new city charter took effect at the turn of the century which authorized San Francisco to own and operate its public utilities, prompting City Engineer C.E. Grunsky to suggest the development of Hetch Hetchy as a municipal water supply.[2] Grunsky, who was familiar with the Tuolumne at least in part because of his service as the Modesto Irrigation District's first engineer, also sketched out a design for the project, including reservoirs at Hetch Hetchy and at Lake Eleanor, which was also in the national park. Once the city's plans became known, it was clear that the Turlock and Modesto districts had a serious competitor for the water they depended upon.

The city's first application to turn a spectacular, if remote, part of a national park into a reservoir was rejected by Secretary of the Interior Ethan A. Hitchcock in early 1903 as being incompatible with the purposes for which the park was set aside and beyond his authority to grant. San Francisco resubmitted the application but in December 1903 suffered a second rejection. During this time there was little public comment on the city's plan from the irrigation districts. The Turlock board resolved in June 1903 to oppose the granting of any rights to store or divert water on federal park or national forest lands, and attorney P. J. Hazen filed legal briefs opposing the project, but otherwise the district did, and could do, little.[3] After the secretary of the interior issued his second rejection, city officials prepared to introduce legislation in Congress to give them what the secretary would not. At the same time there seems to have been an effort to somehow placate the districts. The San Francisco Board of Supervisors suggested that a committee of its members meet with similar committees from the Modesto and Turlock irrigation districts. While it appointed itself and its attorney to be such a committee and resolved to avoid "all unnecessary antagonism with San Francisco," the TID board expressed its doubt that it could agree to any plan that did not fully guarantee enough water to the districts.[4] The Modesto District did not appoint a similar committee, but by that time intense public opposition was developing among the irrigators of the two districts and it seems unlikely that either side would have been willing or able to accept a compromise at that point.

Some San Franciscans grumbled that the hated Spring Valley Water Company, whose monopoly control of the city's water supply would cease if the municipally-owned Hetch Hetchy system were built, was surreptiously stirring up the farmers. Certainly Spring Valley was doing all it could to undermine the Tuolumne project, but the local opponents of the plan were probably motivated by their own vital interest in their water supply rather than by any prompting from the company. Most citizens of the irrigation districts probably agreed with the *Stanislaus County Weekly News* that there was a "beneficent and Divine plan" that water from the

Sierra was ordained to irrigate the dry plains below.[5] There was a common conviction that the river belonged to the territory through which it ran and any other use would be tantamount to theft. As one man put it years later when the issue was about to be decided, "God gave to San Francisco the Pacific Ocean; the rest of the world to the Standard Oil Company, but the waters of the Tuolumne River belong to Stanislaus County."[6] The districts and their residents seemed adamant; San Francisco had to be kept off the Tuolumne watershed.

The legislation introduced in 1904 failed to pass, so the city tried again in 1905. While it was making no apparent headway against either Secretary Hitchcock, who rejected the proposal again in 1905, or Congress, San Francisco's lobbyists were winning important friends in the nation's capitol. San Francisco engineer Marsden Manson managed to win over the man most responsible for the shape of President Theodore Roosevelt's natural resource policies, Chief Forester Gifford Pinchot. As the leading advocate of the doctrine that became known as utilitarian conservation, Pinchot believed that resources like water or timber should be conserved, not to prevent their use but to insure that they were managed as efficiently as possible to bring the greatest good to the greatest number of people. Although not insensitive to natural beauty, Pinchot came to the conclusion that Hetch Hetchy would serve a greater good as a reservoir than as a campground or scenic attraction. Pinchot told the president in early 1905 that he could see no objection to San Francisco's use of Hetch Hetchy and Lake Eleanor as reservoirs. Congressman J.C. Needham, who represented the districts, promptly talked with Hitchcock and found him unwavering, but Needham warned that Pinchot and others were trying to convince Roosevelt that Hitchcock had erred and ought to be overruled. Word that the president was involved brought a barrage of letters to him from the Turlock and Modesto districts urging him to do nothing to aid San Francisco.[7] Whatever his feeling at the time, Roosevelt took no action. In Congress, the city's bill died, in part perhaps because of Congressman Needham's position on the Public Lands Committee. San Francisco had gained some important allies but by the end of 1905 any hope of early action on Hetch Hetchy seemed remote.

In January 1906 word came that San Francisco had decided to abandon the fight. The city's Board of Supervisors, citing the failure of previous efforts and a recent letter from Congressman Needham as reasons to believe that the "Tuolumne supply cannot be acquired for years to come, if at all, and probably not at all," resolved to "refrain from expending further money, energy or time in the futile attempt to acquire the so-called Tuolumne system."[8] The terms of the resolution were deceptive. The abandonment of Hetch Hetchy was prompted not by the evident difficulty of obtaining a Tuolumne water supply but by a classic case of municipal corruption, as artfully practiced by Abe Ruef, boss of the city's Union Labor Party. Ruef, an attorney who held no public office himself, guided city policy and channelled payoffs to Mayor Eugene Schmitz and members of the Board of Supervisors. In January 1906 Ruef had been approached by William S. Tevis, a wealthy San Franciscan whose Bay Cities Water Company held water rights in the Lake Tahoe area and on the American and Cosumnes rivers. Tevis wanted the city to buy his proposed water system rather than build one at Hetch Hetchy, and he was willing to pay to see that it was done. He planned to offer the city a system costing $7,500,000 for a price of $10,000,000, and he offered Ruef over one-third of the profit, a million dollars, for his help in putting the scheme over. Ruef, in turn, planned to split half the sum evenly between himself and Mayor Schmitz and divide the remainder among the supervisors. To clear the way for the scheme, Ruef had the Board of Supervisors pass the resolution abandoning Hetch Hetchy. A special supervisor's committee was then formed to study alternative plans and it selected five, including Tevis's Bay Cities company, for further study by an impartial board of engineers. Ruef, who had written the committee report himself, made sure the other four proposals were so unsatisfactory that the Bay Cities plan would be the only possible choice. When the engineers balked at the limitations imposed on them, the supervisors went ahead regardless. The whole scheme collapsed when a vigorous graft prosecution began in late 1906 which toppled the Union Labor Party from power and landed Ruef in prison.[9] If anything, the Bay Cities episode added a new luster to Hetch Hetchy as a municipal supply. Since it was on public land and depended on city-owned water rights, there was no chance it could turn into the kind of swindle Tevis and Ruef had attempted. The Tuolumne had long been described as a source of pure water; in the wake of the graft prosecution the river took on the appearance of civic purity as well.

In 1907, San Francisco renewed its application for the Hetch Hetchy project and this time it had substantial advantages it had not had before. Ethan A. Hitchcock had been replaced as Secretary of the Interior by James R. Garfield, a protege of Gifford Pinchot and a friend of Marsden Manson. Garfield was clearly someone more likely to give favorable consideration to the city's request.[10] The earthquake and fire that ravaged San Francisco in April 1906 also resulted in a certain sympathy for the city and its need for an improved water supply. After Garfield held a short hearing in San Francisco in July 1907, it became apparent that this time the city was likely to get the reservoir sites it sought. Citizens of the irrigation districts rallied, as they had before, against any threat to their water.[11] More significantly, John Muir began his crusade to save Hetch Hetchy in September 1907 with a letter to President Roosevelt urging that the national park not be violated by the building of a lake in one of its finest valleys. Eventually, the battle over Hetch Hetchy became, in terms of public awareness then and now, a struggle between Muir and his allies and the city – between the value of preserving wild places and the utilitarian concept of managing land and water. Roosevelt found the choice difficult but warned Muir that the parks must not interfere with necessary development, and in the end he reluctantly supported the city's project.[12] Accordingly, Secretary Garfield granted San Francisco a permit to develop a Tuolumne River water system on May 11, 1908.

The Garfield Permit gave San Francisco reservoir sites and rights-of-way over federal lands for its aqueducts, power lines and roads but contained a number of conditions designed to protect the irrigation districts and even Hetch Hetchy Valley itself. Although Lake Eleanor was a smaller, less advantageous site, the permit required that the city develop it to its full capacity before damming Hetch Hetchy. It was widely believed that it would be many years before the city's needs would exceed what it could derive from Lake Eleanor and for at least that long the "Tuolumne Yosemite" would be spared destruction. The permit also required that the city could not "interfere in the slightest particular with the rights of the Modesto irrigation district and the Turlock irrigation district to use the natural flow of the Tuolumne River and its branches to the full extent of their claims, as follows: Turlock irrigation district, 1,500 second-feet; Modesto irrigation district, 850 second-feet."[13] San Francisco was prohibited from storing or diverting any of the river's natural flow until the districts' requirements were satisfied, up to the limits specified. In their original filings in

1889 and 1890, the two districts had claimed much higher flows, totalling 9,500 second-feet. By 1907, however, the combined capacity of their main canals was still under 1,600 second-feet, though they planned to increase it to the total used in the Garfield Permit.[14] Since water rights were a matter of state rather than federal law, the permit effectively kept the city from challenging the districts' rights to a total of 2,350 second-feet but did nothing to prevent the districts from asserting claims to a greater amount.

Besides protecting what were considered the districts' fundamental rights on the Tuolumne, the Garfield Permit offered them other benefits as well. Any surplus stored water in the city's reservoirs was to be made available to the districts at cost, providing them with at least some storage at a reasonable expense. The city was also prohibited from interfering with any storage projects undertaken by the districts anywhere on the watershed except Hetch Hetchy and Lake Eleanor. It was this provision that enabled, and perhaps encouraged, the districts to pursue their own high Sierra surveys beginning in the summer of 1908. Finally, the city was required to sell any surplus electric power generated at its facilities to the irrigation districts at cost for use in pumping for irrigation or drainage. Any disputes over the price to be paid for surplus water or power sold to the districts were to be settled by the secretary of the interior.

Although the permit was not formally issued until May 11, 1908, a United Press International dispatch dated April 30 announced that terms of the grant had been decided and revealed that those terms were the result of negotiations between the districts, represented by Congressman Needham, city officials and Secretary Garfield. Needham had attempted to secure a guarantee that the city would release water, from storage if necessary, to maintain a minimum of 2,350 second-feet at La Grange Dam at all times, even when the river's natural flow fell below that level. The city declined to make that commitment, claiming that it would be tantamount to requiring them to build storage for irrigation. On other points, Needham was more successful, and in the end, the report noted, each side thought Garfield favored its position.[15] Needham considered the conditions attached to the permit a victory, and it appears that the people of the districts agreed with him. The TID board, as well as the businessmen's Board of Trade, sent the congressman their thanks for a job well done.[16]

The San Francisco Board of Supervisors quickly ratified the

Garfield Permit only to have second thoughts a few months later. They worried that they had given up too much to the irrigation districts and their suspicions were further aroused by the districts' willingness to accept the conditions.[17] Whatever their misgivings, they pressed ahead with the project, putting the first bonds needed to pay for it before the city's voters in November 1908. The bonds were overwhelmingly approved and the city set about acquiring the isolated parcels of private land on the upper watershed it still needed. Meanwhile, John Muir and others dedicated to the preservation of Hetch Hetchy in its natural state began a campaign to rally the public to their cause and obstruct San Francisco in any way they could, including preventing congressional action sought by the city.[18]

A new president, William Howard Taft, entered the White House in early 1909 and he chose as his secretary of the interior Richard Ballinger. In October 1909 Taft and Ballinger visited John Muir in Yosemite, as Theodore Roosevelt had done six years earlier. When Taft left the park, Ballinger accompanied Muir on a visit to Hetch Hetchy. The new secretary was locked in a bitter political struggle with Gifford Pinchot over the course of national conservation policy and his dislike for Pinchot, who was closely associated with the Hetch Hetchy grant, may have made him more receptive to Muir's "Tuolumne gospel."[19] At any rate, on February 25, 1910, Ballinger notified San Francisco that it must show cause why Hetch Hetchy should not be eliminated from the permit granted by Garfield, which would leave the city only Lake Eleanor. A hearing held on May 25, 1910, did not produce a decision on Ballinger's order. Instead, a three-man advisory board of Army Engineers was named to review the alternatives to Hetch Hetchy and report to the secretary of the interior. The city's grip on Hetch Hetchy seemed to be slipping and an effort was made to get President Taft to intervene on the city's behalf. He declined to do so, but he did advise city officials to hire John R. Freeman, an eminent engineer from Providence, Rhode Island, to help prepare their case.[20]

San Francisco took Taft's advice and hired Freeman, along with a corps of engineering consultants, who labored until July 1912 to produce a detailed report on the city's plans to develop the upper Tuolumne watershed. The result was brilliant. Originally ordered only to show cause why Hetch Hetchy, which Garfield had decreed should be dammed only when Lake Eleanor was no longer sufficient, should not be dropped entirely from the permit, Freeman fundamentally redesigned the project to make Hetch Hetchy reservoir an indispensable centerpiece of the plan. Instead of the 60 million gallon per day aqueduct system, with provisions for only modest future expansion, planned by previous city engineers, Freeman proposed to ultimately draw 400 million gallons per day from the Tuolumne. Instead of a combination of canals and pipelines to carry water from the high mountains, he suggested a tunnel bored through solid rock. And instead of using a major part of the electricity the water could generate as it flowed down the Sierra slope to pump it over the Coast Ranges, Freeman designed a closed gravity system capable of generating immense amounts of power but requiring no pumping whatsoever to bring the water to the Bay Area. "The Garfield Permit," Freeman wrote, "is not broad enough for present needs."[21] He argued that the old permit had simply been outgrown by the increasing population of San Francisco and the other Bay Area cities that wished to join in developing a water supply, and by advances in the art of water supply engineering. The city was no longer satisfied to defend what Secretary Garfield had given it but wanted a great deal more. Freeman's report dramaticaly redefined the issues surrounding Hetch Hetchy and talk of the Garfield Permit virtually vanished.

Freeman recognized that three groups were likely to find fault with the city's plans; those who desired to sell the city some other source of water, the nature lovers, and those with rival claims to the Tuolumne River. The report therefore dealt not only with what the city wished to do but attempted to discredit arguments that might be raised against it. Proposals to use other rivers, including the Stanislaus, Mokelumne, Yuba, McCloud, Eel and Sacramento, were all shown to compare unfavorably with the Tuolumne. In response to assertions that a dam and reservoir would be blight on one of the Sierra's most spectacular landscapes, Freeman replied with cleverly retouched photographs showing the valley's cliffs and waterfalls mirrored in the calm waters of a reservoir.[22] Roads built by the city would, in Freeman's opinion, permit more than the handful of hardy tourists then using the valley to visit the area. He promised that the city would also build a scenic road around the edge of the lake, and he illustrated the point with photographs of similar lakeside roads in Europe as well as retouched pictures of Hetch Hetchy landmarks like Wapama Falls with the road drawn in.[23] Freeman's report was not just good engineering, it was superb propoganda as well.

Not surprisingly, Freeman also challenged the claims of the Turlock and Modesto irrigation districts. The districts had, of course, acquiesed in the Garfield Permit, but Freeman had no use for the conditions attached to it. He disliked the districts' right to invade reservoir sites that San Francisco might someday want for itself, opposed the mandatory sale of surplus water and power, and felt that the state courts rather than a federal permit should be the arbiter of water rights on the Tuolumne. Still, the districts did have prior rights that had to be respected and unless it was shown that those rights were limited in some way, difficult questions might be raised about whether there would be enough water to satisfy irrigation rights and still divert 400 million gallons per day to the city. To prove that the districts actually needed much less water than they claimed, engineer J.H. Dockweiler was dispatched to the area in 1911 to gather data on the districts' water requirements.

Dockweiler's research was nothing if not thorough. He reviewed diversion records and river flows, studied soils and crops and meticulously examined the irrigation system, from the amount of seepage from the canals to the practices of individual irrigators. Naturally, he gave particular attention to the drainage problem and even drilled his own test wells to measure the water table. His report contended that excessive amounts of water were being used by irrigators, often as a result of poorly levelled checks and wasteful methods, and for proof he pointed to the rising water table and the appearance of lakes, particularly in the Turlock District. Overall, he claimed that 40 percent of the water used by irrigators was actually being wasted.[24] On the basis of experience in other irrigated regions of the West, he concluded that applying two and half acre-feet annually per acre of irrigated land would be a reasonable and economical use of water. Since the districts were still far from fully developed, it was necessary to determine how much water they would need in the future. Here, Dockweiler had to make a set of assumptions. He postulated that at most 80 percent of the 257,353 total acres in the two districts would be irrigated in a single season and he further assumed that a quarter acre-foot of water per acre would be supplied from pumps used to control the water table. These figures, along with an assumed conveyance loss of 15 percent, due mainly to unlined canals, resulted in an estimate that the districts would never need more than 533,050 acre-feet from the Tuolumne River in any one year.[25] That amount was only about 100,000 acre-feet more than the city planned on exporting, and at those levels both irrigation and municipal use combined required only about half the river's annual flow.

Since 1908, events surrounding Hetch Hetchy had been viewed with interest but no great alarm in the Turlock and Modesto districts. In fact, Ballinger's challenge to the use of Hetch Hetchy threatened to emasculate San Francisco's plans for the Tuolumne to the obvious benefit of the irrigation districts. District officials certainly should have known that Dockweiler was making his detailed investigations on behalf of the city, but that in itself was not enough to create any concern. Freeman's report, however, suddenly made it plain that San Francisco coveted much more of the Tuolumne than had previously been imagined and would take a more belligerent attitude toward the districts' claims. The city's new plans became available in early August 1912, and on August 19, the Turlock and Modesto boards ordered their engineers and attorneys to gather information to respond to the assumptions and allegations contained in the Freeman and Dockweiler reports.[26] Time was short, for a hearing before the secretary of the interior was scheduled for November. The Turlock District appealed to Frank Adams of the Department of Agriculture for aid. Adams, who had already turned down a similar request for assistance from San Francisco, declined to become personally involved in the controversy, but he did let Turlock have one of his young engineers, Roy V. Meikle, who had been conducting field studies on irrigation from the Feather River.[27] By the beginning of October, the TID had four engineering crews in the field, three experts in its office and four stenographers, all working on a report to rebut Dockweiler's claims.[28]

Their work was completed by October 31. A cover letter by TID attorney P.H. Griffin and MID attorney E.R. Jones stated simply that,

> We strenuously oppose any further grant to any of the waters upon the Tuolumne River to San Francisco, or its neighboring cities, and believe that there is not a sufficient supply of water to irrigate the lands that can be irrigated from the Tuolumne River in the San Joaquin Valley... The amount of water to be used beneficially when all the land is under irrigation will take practically all of the water, both normal, flood and storm, that can be conserved for the purpose of both of these Districts.[29]

Just as San Francisco had hardened its position, the districts had apparently hardened theirs. They had been willing to share the river with San Francisco under the Garfield Permit when the city's total diversions were expected to be less than 70,000 acre-feet a year, but in the face of Freeman's plan to take over six times that amount, they opposed "any further grant" and implied there would be enough demand for irrigation under existing rights to preclude any use of the Tuolumne by San Francisco.

The district's hastily assembled report attacked Dockweiler's assumptions one by one. At times, their zeal to discredit everything he had written led the engineers to make excessive claims of their own. In discussing the high water table, for example, they claimed that not only was it not particularly harmful but that "these subirrigated portions are and always have been the show places of the districts."[30] Taking issue with Dockweiler's assumptions on future use, the report suggested that up to 90 percent of the districts' total acreage would someday by irrigated and that each acre would need two and three-quarters acre-feet annually, all of it from the river. They charged, probably correctly, that Dockweiler had vastly underestimated the amount of water lost in the canals, laterals and community ditches before it could reach the land, and had not understood how much would be lost to evaporation and seepage from the shallow foothill reservoirs. Using their own estimates of reservoir losses and assuming that 35 percent of the water turned into the canals at La Grange would be lost, they found that the districts would need 1,042,043 acre-feet annually, or almost double the amount Freeman and Dockweiler had contended they could ever reasonably use. They also announced that their projected maximum diversion capacity now totalled 3,100 second-feet rather than the 2,350 second-feet contained in the Garfield Permit.[31]

The first showdown after the Freeman report was made public came at hearings held by Secretary of the Interior Walter L. Fisher in November 1912. They lasted six days but were inconclusive because the Army Engineers board had still not finished its report. TID attorney Griffin and Modesto attorney L.L. Dennett, along with engineers and other advisors, represented the districts and returned generally satisfied that, while San Francisco's new plans still posed a serious threat, Secretary Fisher was not likely to grant a new permit without including provisions protecting the districts. For example, the districts at one point suggested they might drop their opposition in exchange for the right to buy as much water, at cost, during the months of July through October as the director of the federal Reclamation Service determined they legitimately needed. The city summarily rejected the idea, but Secretary Fisher told them it was a fair offer. L.L. Dennett reported that a few days later he met some of San Francisco's representatives in Fisher's office and they told him "that they had talked the matter over and were willing to make such a concession."[32] Nothing was actually conceded, however, by either side.

Even at the time of the hearing in late November, Walter Fisher was a lame duck secretary of the interior, his term in office due to expire in March 1913, when Woodrow Wilson replaced Taft in the White House. San Francisco still hoped to persuade him to issue a permit before he left office and accordingly sent Fisher a draft permit for his approval. In a letter to the secretary dated January 18, 1913, City Engineer M. M. O'Shaughnessy commented on the city's proposed treatment of the irrigation districts:

> Fourth: You will notice that we have respected your suggestion to consider well the application of the farmers and irrigationists for storage water, as requested by you at the hearing in Washington. If you remember, at the time I requested that you grant us opportunity to reflect over the matter. We have done so and I believe have met their demands generously and fairly by making extra storage so that there will be surplus water for them when they need it, at a fair compensation similar to what they have already agreed to pay the Yosemite Power Company for water from the south branch of the Tuolumne. Please do not take our attitude in this matter as being in any way adverse to the strong opinions held by Mr. Freeman as to the undesirability of any relations whatever with the farmers, as we believe he is acting from a thoroughly conscientious standpoint, and fully believes in the correctness of the opinion which he possesses on this subject.[33]

O'Shaughnessy erred when he claimed that the districts had agreed to pay the Yosemite Power Company for water released from the company's proposed reservoirs on the South Fork. Although negotiations were continuing at that time, there was still wide disagreement over what price if any, the districts might be willing to pay for the release of stored water to them.[34] The city's proposal was not acceptable to the districts, and their attorneys soon submitted an alternative stipulation to Secretary Fisher. By that time the Turlock District was doing most of the work associated with the Hetch Hetchy matter, just as it had the previous summer.

The *Modesto Morning Herald* reported that the "lack of interest shown by the Modesto district during the entire Hetch Hetchy business has created a great deal of unfavorable comment in the Turlock district," and since the resignation of their engineer, "the directors of the Modesto district have considered it cheaper to be without an engineer and trust to luck and the Turlock district in case of necessity."[35] The Modesto District's attorney, however, remained active, and with Turlock's P.H. Griffin kept a watchful eye on events in Washington.

The Advisory Board of Army Engineers submitted its long awaited findings in February 1913. It concluded that the Tuolumne offered the least expensive source of water available to the city, and if the river was to be developed they could see no sense in not building the reservoir at Hetch Hetchy first. In terms of the dis tricts' rights and the ability of the Tuolumne to supply all the uses that had been suggested for it, the engineers asserted their belief" . . . that there will be sufficient water if adequately stored and economically used to supply both the reasonable demand of the bay communities and the reasonable needs of the Turlock-Modesto Irrigation District for the remainder of this century."[36] Bolstered by the board's endorsement, San Francisco officials made a last ditch effort to win a permit from the outgoing secretary, but Fisher, and President Taft, declined to act on the matter. Fisher announced in March that congressional action would be necessary to grant the city rights in Hetch Hetchy.[37] When asked later by a congressional committee why Fisher had held hearings and then failed to take action, San Francisco city attorney Percy Long said, "He was also doubtful of his power to enforce conditions; that seemed to be the greatest doubt in his mind. Particularly, he doubted somewhat whether the rights of the Turlock and Modesto irrigation districts could be fully protected by such conditions as he might make."[38] If Percy Long correctly understood Fisher's attitude, one of the primary reasons for his deference to Congress was a concern for the rights of the irrigation districts. Whether Congress could do any better was, of course, far from certain.

San Francisco's positon seemed much improved when Franklin K. Lane, a former San Francisco city attorney and Hetch Hetchy supporter, was appointed secretary of the interior by President Wilson. Lane, however, quickly made it clear that, like his predecessor, he, too, would adhere to the theory that Congress must authorize any use of federal lands for San Francisco's water supply. Legislation for that purpose was introduced by California congressman John E. Raker and, on the flimsy pretext that a water famine was imminent, it was pushed ahead in a special session devoted only to certain issues such as tariff and banking reform and emergency measures. The irrigation districts reacted quickly and at a joint meeting of the two boards on May 30, former Congressman Needham was appointed to represent the districts in Washington.[39] The following afternoon a mass meeting was held at the Turlock Opera House under the chairmanship of TID engineer Burton Smith where petitions were circulated against the bill, and recent settlers in the districts were urged to write their former congressmen protesting the city's plans.[40] On June 5 the chairman of the House Committee on Public Lands announced that hearings on the bill would be scheduled for June 23, less than three weeks away.[41] An emergency meeting of the Turlock and Modesto boards was held June 7, where it was decided that representatives of the districts must go to Washington immediately to present their case at the hearing. Superior Court Judge L.W. Fulkerth, attorneys Griffin and Jones and Burton Smith were chosen to go, accompanied by W.H. Langdon and consulting engineer H.T. Cory.[42] By the time they left Modesto on June 10, James W. Corson of the Modesto Chamber of Commerce and attorney L.L. Dennett, now representing the fledgling Waterford Irrigation District located along the river just east of the Modesto District, had also decided to go to Washington.[43]

On their arrival in the capitol they found the situation grim indeed. Except for the districts' representative, Congressman Denver Church, the entire California congressional delegation supported the city, as did Secretary of the Interior Lane and other prominent members of the Wilson administration. The only substantial opposition came from the so-called nature lovers, who defended the wilderness values of an untouched Hetch Hetchy. More than anyone else, John Muir set the tone for their battle, declaring in 1912 that,

> The proponents of the dam scheme bring forward a lot of bad arguments to prove that the only righteous thing to do with the people's parks is to destroy them bit by bit as they are able . . .
>
> These temple destroyers, devotees of ravaging commercialism, seemed to have a perfect contempt for Nature, and, instead of lifting their eyes to the God of the mountains, lift them to the Almighty Dollar.

The Hetch Hetchy reservoir.

> Dam Hetch Hetchy! As well dam for water tanks the people's cathedrals and churches, for no holier temple has ever been consecrated by the heart of man.[44]

Despite their well-publicized opposition and the lesser known objections of the irrigation districts, it appeared that Raker's bill stood a good chance of passing. The committee from the irrigation districts had been sent to Washington to fight the bill, but when they discovered that sentiment in Congress and the Wilson administration was overwhelmingly in favor of San Francisco, they quickly sought to open negotiations. At first, city officials, confident that concessions were no longer necessary, refused to discuss changes in the bill, but they soon relented.[45] By June 18 the committee was able to wire, "Favorable agreement about closed. Hope to finish and have bill introduced this week."[46] On Sunday, the 22nd, all sides met at the Interior Department to finalize the agreement, which was to be implemented by amending the Raker bill.

When they considered their committee's work almost two months later, the districts' boards asserted that, "the committee has succeeded in forcing the City and County of San Francisco to recognize certain rights of said Districts and also to make concessions for the benefit of said Districts," but the rights and concessions were actually a far cry from the kind of protection the districts wanted.[47] As amended, the Raker bill, like the Garfield Permit, guaranteed that San Francisco would recognize the districts' prior rights to the natural flow, as opposed to any releases from storage, only up to 2,350 second-feet, a number based on now antiquated projections of future canal capacities. The city did agree to raise that figure to 4,000 second-feet for a sixty day period beginning on April 15 of each year, when the river was full of snow-melt. The districts were to be allowed to use the extra water directly or store it in reservoirs of their own below Jawbone Creek, a tributary stream a short distance below the point where San Francisco planned to divert water into its tunnel and aqueduct system.[48] Freeman disliked the concession of extra water to the districts, but in a letter written the following year, City Engineer O'Shaughnessy reminded Freeman that under the Garfield Permit the districts had the right to leapfrog above the city's dams to store water in places like Benson Lake. O'Shaughnessy told Freeman:

> The waiving of this right by the irrigationists could not be procured without a counter concession on the part of the City, which was

> made by Mr. Long and myself in Washington, after very due and mature consideration, and from this study of the possibilities of the watersheds and practicability of making cheap rock-filled dams at the higher levels, I believed the exchange was a very desirable one for San Francisco and of great future benefit to the City, as there cannot be in the future any interlocking of interests in construction work between the irrigationists and ourselves.[49]

In other words, in exchange for abandoning projects like Benson Lake that could complicate San Francisco's control of the watershed above its reservoirs, the districts would receive an extra apportionment amounting to a little under 200,000 acre-feet that they could store further down the river. Although it was only implied in the bill, the result was a division of the river at Jawbone Creek, with San Francisco building its project above it and the districts constructing any dams they might want below it.

In previous negotiations, the districts had consistently contended that in exchange for reducing the quantity of water reaching La Grange Dam, San Francisco should agree to sell the districts enough water at cost for late summer and fall irrigation. The city had just as consistently resisted guaranteeing anything beyond a minimal recognition of the districts' prior rights. The compromise agreement settled the matter by requiring the city to sell water to the districts if they requested, the amounts to be provided and the price to be determined by the secretary of the interior, but with provisions that essentially required the districts to first build or expand their own reservoirs and reduce waste. The rate to be paid would include a portion of the cost of the entire water supply system rather than just the reservoirs used to store it, which could make Hetch Hetchy an expensive source of supply. The city could also force the districts to pay for a minimum amount of water each year whether they needed it or not. Once the districts notified the secretary of the interior that they had enough storage of their own, both sides were released from all obligations to provide or pay for stored water. The formula gave the districts an opportunity to buy water but on terms that would hardly encourage them to do so.[50] Finally, the districts retained the right to buy surplus electric power at cost for irrigation and drainage pumping and for municipal purposes.[51] In general, the compromise gave the districts the same degree of protection as the Garfield Permit, though in more detailed and restrictive terms.

Once the agreement had been reached, the districts ceased all

San Francisco's Moccasin powerhouse went into operation in 1925.
Source: City and County of San Francisco.

activity in opposition to the Raker bill. At hearings before the House committee in June and July, the only protest on behalf of irrigation came from L.L. Dennett. The Waterford Irrigation District, which he represented, was not a beneficiary of the protections granted the older districts because it had no prior rights to assert. Directors of the Turlock and Modesto districts formally considered the agreement at a joint meeting on August 13, 1913, resolving that, "We endorse and ratify the amendments secured by our said committee and consider the same to be the best that could be done for said Districts owing to the conditions existing at Washington."[52] They also sent a telegram to Congressman Church informing him of their sentiments;

> At a joint meeting of the board of directors of the Modesto and Turlock Irrigation Districts held in Modesto this day, the action of the committee sent to Washington to represent the district was fully indorsed, and the Raker Bill, as recommended by the House Committee, was approved. The boards also passed resolutions requesting our Representatives in Congress to use their best efforts to pass such bill and oppose the passage of any bill granting San Francisco the Hetch Hetchy which does not contain provisions recognizing and protecting the rights of the districts in the Tuolumne watershed, as provided in the bill.
>
> Stanislaus County Board of Trade passed resolutions on Monday night in effect that no further opposition would be made to the Raker Bill. Some little opposition to the bill had been engendered by persons having special interests outside of the districts and by a few others who feel that the waters of the river should never be taken from the valley. People generally of the irrigation districts believe that under all the circumstances the Raker Bill should be adopted without material amendment and that the strongest opposition should be made to any change in the bill which would eliminate any of the conditions in favor of the districts.[53]

Although other telegrams from meetings of irrigators in Hughson and Turlock endorsed the amended Raker bill, the "few others" still opposed to the grant refused to surrender. Their most vocal spokesman was attorney W.C. LeHane, who had tried and failed to convince the joint board to reject their committee's work in Washington. The Raker bill passed the House easily in early September and won the unanimous approval of the Senate Public Lands Committee a few weeks later, despite the protests of Mr. LeHane, appearing, officially at least, on behalf of the Waterford Irrigation District.[54] When it reached the full Senate, the bill hit its first substantial delay. Several senators threatened a filibuster, and California senator John D. Works called for further investigation, alleging that 99 percent of the water users in the Turlock and Modesto districts felt the compromise had betrayed their interests.[55] Senator Works had already expressed his opposition to the Raker bill on grounds that its provisions regarding the water rights of the districts constituted an infringement of the state's control of its water resources. In the face of such determined opposition, Senate consideration of the bill was put off until early December.

The intervening two months saw intense activity in regard to Hetch Hetchy. the campaign of the nature lovers turned Hetch Hetchy into a national issue. It was discussed in leading newspapers and journals across the country and its opponents unleashed an unprecedented torrent of letters, telegrams and resolutions urging the Senate to prevent the destruction of the valley.[56] In the Turlock and Modesto irrigation districts, support for the compromise began to crumble rapidly when it appeared that the Senate might still reject any grant to San Francisco. When the Hughson Board of Trade announced in mid-October that it still supported the Raker bill, the *Modesto Morning Herald* editorialized that, "From all that can be learned public sentiment in the two districts is absolutely against the Raker bill on the general principal that water should remain in the valley."[57] A Water Users Association was formed to oppose the bill, and in late October the Modesto Irrigation Board capitulated to the rising clamor and gave the group $2,500 to continue the fight. The Stanislaus County Board of Supervisors soon matched the Modesto District's initial contribution and in early December the district gave another $1,000.[58]

Pressure was building by early November for the Turlock board to join the Modesto District in repudiating their endorsement of the Raker bill. At the end of October, engineer Burton Smith announced that he would soon submit his resignation, a move thought by many to be connected to his unpopular stand in favor of the compromise.[59] Since the formation of the irrigation districts, probably no single issue had excited the area as much as the debate over the Raker bill. On November 10 the TID board was handed petitions bearing 475 signatures asking for money to fight the bill, but action was deferred until the following week.[60] By that time the petitions carried 610 names, and petitions submitted to the county Board of Supervisors were said to have the names of 60 percent of the water users in the Turlock District.[61] Despite the hue and cry against it, the Turlock board refused to retreat from its endorsement of the amended bill. If the Raker bill were defeated, however, the board pledged to spend the entire annual tax assessment if necessary to prevent any unconditional grant of reservoir sites to San Francisco.[62]

When Senate debate on the Raker bill began on the first of December, Congressman Church was still working on behalf of the bill largely because of the Turlock board's continued support for the legislation. Senator Works, however, repeated his charge that the people of the districts had repudiated the bill. Mayor Rolph of San Francisco wired Judge Fulkerth and others asking for information regarding these allegations. Fulkerth replied,

> Public sentiment has greatly changed in the irrigation districts since the committee was in Washington in July. The Modesto board had been forced to pass resolutions receding from the original position approving the bill. The people of these two districts are unquestionably against the bill or any bill permitting water to be taken out of the valley.[63]

In a very similar statement, the Turlock Board of Directors wrote Rolph that, "the people are greatly agitated" and vehemently opposed to taking water from the San Joaquin Valley. At the same time, they telegraphed Church that he was no longer bound by any obligation to support the Raker bill.[64] That was as close as the TID board would go toward joining the chorus of opposition. It was reported the next day that the congressman had quickly recanted his previous stand and told Senator Works that his support for the bill had always been reluctant and only because the districts had told him that was what they wanted, but now the irrigators themselves had rejected the bargain.[65]

It is difficult to judge exactly how much impact was had by the late blooming but intense opposition effort, which lacked only the official support of the Turlock Irrigation District. Richard Waterous of the American Civic Association, a group allied with Muir in the fight, wrote to the organization's president, J. Horace McFarland, that,

> You remember I wrote you in June that there had been some kind of an agreement between the Irrigationists and San Francisco, whereby the San Francisco people had promised the Turlock-

> Modesto people an amount of water which *at that time* the representatives accepted and thereupon they ceased their opposition. It developed during the month following that those representatives entered into an agreement which did not meet with the approval of the great body of the irrigationists and when they woke up to the situation it was with the determination to fight harder than ever to save the water for their use. Thus, it was that their fight during the big week in December was an important phase of the matter.[66]

In the end, the efforts of the wilderness preservationists and the last ditch intervention of the irrigationists were no match for San Francisco's skilled lobyists. The Raker bill passed the Senate at midnight, December 6, 1913, and a little less than two weeks later was signed into law by President Wilson.

The battle over Hetch Hetchy and its outcome was in many ways more important to the city's opponents than it was to San Francisco. It became one of the landmarks of the American conservation movement. The wilderness preservationists had lost a hard-fought struggle but had nevertheless managed to preach their gospel of wilderness values to a national audience. For them, the loss of Hetch Hetchy was not an end, but a beginning. It is in this context of the wilderness movement that the story of Hetch Hetchy is most often told. For the irrigation districts, the results were equally dramatic; they would finally be forced to share their greatest asset, the Tuolumne River. Also, for the first time they had had to involve themselves in the national political arena. They would have to continue to do so, for having established its authority to oversee the development of much of the Tuolumne watershed, the river's future would involve the federal government almost as much as it involved San Francisco. In terms of the principles it embodies, the Raker Act was a notable piece of national legislation. Its central precepts were those associated with the Progressive movement – a complex reform impulse that generally called for honest government, restrictions on the power of monopolies and other special interests, and even for the public ownership of certain utilities. Progressivism was identified with utilitarian conservation and its emphasis on the rational and efficient management of water, minerals, land and timber. For water development, the epitomy of the Progressive conservation program was coordinated, multi-purpose river basin planning; a goal that was not generally achieved until years later.[67] In a sense, however, that is what the Raker Act tried to do. It attempted to strike a balance between domestic and irrigation water uses, and it authorized hydroelectric power development. It also mandated the public ownership of the water and power produced by the city's project, and prohibited any private company from profiting from the development of the upper Túolumne. It was a far-reaching plan for the use of the river. The Raker Act defined the Tuolumne's future and touched, directly or indirectly, almost everything the Turlock Irrigation District did after 1913. Its influence was so pervasive that it would hardly be an exaggeration to say that, for the TID, it was second in importance only to the Wright Act itself.

CHAPTER 7 – FOOTNOTES

[1] John Muir, *The Yosemite,* (Natural History Library edition, Garden City, New York, 1962), pp. 192-196.

[2] Ray. W. Taylor, *Hetch Hetchy: The Story of San Francisco's Struggle to Provide a Water Supply for Her Future Needs,* (San Francisco, 1926), pp. 27-47.

[3] Board minutes, vol. 2, p. 297 (June 2, 1903); *Stanislaus County Weekly News,* Jan. 27, 1905.

[4] Board minutes, vol. 2, pp. 352-354 (Apr. 5, 1904).

[5] *Stanislaus County Weekly News,* June 3, 1904. The quotation is from the Aug. 5, 1904 issue.

[6] U. S. Congress, House Committee on Public Lands, *Hetch Hetchy Dam Site, Hearing before the Committee on Public Lands, House of Representatives,* 63rd Cong.,1st Sess; H. R. 6281, June 25, 1913, vol. 1, p. 112.

[7] *Stanislaus County Weekly News,* Feb. 24, 1905.

[8] Quoted in M. M. O'Shaughnessy, *Hetch Hetchy: Its Origins and History,* (San Francisco, 1934), p. 32.

[9] Walton Bean, *Boss Ruef's San Francisco,* (Berkeley, 1952), especially pp. 140-144.

[10] On Garfield see Samuel P. Hays, *Conservation and the Gospel of Efficiency: The Progressive Conservation Movement, 1890-1920,* (Atheneum edition, New York, 1974), pp. 72-73. Kendrick A. Clements, "Politics and the Park: San Francisco's Fight for Hetch Hetchy, 1908-1913," *Pacific Hist.Rev;* 48 (May 1979), pp. 188-189 suggests Garfield had been won over by San Francisco City Engineer Marsden Manson before becoming Secretary of the Interior.

[11] *Stanislaus County Weekly News,* Aug. 9, 1907.

[12] Roderick Nash, *Wilderness and the American Mind,* (revised edition, New Haven, 1973), pp. 163-164.

[13] The permit is found in *Hetch Hetchy Valley: Report of Advisory Board of Army Engineers,* House Doc. No. 54, 63rd Cong. 1st Sess. (1913), pp. 7-8.

[14] U. S. Geological Survey, *Surface Water Supply of the United States, 1907-8, Part XI: California,* Water Supply Paper 251 (1910), pp. 274, 277.

[15] *Stanislaus County Weekly News,* May 1, 1908.

[16] *Ibid.,* May 15, 1908; Board minutes, vol. 2, pp. 708, 711 (June 2, 1908).

[17] *Stanislaus County Weekly News,* Sept. 18, 1908.

[18] Nash, pp. 164-169.

[19] For background on the Pinchot-Ballinger feud, see Elmo Richardson, *The Politics of Conservation: Crusades and Controversies, 1897-1913,* (Berkeley, 1962).

[20] Clements, p. 200.

[21] John R. Freeman, *On the Proposed Use of a Portion of the Hetch Hetchy, Eleanor and Cherry Valleys . . . for the Water Supply of San Francisco,* (San Francisco, 1912), p. 140. [22] Freeman, p. 10. [23] *Ibid.,* p. 16.
[24] See Dockweiler, *Water Needs of the Turlock and Modesto Irrigation Districts,* (1912).
[25] Freeman, Appendix No. 17, "Abstract of Report on Modesto and Turlock Irrigation Districts . . .," " by J. H. Dockweiler, pp. 358-359.
[26] Board minutes, vol. 4, pp. 152 (Aug. 19, 1912).
[27] Frank Adams oral history transcript, Bancroft Library, Berkeley.
[28] *Modesto Morning Herald,* Oct. 1, 1912.
[29] *Report of H.S.Crowe, Engineer of Modesto Irrigation District and Burton Smith, Engineer of Turlock Irrigation District, In Answer to Reports on Behalf of San Francisco . . .,* Vol. 1, (Turlock, Oct. 31, 1912), cover letter by P. H. Griffin and E. R. Jones.
[30] Crowe and Smith, vol. 1, p. 92. [31] *Ibid.,* pp. 125-126.
[32] *Modesto Morning Herald,* Dec. 10, 1912. [33] O'Shaughnessy, p. 43.
[34] *Modesto Morning Herald,* Jan. 14, 1913. [35] *Ibid.,* Feb. 6, 1913.
[36] House Doc. No. 54, p. 50. [37] Clements, p. 206.
[38] House Committee on Public Lands, *Hetch Hetchy Dam Site,* p. 107.
[39] Board minutes, vol. 4, p. 415 (May 30, 1913).
[40] *Modesto Morning Herald,* June 1, 1913. [41] *Ibid.,* June 6, 1913.
[42] Board minutes, vol. 4, p. 416 (June 7, 1913): *Modesto Morning Herald,* June 8, 1913.
[43] *Modesto Morning Herald,* June 10, 1913. [44] Muir, pp. 200-202.
[45] *Turlock Weekly Journal,* July 10, 1913. [46] *Ibid.,* June 19, 1913.
[47] Board minutes, vol. 4, pp. 477-478 (Aug. 13, 1913).
[48] Raker Act (H. R. 7207), section 9c.
[49] O'Shaughnessy, p. 49. [50] Raker Act, sections 9d, 9e, 9f, 9g.
[51] Raker Act, section 9(1). [52] Board minutes, vol. 4, p. 478 (Aug. 13, 1913).
[53] O'Shaughnessy, p. 49.
[54] *Turlock Daily Journal,* Sept. 25, 1913, Oct. 7, 1913.
[55] *Ibid.,* Oct. 6, 1913.
[56] Nash, pp. 170-179. [57] *Modesto Morning Herald,* Oct. 17, 1913.
[58] *Ibid.,* Oct. 29, Nov. 11, Dec. 3, 1913.
[59] *Modesto Morning Herald,* Oct. 3, 1913.
[60] Board minutes, vol. 4, p. 510 (Nov. 10, 1913).
[61] *Modesto Morning Herald,* Nov. 11, 1913.
[62] Board minutes, vol. 4, p. 524 (Nov. 18, 1913).
[63] *Modesto Morning Herald,* Dec. 4, 1913. [64] *Ibid.,* Dec. 5, 1913.
[65] *Turlock Daily Journal,* December 5, 1913.
[66] Richard Watrous to J. Horace McFarland, Dec. 15, 1913, in Holway Jones, *John Muir and the Sierra Club,* (San Francisco, 1965), pp. 165-166.
[67] This brief description hardly does justice to Progressivism and the conservation movement. See Hays, *Conservation and the Gospel of Efficiency* for a detailed discussion. Nash, *Wilderness and the American Mind* discusses the significance of Hetch Hetchy to the wilderness movement. While Hetch Hetchy has been examined in detail as an environmental issue, much more work needs to be done on the Raker Act and the history of the Hetch Hetchy water and power project.

8

Crisis and Progress
1914-1918

The TID board's decision to stick to its bargain with San Francisco despite overwhelming local opposition to the Raker bill soon resulted in an attempt to remove some or all of the directors from office. Even before the final Senate vote on the bill, the Hickman Board of Trade had issued a fierce denunciation of the board that ranged from its failure to do battle against the Hetch Hetchy grant, to allegations of improper handling of finances and contracts, including rumored defects in the construction of the new Davis Reservoir outlet gate.[1] An angry meeting was held in Ceres at about the same time and a petition for the recall of Division 2 director T.A. Owen of Hughson began circulating.[2] The petition, which ultimately carried at least 236 names, complained primarily that Owen had favored the Raker Act, had been party to the hiring of former Congressman Needham to, in effect, assist in its passage, and had endeavored to retain engineer Burton Smith, who was called "incompetent, unreliable and inefficient." The circulators claimed that unless the board was replaced, there could be no cooperation with the Modesto board in defending the rights of the two districts on the Tuolumne.[3] A move to recall N.J. Witmer of Denair, representing Division 1, for the same reasons was also started but technical flaws in the petitions delayed the calling of a recall election in that division.

Interest focused on the March 10, 1914, election in Division 2 that would decide the fate of T. A. Owen, the first TID director ever subjected to a recall. The Keyes neighborhood was reportedly heavily against him, with Ceres leaning toward his removal and Hughson solidly in favor of its hometown director. A mass meeting was scheduled for Ceres the day before the election, but backers of the recall found Owen's supporters had managed to get control of the gathering and Owen himself was one of the major speakers.[4] Both sides worked hard to bring their partisans to the polls on election day, but Owen emerged victorious by a margin of 449 to 345.[5]

Following the failure of the effort to recall Owen, the movement to replace the board members collapsed and there was so little interest in the recall election in Division 1 that no candidate came forward to oppose Director Witmer, who won by a better than nine-to-one margin.[6] The board had survived its first, and last, recall challenge. It ended, on a dramatic note, the kind of agitation that had been all too commonplace since irrigation began. The recall episode was apparently not much more than an angry reaction to district policy in the Hetch Hetchy battle. Certainly, if fundamental issues of district governance had been the cause of it, the dissident movement would not have disappeared so rapidly or thoroughly. As it was, the events of early 1914 indicated that the district as an irrigation community was finally at peace with itself. In the following months that peace would be severely tested and proven sound.

One of the charges levelled against the directors by their detractors was that they had endeavored to retain their controversial engineer, Burton Smith. Smith had submitted his formal resignation in November 1913, at the height of the Hetch Hetchy furor, but in December he withdrew it, with the comment that the large amount of construction work then underway would make it difficult for a new man to take over the job.[7] If the board had wanted to retain Smith, the withdrawl of his resignation gave them that opportunity. Instead, on January 30, 1914, they declared the office of chief engineer vacant as of March 1 and named Roy V. Meikle to be Smith's successor.[8] A native of Ohio, Meikle attended Stanford University but did not graduate after completing his engineering program, later commenting, "I couldn't wait another year for a degree. The world needed engineers."[9] He worked first for municipal water systems in Portland and Tacoma and then for an irrigation project in Medford,Oregon. In 1912 he joined the U.S. Department of Agriculture's Irrigation Investigation Division in Berkeley and was sent to the Feather River to study its irrigation potential. Later that year Frank Adams sent him to Turlock to help prepare the districts' defense against San Francisco, and on October 17, 1912, he became a TID engineer.[10]

Meikle's tenure as chief engineer got off to a disheartening start. He was forced to shut off water in the main canal in early June when a section of concrete lining broke loose at a weak spot a short distance below the new reservoir. Prompt repair was needed to prevent the seepage that could lead to a much more serious washout. Meikle soon had fifty men and thirty horses and mules working in two shifts to fill the eroded spots and install a new lining, but water was still out of the system for about ten days. During the time the repairs were being made, a note was found pinned to the door of the F.M. Hudelson house a short distance away which said, "Have inspected the canal and works hereabouts and hereby condemn them and advise you to move before long."[11] The board was reportedly amused by the message, but the good humor was destined to be short-lived.

At three o'clock on the morning of June 27, 1914, a section of concrete-covered fill at the reservoir about fifty feet south of the outlet gate suddenly gave away. At first, the torrent of water that poured out of the reservoir through the broken wing wall followed the canal. A night watchman on the Nels Johnson place, two miles below the outlet gate, heard the roar of the oncoming flood and saw it spilling over the sides of the canal bank. He quickly telephoned La Grange to have the water shut off at the dam, and continued to spread the alarm until the canal-side telephone poles were washed out. Some of the water went into the Dickinson Lake basin,but before long, a bank gave way in another spot and the full force of the escaping water went down the bluff to the Tuolumne near Roberts Ferry. The cliff, composed of sand and hardpan, was quickly eroded by the rushing stream, cutting an immense gully back 250 feet to the canal, which itself slowly became part of the growing ravine. It took nearly three days to empty the reservoir, and sightseers by the score journeyed out to see the broken wall and the devastation it had caused. Below the bluff, the George Nelson ranch was nearly covered by sand and mud, and the approaches to Roberts Ferry were clogged.[12]

The failure of the reservoir was perhaps not entirely unexpected. In December 1913, the Hickman Board of trade, as part of their general denunciation of the board at that highly charged time, had asserted that "the fill on either side of the outlet gates of the Davis reservoir is not heavy enough and the concrete work at this point is very defective and part of it is broken and only sustained by the reinforcing."[13] In response, over fifty representatives of local organizations, district officials and newspapermen toured the main canal and reservoir in early January. Charges that there were cavities in the puddled fill behind the concrete facing were investigated by hammering holes in the four-inch thick concrete. No cavities were found, although some deficiencies in the reinforcing on the south side of the outlet gate were later discovered and corrected by contractor T.K. Beard before the district accepted the work in March. The reinforced section was still standing after the washout.[14] There was properly still a nagging doubt that the fill and concrete work had been properly done, and stories persisted that weak spots had been left in the wall.[15] Assistant State Engineer P.M. Norboe had inspected the district's works in December 1913, on behalf of the Irrigation District Bond Commission and he, too, seemed to have had misgivings about the new reservoir. His report, issued by the commision only two days before the break, said:

> The ground upon which the dams surrounding this reservoir rest appears to be firm, but as it is underlain a few feet beneath the surface with hardpan which in turn frequently rests upon sandy or other pervious earth, the concrete cut-off walls are a necessity to prevent dangerous seepage beneath the dam.
>
> The engineer in charge, Mr. Burton Smith, conducted quite extensive exploration at each site to ascertain the depth necessary to carry the walls so as to cut off all pervious strata.
>
> Considerable seepage below some of the dams indicates that not quite all loose strata were intercepted. However, should a dam fail at any point but little damage is likely to result excepting expense of repairs, and the loss of stored water pending the repairs.
>
> The outlet, situated in the line of the canal, is a concrete structure with counter balanced regulating gates. With 37 feet of water against the gates considerable leakage was observed. This does not indicate that any danger of collapse necessarily exists, and it is not probable that absolute safety can be felt until the gates have stood some years. As stated above in regard to the dams, should the gate structure collapse, no great damage would result as the flood pouring down the canal would be turned into the river through wasteways provided for that purpose.[16]

The repeated assurances that everything would be all right if the reservoir went out underscored not the assurances but the threat. The note found on the Hudelsons' door added an element of mystery as well until it was learned that friends of the family had concocted the warning as a practical joke.[17]

The last of the 40,000 acre-feet stored in the reservoir were still draining into the Tuolumne when the district began the job of getting the damaged canal back into service. Water could still be run through the section innundated by the reservoir and through the outlet gates, but below the reservoir a short section had been destroyed by the break. A temporary "shoofly" canal was planned

The Davis Reservoir break.

Rebuilding the main canal below Davis Reservoir, 1914.

and approved by the board on Monday, June 29, 1914. It would run from a half mile below the outlet gate to a quarter mile below the break at the bluff. One section of the work was to be part of a long contemplated permanent improvement to eliminate the weak spot that had been repaired just before the reservoir broke. The board passed a resolution giving a committee of its members the authority to award contracts without waiting for bids, and decided to levy a $40,000 emergency tax assesment without an election to pay some of the repair costs.[18] Following the formal meeting in Turlock, the directors went to the reservoir where preparations were already underway to begin construction. The following day, A.G. Chatom was given a contract to construct part of the new canal, while the remainder would be built by the district's own forces under the direction of Superintendent Milo J. Caton. Anyone with teams and scrapers was offered work.[19] A week later, Caton's construction camp near Dickinson Lake housed one hundred men and the three hundred animals needed to pull over fifty of the four-horse Fresno scrapers and several plows. Chatom had a smaller number of scrapers on the job but soon put a clamshell dredge to work, which was better able to deal with the hardpan in that section. By July 15 Caton's district crews had completed their section and moved onto Chatom's in order to speed the work. Despite optimistic reports that the bypass canal would be ready in a month or less, it was not until August 15, almost seven weeks after the reservoir disaster, that water was once again flowing down the main canal.[20]

The calamities that befell the district were not limited to structural failures. What seemed to be a jinx haunted the Board of Directors. The recall elections early in 1914 had been a decided strain, but with the failure of that effort and the completion of work on the reservoir and main canal improvements, nothing out of the ordinary was anticipated. S.A. Hultman, the director from Division 5, took a leave of absence in the spring for a vacation in Europe. Then, in early June, Division 3 director E.P. McCabe mysteriously disappeared one evening after saying that he was going to bed. Instead, he was seen boarding a southbound train for points unknown. He was found a few days later in a Merced hotel, where he had suffered a stroke.[21]

That left the board with only three active members when the reservoir broke. They were frequently at the scene of the disaster and the canal construction site, and on July 21 were due to make another trip to the area. The three men left Turlock in a TID Ford with T.A. Owen at the wheel sometime after 9:00 a.m. They were on the Denair road a little way past the Hawkeye dairy when the car, travelling about fifteen miles per hour, overtook a wagon. Owen tried to pass, but the car swerved in the sandy road and headed toward a fence. When Owen tried to turn it back onto the roadway, the vehicle flipped over. Director Witmer was thrown clear but the other two directors were trapped under the overturned car, which was across Owen's head and Edward Kiernan's chest. Although he had a broken rib and shoulder, Witmer helped the wagon driver lift the car off the two men. Kiernan was reportedly "badly squeezed" in the effort to move the car but was seen around Turlock that afternoon. Owen, however, was suffering from a fractured skull and was taken unconscious to the Turlock sanitarium.[22]

With only Witmer and Kiernan still able to work, the board lacked the quorum necessary to do business. The outbreak of the war in Europe disrupted travel there and delayed Hultman's return from his Swedish holiday. McCabe was recovering from his stroke and some hope was expressed that he might be able to rejoin the board, but in early August, he resigned due to continuing ill health. A few days after McCabe announced his resignation, Kiernan underwent an operation for a srangulated hernia, an old injury that had been aggravated by the auto accident some weeks earlier. The operation came too late, and he died without regaining consciousness. With Owen still recovering from his injuries, only N.J. Witmer was able to carry on his duties as a director, and even he had already planned to resign that fall in order to move to another community. Informal advisory elections were promptly held to choose directors to replace McCabe and Kiernan, and the county Board of Supervisors appointed the winners. When John A. Sisk and Claus Johnson joined Witmer on the board, it was at last able to resume normal operations. In October, Witmer handed in the resignation he had planned to submit earlier, and T.A. Owen also resigned, even though he had largely recovered from the accident.[23] By November, Hultman had returned to join the four new members who had been appointed in his absence.

The year had, of course, been a disasterous one for the district's farmers. Asked years later what hard times he could remember, long-time Hughson resident I.W. Swagerty named 1914 because the reservoir break cut off the water in early summer and the alfalfa dried up.[24] When the water was turned off in June for main canal repairs, it was reported that most farmers had already had one irrigation, so a brief interuption was not especially serious.[25] After less than two weeks of full operation at the height of the irrigation season, the supply was cut off again by the failure of the reservoir. It seems unlikely that during those two weeks any more than half the district received an irrigation, and probably much less. By the time the bypass canal was in operation, it was too late to salvage the season. U.S. Geological Survey measurements of the TID canal at La Grange show that water was run into the system for a few days before regular service was resumed in mid-August, but by then the river was falling rapidly. Diversions of 500 second-feet or more were possible for only three days, and by August 27 the flow through the headworks had dropped to less than 200 second-feet.[26] Even with 600 second-feet running, it would not have been possible to irrigate more than 1,000 acres a day, and with far less than that, only a few hundred acres a day could expect to receive water.[27] During all of August and September, the district took only about 21,000 acre-feet out of the Tuolumne River, which could not even cover a quarter of the district's irrigated area.[28] The results could be seen in the interviews with farmers conducted by Wells Hutchins the following year. Henry Thake, for example, took in no money at all in 1914, while A. W. Johnson saw his $600 annual gross income cut in half by the break.[29] Some farmers, who were fortunate enough to get some of the scarce water, may have fared slightly better, but few would have viewed 1914 as anything less than a catastrophe.

Considering the hardships caused by the loss of the resevoir and the recent history of turmoil and dissension among the irrigators, it seems surprising that outrage over the disaster was nearly nonexistent. Some people remembered that Chief Engineer Meikle had been an assistant to the widely disliked Burton Smith when the defective work was done and petitions were circulated asking that Meikle be replaced by B. W. Child, the district's first superintendent.[30] The board declined to make the change and the matter died. The only other organized criticism came from a committee appointed by the Turlock Board of Trade and other local organizations to investigate the causes of the reservoir break. They condemned the poor construction of the walls on either side of the outlet gate and severely criticized the board for continuing to fill the reservoir even after signs of leakage occured. Although they did not suggest that Meikle must necessarily be replaced, they asked that a consulting engineer be hired to redesign the outlet structure.[31] The board – what little there was of it at the time – took no action on the committee's report. Instead, Meikle drew up plans for a new and stronger outlet gate, which were then approved in detail by State Engineer W. F. McClure.[32] General support for the district was reflected in the election held on assessments covering ordinary repairs and operations as well as special levies to pay for the new outlet gate and buy the Nelson ranch, which had been damaged to the extent of over half its value by the break. All amounts were approved in a heavy voter turn-out by a margin of over 80 percent.[33] The comparative absence of bitterness and complaint in the wake of the worst misfortune in the district's history was, like the failure of the recall effort, a sign that the community had reached a state of maturity. That was undoubtedly the most significant thing about the tumultuous events of 1914; they proved that the long era of controversy that dated back to the Wright Act itself was over.

The TID, too, had matured by 1915. The water distribution system that a decade before had been run without instruments or records now had an abundance of both. The chief engineer assumed full authority for the operation of the canal system following the resignation of Superintendent Caton in 1914. Meikle had an assistant engineer to help in the work and kept two men with current meters in the field measuring flows to determine losses and monitor operation of the canal and lateral system. The district's thirty-one ditchtenders also used various gauging devices to insure that the heads – twenty second-feet for alfalfa and half that for orchard and row-crops – delivered to irrigators were uniform and that each farmer received his fair share of the water. Applications for water, which included the acreage in each crop, were collected from all irrigators before the season started, and record-keeping extended to issuing receipts to irrigators for the water used in each irrigation. Accurate water measurements and complete records improved the management of the distribution system to the point that complaints about water service were no longer a problem.[34]

In 1915, the district's operations were run from its building on Turlock's Main Street. It housed the board room and offices of the chief engineer and his assistants, Secretary Anna Sorensen, the assessor-collector-treasurer (consolidated into a single office after 1914), a bookkeeper and the clerical staff. On the outskirts of town, at Main and Pioneer streets just north of Lateral 4, the district had its machine and blacksmith shops and corrals where its animals were kept when they were not assigned to outlying "horse camps" at the scene of major repairs.[35] In 1915 there was a repair camp manned by a foreman, two helpers, eight day-workers and a cook, and a ditch maintenance camp with fifteen men and a foreman.[36] The district paid its mule-skinners, teamsters and laborers about $2.50 per day with deductions of 25¢ per meal for those living in the camps.[37] During the winter months when water was out of the canals, the ditchtenders joined the maintenance crews and additional men and teams were hired to do all major work before the water was turned in again.

The district faced the same problems in maintaining its dirt canals that farmers faced on their farm and community ditches, but on a larger sacle. Silt could form berms where grass and tules took root, and aquatic plants collectively referred to as moss were a problem throughout the system. Plowing out the canals during the off-season was the best way to deal with the buildup of vegetation, but at $250 to $500 per mile for canals with a twenty-foot bottom, it was expensive. Other ways of dealing with weeds in the canals included dragging heavy chains along the bottom to break off all but the smaller, more supple weeds and moss, or using a cutter similar to a two-man saw, but neither of these was very effective.[38] Weeds above the water line could also pose a problem when they spread or drooped into the water. Johnson grass, bermuda grass and Russian thistle were among the worst common weeds on canal banks and their presence was not only a nuisance for the district but a source of infestation for the entire community since it was easy for their seeds to float into fields with the water. Weeds growing on the interior banks were plowed out along with the other vegetation, while those on the outside could be controlled to some extent by burning. The fires sometimes got out of hand, as evidenced by a damage claim for raisins and trays that fell victim to a TID burning crew along Lateral 6 in 1915.[39] Aside from weeds, the most persistent problem was burrowing animals such as squirrels, muskrats and gophers, whose tunnels could weaken banks and cause washouts. Ground squirrels, because of their numbers and the damage they could do to crops, seem to have received the most attention. The district mixed its own poison baits or used carbon bisulfide gas in a effort to rid the canal banks of the pests.[40]

Winter work on the irrigation system extended beyond normal maintenance to an on-going program of modifications and improvements. The capacity of the main canal above Davis Reservoir (renamed T. A. Owen Reservoir in 1917) was gradually increased from 1,560 second-feet in 1914 to 1,820 second-feet in 1921, and an ambitious program of canal lining was begun about 1917.[41] The lining, of two-inch thick concrete in most cases, or one and a half-inch thick gunite over chicken wire in others, had several benefits.[42] It eliminated seepage and reduced the possibility of washouts, allowed water to flow freely and sharply reduced maintenance. It was costly and time consuming, so the first work was restricted to places where canals were built in fills or in spots where excessive seepage required corrective action. Concrete lining, like the growing number of concrete drops, waste-gates and bridges, was an important permanent improvement. The policy begun years before of replacing wooden drops with concrete ones had largely eliminated the failure-prone old structures. What's more, many of the new drops had automatic regulator gates, which added a margin of safety to the canal system, as did automatic alarms at critical places in the system. If the water level rose or fell too far, an alarm would sound in the nearest ditchtender's house.[43]

Meanwhile, the drainage problem worsened. It was estimated that 13 percent of the district had water at four feet or less below the surface in 1915. The following year, 33.7 percent of the area had a water table at that level and in 1917 the proportion rose to 42.3 percent before falling to 35 percent for the next two years.[44] A survey in 1918 showed almost 2,000 acres covered by lakes and ponds, and banks were beginning to refuse loans to farmers in the waterlogged regions.[45] In 1916, test wells, some 200 in all, were drilled at section corners throughout the district and checked monthly to monitor fluctuations in the water table. Doing something about the problem was more difficult than studying it. The open drains were a poor expedient and no new ones had been dug since 1910. Despite the poor results achieved by early drainage pumping experiments, the engineers appear to have been convinced that pumping remained the best solution to the drainage problem, but only if a reliable and economical source of power was available to run the pumps. Balky gasoline engines were blamed for the failure of the first experiments, so by 1910 Burton Smith had begun advocating electric pumps. In that connection, a board resolution that year cited the potential for hydroelectric generation at La Grange Dam and Hickman drop on the main canal. Estimates for a complete electrical system were included in initial plans for the 1910 bond issue, but were later dropped without comment.[46] The Raker Act recognized the need for drainage and other pumping by granting the districts the right to buy power at cost for that purpose, but it would clearly be many years before San Francisco would have its dams and powerplants in operation.

Pumping tests were resumed in 1915 and were apparently successful enough to prompt Meikle to write "A Consideration of Power Development in the Turlock Irrigation District" in February 1916. The board authorized him to pursue his investigations and he soon outlined plans for a powerplant at Hickman drop with an oil-fired auxillary plant to take over the load in emergencies and when water was out of the canals. Fifty electric drainage pumps were included in the proposal, and Meikle indicated power would be for sale to district residents as well.[47] For reasons never given, the board decided not to go forward with the plans at that time. Instead, open drain construction was resumed. By 1918, five dredgers were busy digging drains from the worst waterlogged areas to the San Joaquin River.[48] There was never any real expectation that the network of open drains, which eventually totalled nearly sixty miles, would solve the water table problem. The drains were dug to a depth of eight to ten feet, but because the nature of the soil at that level did not favor the lateral movement of water through it, only the area directly adjacent to the drain ditch was affected. The first underground strata where water moved freely was found at about forty feet.[49] Tapping that level would have been more effective in draining large areas but that was something pumps could do and open drains could not. Until an economical source of power for pumping could be found, however, the drainage ditches were the only alternative.

Besides the drainage situation, the prinicpal unresolved problem facing the Turlock District after 1914 was water storage. Owen Reservoir added up to a month to the irrigation season, but irrigated acreage did not exceed 100,000 out of the district's 176,000 acres until 1918. In order to serve the remaining acreage, or extend the irrigation season into September or October to match the growing season, the district would have to have more water, and the only way to get it was to store more of the spring snowmelt. Furthermore, if the district was limited to only the natural flows stipulated in the Raker Act, additional storage was even more necessary. By 1918, it appeared that a single project could solve both of the district's most pressing problems by providing water storage and at the same time the power needed to run a network of drainage pumps. That project was the Don Pedro dam and powerhouse.

CHAPTER 8 – FOOTNOTES

[1] *Turlock Daily Journal,* and *Modesto Morning Herald,* Dec. 3, 1913.
[2] *Turlock Daily Journal,* Dec. 4, 1913, Jan. 5, 1914.
[3] *Ibid.,* Jan. 20, 1914; *Modesto Morning Herald,* Jan. 22, 1914.
[4] *Modesto Morning Herald,* Mar. 8, 10, 1914.
[5] *Ibid.,* Mar. 11, 1914; Board minutes, vol. 4, p. 636 (Mar. 16, 1914).
[6] *Modesto Morning Herald,* Apr. 11, 1914; Board minutes, vol. 4, pp. 661-662 (Apr. 20, 1914).
[7] Board minutes, vol. 4, p. 511 (Nov. 10, 1913), vol. 4, p. 558 (Dec. 22, 1913).
[8] Board minutes, vol. 4, p. 595 (Jan. 30, 1914).
[9] *Modesto Bee,* June 16, 1975.
[10] "Testimony of R. V. Meikle," before Federal Power Commission (1962), pp. 1-2.
[11] *Turlock Daily Journal,* June 15, 1914.
[12] *Ibid.,* June 27, 1914; *Modesto Morning Herald,* June 28, 1914.
[13] *Turlock Daily Journal,* Dec. 3, 1913.
[14] *Ibid.,* June 28, 1914; *Modesto Morning Herald,* June 28, 1914.
[15] Author's interview with Ray Knowles, July 11, 1984.
[16] Irrigation District Bond Commission, *Report on Turlock Irrigation District,* p. 15.
[17] *Turlock Daily Journal,* July 1, 1914.
[18] Board minutes, vol. 4, pp. 703-706 (June 29, 1914).
[19] *Modesto Morning Herald,* June 30, July 1, 1914.
[20] *Turlock Daily Journal,* July 8, 15, Aug. 17, 1914.
[21] *Turlock Daily Journal,* June 1, 2, 5, 1914.
[22] *Ibid.,* July 21, 1914.
[23] Board minutes, vol. 5, p. 11 (Oct. 5, 1914), p. 25 (Oct. 14, 1914).
[24] I. W. Swagerty interviewed by Bill Hurd, Jan. 1967.
[25] *Turlock Daily Journal,* June 5, 1914.
[26] U.S. Geological Survey, *Surface Water Supply of the United States, 1914, Part XI: Pacific Slope Basins in California,* Water Supply Paper 391 (1917), p. 179.
[27] *Turlock Daily Journal,* Aug. 17, 1914.
[28] Water Supply Paper 391, p. 179.
[29] Hutchins, *The Settlement of Small Farms,* pp. 29-30.
[30] *Modesto Morning Herald,* July 14, 1914.
[31] *Ibid.,* Aug. 12, 1914.
[32] Board minutes, vol. 5, p. 72 (Nov. 30, 1914), pp. 76-77 (Dec. 8, 1914).
[33] Board minutes, vol. 4, pp. 779-785 (Sept. 16, 1914), vol. 5, p. 20 (Oct. 13, 1914).
[34] Adams, 1916, p. 83.
[35] Author's interview with Charles O. Read, Apr. 25, 1983.
[36] Adams, 1916, p. 83.
[37] R. V. Meikle to Miss E. D. Bishop, Apr. 5, 1917.
[38] Harding, *Operation and Maintenance of Irrigation Systems,* pp. 18-19.
[39] Board minutes, vol. 5, p. 391 (Dec. 14, 1915).
[40] Read Interview.
[41] R. V. Meikle to A. Griffin, June 4, 1918; R. V. Meikle to Samuel Fortier, Mar. 21, 1919. Canal capacities and diversions are shown in annual U. S. Geological Survey Water Supply Papers.

[42] R. V. Meikle to S. O. Harper, Nov. 13, 1940.
[43] Irrigation District Bond Commission, pp. 16-17.
[44] Excerpts from reports of B. A. Etcheverry and Thomas H. Means in Meikle files, vol. 3, item 4.
[45] "History of the Hilmar Drain Area," memorandum by R. V. Meikle, Oct. 16, 1963, in Meikle files, vol. 36, item 25.
[46] Board minutes, vol. 3, p. 192 (Feb. 14, 1910), p. 262 (June 7, 1910), pp. 304-305 (July 28,1910).
[47] *Ibid.*, vol. 5, pp. 457-460 (Feb. 28, 1916), p. 462 (Mar. 13, 1916); R. V. Meikle to Allis-Chalmers Mfg. Co. Mar. 3, 1916; R. V. Meikle to C. S. Northcutt, May 26 1916.
[48] R. V. Meikle to A. Griffin, June 4, 1918.
[49] Report on TID irrigation, drainage and weed control, 1936, in Meikle files, vol. 7, item 53.

9

Don Pedro

The first gold seekers came to the Tuolumne River where it passes through the oak-and grass-covered hills in 1848, not long after news of the first discovery of the precious metal at Sutter's mill. Among the earliest was the Stockton Mining Company. One of its members, a Frenchman named Pierre Sainsevein but known as Don Pedro to the Mexicans, established himself in a remote part of the canyon, and left the following year with what was rumored to be a considerable fortune in gold. Don Pedro's Bar, as the place on the river became known, was soon bustling and second in population only to Jacksonville among the Tuolumne River mining camps. The settlement was built on both sides of the river with a ferry, and after 1859 a bridge, connecting the two. It had a restaurant, a brewery, a notorious fandango hall, a hotel on each side of the Tuolumne and, at least for a time, no church. It was said that Wells Fargo shipped $13,000,000 in gold from Don Pedro's Bar and still more left in other ways. The great flood of 1862 washed away the town but it was soon rebuilt, only to burn down in 1864. By then, the gold had almost run out and only a few Chinese miners remained to get what little was left. In 1871, a mining ditch beginning at Indian Bar, upstream from Don Pedro's Bar, was carried in ditches and precarious wooden flumes along the side of the canyon past the old Don Pedro cemetary and remains of the abandoned town to La Grange, where there was still gold to be mined.[1]

Except for cattlemen and a dwindling population of Chinese miners, the canyon around old Don Pedro's Bar was deserted until sometime shortly after the turn of the century when surveyors arrived looking for places to put dams. Just when the first reconnaissance was made is unknown but it was probably conducted on behalf of one of the many companies prospecting for places to generate eletricity. The ability of falling water to turn generators had been proven by 1890, and once problems associated with the long distance transmission of electricity had been solved, efforts were begun in earnest to harness the potential power in the streams coming out of the Sierra. On the Tuolumne, the first powerhouse was a small one in the mountains near the mouth of the Clavey River tributary, built about 1904 to serve the Tuolumne City area. About 1906 the La Grange Ditch and Hydraulic Mining Company decided to use its old mining ditch from Indian Bar to supply a powerplant, which would in turn run a gold dredger and provide power to sell in surrounding communities, including Turlock. That year they began construction of a powerhouse along the river a short distance below La Grange Dam, and were reportedly considering joining with another company to develop a much larger project further up the river.[2] By the time the irrigation districts sent their engineers to the mountains for the first time in 1908, most of the watershed had been surveyed at least in a general way for reservoir sites and, where land could be purchased, property at the likely spots was being acquired by private companies. Whether the districts' engineers even looked at the lower elevations closer to La Grange is doubtful. The evidence suggests they headed directly for the upper watershed, where favorable sites for silt-free, inexpensive dams could be found on federal forest or park land. But while the irrigation engineers were in the high country, others were at work closer to the valley in the area around Don Pedro's Bar.

On June 1, 1909, a Mr. A. Scott of the La Grange Water and Power Company, the successor to the old hydraulic mining company, spoke at length to the TID board about the joint construction of a reservoir at Indian Gulch Bar in the Don Pedro area and asked their permission to make a more definite proposal at some later time.[3] That time came in early December 1909, when E. A. Wiltse of New York, a representative of John Hays Hammond, a well-known mining entrepreneur and the principal backer of the La Grange Water and Power Company, addressed a meeting of the Turlock and Modesto boards. He announced that the company owned several reservoir sites on the Tuolumne, including one about seven miles above La Grange dam. A 160-foot dam at that point could store 41,000 acre-feet and produce 10,000 horsepower. The company had already acquired mining claims and other property in the area, and apparently could build the project by itself if it wished. Instead, it offered to cooperate with the irrigation districts. Its proposal called for the districts to pay half the cost of construction with their bonds in exchange for a guarantee that water released from the dam's powerhouse would always be theirs to use for irrigation. Wiltse claimed that the company's plan would give the two districts the additional storage they wanted at a quarter of what it would cost them to buy the site and construct the dam themselves. As an added inducement, he suggested that the existing La Grange powerplant, which was bypassing water around La Grange Dam by way of the old mining ditch, would be used only in emergencies, which would mean more water at La Grange Dam for the districts.[4] Further meetings were held in 1910, though the details of any cooperative arrangement were still hazy. One thing that was clear was that the company expected to control the reservoir, and releases would be made when needed for power, which was not necessarily when the districts might need the water they had helped pay to store.[5]

Negotiations continued off and on until early 1913 without producing an agreement. The districts made it abundantly clear that they and not the company should build, own and operate what was sometimes referred to as "dam number two."[6] The company, reorganized as the Yosemite Power Company in 1911, still wanted to control the output of the powerplant at the dam, but there, too, the districts refused to surrender any of what they considered their rights to control the facility. The Turlock District was more disposed than the Modesto District to come to terms with the company and expedite the project, but in the end the TID had little incentive to do so.[7] The only real leverage the power company had was its ownership of property in the reservoir site and, of course, the mining ditch and powerplant the districts wanted to put out of service. As the company's bargaining position at Don Pedro was eroded by district intransigence, it offered additional storage at reservoirs it hoped to build on the South Fork, but the two sides remained far apart on how much the extra storage would be worth.

Since a cooperative venture seemed unlikely, the Turlock District went forward by itself. During 1913, TID engineer Roy V. Meikle and a crew of eight men were sent to a reservoir site somewhat below the one first proposed by the power company for a thorough five month survey. Their isolated camp was supplied by mule teams from La Grange.[8] Using data from the survey, Burton Smith drew up plans for a 283-foot-high dam what would cost about $2,000,000 and impound 259,000 acre-feet. At its December 22, 1913, meeting, the board adopted those plans and ordered that steps be taken to acquire federal land at the site known as Don Pedro reservoir. A few days later, Smith and attorney P.H. Griffin left for the General Land Office in Sacramento to file a formal request that land in the reservoir site be reserved for the Turlock Irrigation District.[9]

In late 1914, the Yosemite Power Company returned to try to strike some sort of deal to salvage its hopes to profit from the dam it first proposed.[10] The company was still pursuing its plans for dams on the South and Middle forks and even at Poopenaut Valley, on the main stem of the Tuolumne below Hetch Hetchy, where San Francisco envisioned a future dam of its own. Both San Francisco and the districts had entered protests against the grant of national forest lands to the company, in part out of a concern that the presence of a third party on the river between Hetch Hetchy and La Grange could complicate compliance with the Raker Act. In November 1915 Henry Floy was sent by the Department of the Interior to investigate the protests and see if some means of accomodating the various conflicting interests could be found. A meeting with all parties did nothing to solve the dilemma. Charles S. Wheeler, an attorney for the company, explained that if matters on the upper reservoir sites were satisfactorily settled, the company would be willing to sell the mining ditch, La Grange powerplant and property in the reservoir site for $800,000, a sum that was said to have staggered the districts. The company would have liked to build Don Pedro itself, but since it could not, it was offering to abandon the area in exchange for a handsome cash reward and cooperation, or at least lack of opposition, to its other plans.[11] Perhaps realizing that once again they had set their sights too high in demanding so much money from the districts, the company retreated and a few weeks later offered to sell its property at Don Pedro for a mere $58,000 and a promise to keep seventy-five second-feet flowing to the La Grange powerhouse. Not only was that a larger flow than the districts were ready to admit the old mining ditch was entitled to, but the $58,000 soon soared to $88,000 when an overlooked mining claim was added. The districts thereupon cancelled all negotiations pending the preparation of a complete property map of the reservoir.[12] Meanwhile, on December 15, 1915, the two districts agreed to construct Don Pedro Dam at some unspecified future time when arrangements could be made.[13]

Although both districts had agreed to build the dam, only Turlock continued actively working on the project. The district began purchasing land at the site, including several parcels from

Albert Chatom, an old-time cattleman and one of the builders of the Highline Canal and the shoofly canal around the 1914 reservoir break. Chatom purchased the property and options at the time when the district itself was not prepared to do so and held them for the district's later use. He recalled that in 1910 or 1911, Edgar Annear had told him that he was surveying a reservoir site for John Hays Hammond's power company, and Chatom learned from George Bates that the company's option on some of Bates' land at the site would soon expire. When it did, Chatom stepped in and took the lands himself. He later acquired over 3,000 acres at Don Pedro, which he resold to the TID for little more than he had paid for it.[14] Chatom was widely revered in the Turlock District for his assistance in securing the reservoir site.

One highly important matter was that of financing the reservoir project. Certainly, the two irrigation districts could not afford to pay for it when it was first proposed; instead they opted for the cheaper foothill reservoirs. Very few local water agencies had the resources to build large dams, and to complicate matters, many streams had a number of competing users, who had to agree on how to share costs and benefits. For these and other reasons, reservoir building in the West became the almost exclusive province of the federal government, primarily through its Reclamation Service (later the Bureau of Reclamation). The Turlock and Modesto districts, however, gave only slight consideration to the idea of making Don Pedro a federal reservoir. At the 1915 meeting presided over by Henry Floy, for example, a suggestion was made that perhaps the government, or even San Francisco, might build Don Pedro and then turn it over to the districts, free of charge.[15] Of course, that was not how the Reclamation Service worked; if it had taken over the project the districts would still have had a repayment obligation and would have had to submit to federal control of their water supply. Despite such drawbacks, the attractive possibility of federal financial aid kept the districts at least mildly interested in Reclamation Service construction, but nothing came of their half-hearted efforts at involving the federal government in Don Pedro.[16]

In 1916, the Turlock District's application to use federal lands for the reservoir was approved and by the beginning of 1918 the district was ready to begin detailed planning. On January 14, 1918, Chief Engineer Meikle formally requested technical assistance from the Reclamation Service. His letter said:

> The Turlock Irrigation District to reach full development must have more storage. To this end surveys have been made and maps filed with the Government and about 3000 acres of land purchased. This Don Pedro site is the only site now available, all other sites having been taken up by San Francisco for the Hetch-Hetchy project. With the proposed dam it will be possible to store sufficient water for the late irrigations of lands in the Waterford, Modesto and Turlock Districts, together with other lands tributary to the Tuolumne River.
>
> We are now preparing to make borings at the proposed dam site, but if it is possible we would prefer to have this work done by the Reclamation Service at our expense.[17]

Copies of the letter were sent to the Modesto and Waterford districts. In response, Director Hilton of the Modesto board prepared a seemingly innocuous resolution that recited Modesto's rights under the 1890 agreement to join the Turlock District in the project and requested a conference with the Turlock board "for the purpose of arranging a plan whereby a thorough investigation may be harmoniously and expeditiously made."[18] Surprisingly, the resolution lost on a three-to-two vote. In a statement explaining his negative vote, MID board chairman Allen Talbot argued that the dam site was unsatisfactory, the whole project would cost much more than estimated and nothing should be done anyhow until legal actions contemplated against San Francisco were completed.[19] Suddenly it appeared that Turlock might be without a partner in the project.

At its next meeting, on February 18, the Turlock board moved to find out what the Modesto District's intentions were by invoking the sixty day clause in the 1890 agreement. To maintain its rights in the project, the MID had only that long to pay its proportional share, or about one-third, of expenses to date, which totalled $66,827.18 for land acquisition and surveys.[20] A letter accompanying the resolution said bluntly that, "The Turlock Irrigation District is unable to proceed further in the construction of the dam without definitely knowing the intention of the Modesto Irrigation District, whether it will participate in building the Don Pedro storage reservoir."[21] The first indication of the Modesto board's probable response came in the form of letters to the local press from Director Talbot, who essentially reiterated his previous statements against Don Pedro, and MID attorney W. C. LeHane, whose remarks were much longer and far less temperate. LeHane had a persistent, all-consuming grudge against the Turlock District, dating back to the district's resistance to demands that it repudiate its support for the Hetch Hetchy compromise. LeHane had been a leader in the last minute attack on the Raker bill, and he apparently blamed the TID for the failure of his efforts. His letter was full of invective against the Don Pedro project, the Turlock Irrigation District and especially P. H. Griffin, the TID's attorney. He concluded that,

> Our board should cancel the old agreement with reference to joint action with the Turlock district. Since that agreement was made, laws have been enacted which take its place and give us more rights than the agreement provides for. We can share in the reservoir whenever we want to.[22]

The Modesto board's formal response read much like LeHane's letter. In a rambling resolution, it claimed that its district lacked enough money to pay its share of past costs and the proposed investigation, that other reservoir sites were available, that Turlock did not help fight the Raker bill, and that laws had, in some unspecified way, changed. After citing all these alleged facts, the Modesto resolution abrogated the clause in the 1890 agreement dealing with the sharing and division of future water rights and projects above La Grange Dam.[23] Having so vehemently denied that they had to do anything to maintain their rights in a project they openly considered to be of questionable merit, the Modesto directors still left the door ajar to a compromise with the TID in a second resolution suggesting an informal meeting with the Turlock board.[24]

The position taken by the Modesto directors created an immediate disturbance in that district. The *Morning Herald* branded the action "A Damned Outrage and Robbery" and wondered aloud, "Why did this unholy bunch abrogate this very, very valuable agreement with Turlock district without suggestion, direct or remote, to the taxpayers?"[25] It was soon clear that those taxpayers were appalled at what their board had done. A heated debate at a meeting a few days later showed LeHane's viewpoint had few defenders and that even his legal arguments regarding the Modesto District's right to force its way into the project whenever it cared to were flawed.[26] While the Modesto District was thrown into chaos by its board's action, the Turlock District was not greatly distressed. Meikle wrote to State Engineer W. F. McClure that,

> The Modesto Directors refused to join the Turlock District and even went so far as to cancel the agreement of 1890 between the two

> districts in regard to future water rights procured by either District. This move is very satisfactory to the Turlock District as we would much rather handle the project alone, as we will have to pay two-thirds of the cost anyway.[27]

Meikle acknowledged, however, that the outcry in the Modesto District would probably force their board to participate in the project, and indeed an informal meeting was soon arranged. Before travelling to Turlock for the meeting on March 11, the Modesto board rescinded its abrogation of the 1890 agreement. For its part, the Turlock board agreed that, for the time being, Modesto should pay only one-third of certain survey and investigation costs, deferring payment for land purchases until later.[28] A week later, the boards met again to formally settle their differences. The Modesto directors accepted Turlock's offer of partial payments and the Turlock board rescinded its resolution giving Modesto the formal sixty day notice.[29] Meanwhile, the MID had taken a major step toward restoring harmony between the two districts by firing attorney LeHane on March 13. The crisis was over and the districts were ready to proceed with Don Pedro Dam.

Apparently, the Reclamation Service declined to do the test borings Meikle had requested, but federal authorities did recommend two engineers with experience in dam design. By July 1918 one of them, A. J. Wiley of Boise, Idaho, had been hired as a consulting engineer and the International Diamond Drill Company had been chosen to bore the test holes.[30] At the end of October, Wiley was able to make a verbal report, and the news was excellent. Twenty-one test holes had been drilled at the site from fifty to two hundred feet deep, and in every case they showed the underlying rock was solid and an excellent foundation for the dam without extensive excavation to bedrock. Directors viewed the core samples on display at the TID shops and heard Wiley describe the construction process, which he said involved "no unusual problems."[31]

Wiley submitted his formal report in December 1918. The site chosen was about four miles above La Grange Dam, at the upper end of the rock gorge that ended near La Grange. The concrete dam he recommended would rise 279 feet above the old river bed, making it the highest dam in the world. The height of the dam and the maximum water level were, according to Wiley, "fixed by the necessity of keeping the flow line beneath a certain level above which valuable property would be flooded, so that the usual investigation of the relation between the water supply and the irrigation requirements is not needed for determining the necessary reservoir capacity."[32] Nevertheless, a great deal of the report did deal with the amount of water needed to meet actual irrigation requirements and the inevitable reservoir and canal losses. Wiley estimated that the two districts would ultimately need to divert over 1,100,000 acre-feet annually, which was not far from the estimate arrived at by the districts' engineers in 1912. Don Pedro reservoir would have a useful capacity of about 250,000 acre-feet, which in most years would extend the irrigation season through September. Critics of the project had often contended that the reservoir would silt up and soon be nearly worthless, but Wiley's report showed that silt, while present, would not have much of an impact on Don Pedro, and sluice gates at the bottom of the dam could be used to periodically flush out whatever did accumulate.[33]

The engineering report confirmed what had been general knowledge for years; that Don Pedro would produce a great deal of electricity. The question, of course, was what to do with it. At least in the Turlock District, the generation and use of electrical power had been considered since 1910, when Burton Smith drew up the first plans for powerplants at La Grange and Hickman drop. Smith was thinking primarily in terms of powering drainage pumps, but plans for district-owned powerhouses soon prompted talk of lighting every farmer's home and even the formal cost estimates were for the vague purpose of "installing power plant and wiring same throughout the district."[34] The plans were dropped until R. V. Meikle revived them in 1916 with a proposal for a powerhouse at Hickman drop and an oil-fired back-up generator. Once again, drainage pumping was at the heart of the idea, but Meikle's correspondence made it clear that he contemplated "furnishing power to the entire district, for all purposes, including lighting of several towns."[35] Although Meikle later claimed that the original plans for Don Pedro contemplated the sale of its power to the power company, the districts' insistence on keeping absolute control of the Don Pedro powerhouse for themselves belies any such assumption, and there was apparently considerable interest in selling it to local residents. A Modesto newspaper, citing information from the *Turlock Tribune,* reported in 1912:

> Although it has never been announced officially, it has been a sort of mutual understanding between the people and the board of directors that advantage of the excellent opportunity be taken to install an electrical power plant at reservoir No. 2 should be taken,

> and there is hardly a question but what such a project will be carried out with the result that all of the people of the Turlock Irrigation District will, in the not far distant future, be enjoying public ownership of a very valuable public utility.
>
> The present plans for the new reservoir make it possible to generate a sufficient amount of electricity to supply the needs of the district and such a project is worthy of the unanimous support of all who cannot help but benefit by it.[36]

There were, however, some legal constraints that had to be dealth with before such plans could be put into effect. There was no doubt that irrigation districts could generate and use power for pumping and other irrigation-related purposes, or even lease power sites to private companies, but despite the consent implied by at least one statute, there was some question of their legal ability to, in effect, become electric utilities by distributing power to homes, farms and businesses.[37] In order to remove these uncertainties and give the districts a free hand in dealing with the power that could be generated at Don Pedro, legislation was introduced in early 1919 by Assemblywoman Esto Broughton and Senator L. L. Dennett, both of Modesto, to permit irrigation districts to build powerplants and sell or distribute the electricity in whatever way they wanted. The bill simply extended the broad authority granted by the irrigation district act to cover power as well as water. The measure met substantial opposition from San Francisco, which apparently feared the Tuolumne River districts would use additional water to make power and thereby enlarge their prior rights to the detriment of the Hetch Hetchy project. The city managed to insert an amendment making the use of water in excess of irrigation requirements subject to the city's own rights. The districts objected strenuously to the limitation but were unable to defeat the city's influence, and the bill became law with the provision intact.[38] Despite the amendment, the bill offered the districts a virtual carte blanche to enter the power business if they decided to do so.

By early 1919 the districts were ready to proceed with financing and construction of the dam. First, however, they needed to formalize their division of the project. As a result, the two districts, or at least their directors, were again involved in a controversy. At a joint meeting on April 24, 1919, the attorneys were ordered to draft a working agreement dividing the project on a two-thirds, one-third basis with Turlock, the larger district, taking the larger share.[39] The 1890 agreement, which covered the division of future projects like Don Pedro, specified that they would be divided in proportion to acreage, which if done accurately would yield a slightly different result, in Turlock's favor, than the approximate two-thirds, one-third ratio that had been applied at times as a matter of convenience. The question came up at a meeting on June 28, when a Turlock director proposed that the attorneys use the exact acreage proportion in the agreement they were writing. A counter-move by the Modesto board to retain the two-thirds, one-third division failed and the acreage ratio was approved, but since a majority of the Modesto board opposed it, fears were expressed that they might not ratify the end result.[40] The two boards met again on July 17, 1919, and once again argued over their respective shares in the project, but the Turlock board remained adamant and the final document divided the project on the basis of 68.46 percent for the TID and 31.54 percent for the MID. There was still some doubt that the Modesto District would accept the agreement, so when Modesto's attorney Walthall asked how soon work would begin, P. H. Griffin made his client's position clear. "Just as soon as you affix your names to that agreement," he replied. "If you decide to sign, well and good. If you decide not to, we'll simply go ahead and build it alone."[41] After a brief discussion of several other, non-controversial points on July 19, 1919, the two districts signed the working agreement.

By the time the project had been formally divided, cost estimates for construction of the dam and powerplant had been completed. On July 24 the boards decided that bonds for the powerhouse should be voted separately from those for the dam and reservoir. The dam was expected to cost a total of $3,750,000 and the powerplant about $609.000.[42] To its share of those amounts, the Turlock board added $510,000 for drainage and canal lining. Before putting the bonds to a vote of the people, the districts had to have their plans approved by the state's bond certification commission. The commission readily endorsed the bonds for the dam, but was reluctant to approve those earmarked for power production without more information on what would be done with the power.[43] Supplemental reports on the power question were called for and were prepared in less than two weeks. Meikle's statement of the Turlock District's intentions assumed that the district would retail power itself. He emphasized the need for drainage and for pumping irrigation water to land higher than the canals. The amount of power sold to consumers was expected to grow slowly, and in the meantime the surplus could be sold to one of the power companies. The report also acknowledged that no formal decision on power disposal had been made, but noted that the long construction period would give the districts ample time to decide whether they would distribute the power themselves or simply wholesale the entire output of the plant to a private company. Meikle estimated that an electrical distribution system for the Turlock District would cost $608,000, and that amount was added to the other bonds the district planned to issue for the project.[44]

The bond commission gave its full approval to the districts' bonds in late November 1919, and petitions calling for an election were soon being circulated. In the Turlock District, they were sent out first to the local Farm Bureau centers, but when the volunteer circulators did not gather signatures fast enough, the board put its thirty-five ditchtenders on the job, and a member of the Board of Trade stationed himself inside the district's office to collect signatures from people coming in to pay their irrigation taxes.[45]

Within a month, the petitions carried more than enough names and a bond election was scheduled for February 10, 1920. Three separate propositions were before the voters of the Turlock Irrigation District; one for $2,570,000 in bonds for the dam itself, one for $1,028,000 to cover the cost of the powerhouse and electrical distribution system and another for $510,000 for canal lining and drainage. There was no organized opposition to the bonds in the Turlock District, although two members of the Modesto board and a handful of others in that district were actively against the project. Their arguments had little effect, and voters in the Turlock District approved all three propositions by a vote of about fifteen-to-one. The election in the Modesto District a week later yielded similar results.[46]

On March 1, 1920, the districts hired A. J. Wiley as consulting engineer for the construction of the dam and ordered their own engineers to begin laying out routes for the railroad lines that would carry gravel and other materials to the construction site.[47] However, when a motion was made at a joint board meeting ten days later to appoint TID chief engineer R. V. Meikle to the post of chief engineer of the Don Pedro project, another interdistrict feud erupted. The motion to appoint Meikle passed, but the Modesto board refused to ratify the action of the joint board. As it developed, the dispute was not over Meikle but over objections by a majority of the Modesto board that a single Modesto director voting with the solid Turlock board at joint meetings could, in effect, decide how to spend MID's share of money for the project.[48] The Modesto board brought in noted San Francisco attorney John Francis Neylan to negotiate with Turlock, and meanwhile rehired W. C. LeHane as a special counsel, ostensibly to conduct a suit against San Francisco over Hetch Hetchy. The nasty little crisis reached its height in a chaotic meeting in Turlock on March 30, 1920. When the Modesto directors arrived for a scheduled joint meeting, they immediately went into a secret session of their own, leaving the Turlock board waiting for over two hours. At last, the Modesto board emerged and headed straight for the door, which led to a heated exchange and then to a no less heated joint meeting, which at one point nearly disintegrated into a fist fight. The Modesto directors offered a resolution that would have required the approval of at least three members of each board for the passage of any motion before the joint board, but a substitute TID motion calling for simple majority approval passed instead.[49] As a result of the bitter procedural impasse, Modesto District voters recalled the three so-called "majority" directors, and in June the new Modesto board agreed that any measure receiving majority approval at a joint board meeting should be ratified by both boards at their regular monthly meetings.[50] The controversy was more a matter of personality and pride than of substance, and once it was resolved the two districts functioned smoothly together throughout the construction period.

No major progress was evident until near the end of 1920. In December the TID sold its $2,570,000 in bonds for the dam at par, with all but $60,000 going to a San Francisco syndicate.[51] Meanwhile, the detailed plans and specifications needed to call for bids had been prepared by Reclamation Service engineers under contract to the districts. Bids were then requested and were opened on February 24, 1921. Two companies submitted bids, but the lowest exceeded the engineers' estimates by nearly $400,000. The boards therefore rejected the bids and announced that they would consider further proposals at their March 1 meeting.[52] On March 1 they received thirteen informal proposals from ten different contractors. The districts' engineers reported that, "The proposals are unusually complete, are from contractors of the very highest standing and include the latest and best types of 'cost plus' contracts."[53] Cost-plus contracts were very common at that time as a result of rapid inflation during and just after World War I. The directors now faced a dilemma. A cost-plus contract meant that the districts assumed a certain amount of risk, since they, not the contractor, would have to pay for higher than estimated costs, but the engineers warned that to readvertise for fixed price bids could delay the project for a year. The final option was for the districts to build the dam themselves without a contractor. A decision was made quickly, and on March 11, 1921, a cost-plus contract for the construction of the dam and a narrow-gauge railroad from Hickman to La Grange was awarded to W. A. Kraner. According to the agreement, Kraner, who had the backing of the syndicate of bond buyers, would receive a flat fee of $185,000 for his work.[54]

There seems to have been some uneasiness among the districts' citizens with the flat fee or cost-plus arrangement, and on March 14, the California State Land Settlement Board, which was putting in a colony at Delhi, complained that the contract with Kraner lacked sufficient guarantees that it would be carried out.[55] Apparently in response to these concerns, the boards rescinded the original contract and considered a new one that reduced the fee but offered to pay a premium of 25 percent of the savings realized if the job was finished for less than the engineers' estimates. The plan had been to rescind one contract so a new one could be substituted, but the directors deadlocked and the revised contract was not approved.[56] This surprising development brought another round of proposals from contractors hoping to take Kraner's place. While the directors tried to sort through their options, they ordered their engineers to proceed with the construction of a railroad under district supervision. Serious consideration had been given to building a narrow-gauge line from Hickman to La Grange, and to the use of trucks to haul gravel from near La Grange to the dam. However, negotiations with the Sierra Railway, which ran from Oakdale to Sonora, resulted in a decision to build a standard gauge railroad from the Sierra line at Rosasco eight miles to the dam site. Except for thirty gravel cars to be purchased by the districts, the Sierra would supply the equipment and operate trains from gravel beds just outside of Oakdale to the dam site, a total distance of thirty-three miles. A contract with the railroad was signed on March 29, and grading for the roadbed of the Don Pedro branch began almost immediately.[57]

Kraner continued to submit modified proposals and even demanded immediate delivery of the TID bonds purchased by the syndicate he was aligned with in order to prove that his financing was solid. Although Kraner and the syndicate did take the bonds on terms that were advantageous to the district, sentiment was growing to dispense with outside contractors altogether and build the dam under district superintendence.[58] On April 26, 1921, the districts made the decision official when they hired D. H. Duncanson as general superintendent for construction and ordered their engineers to proceed at once to build the dam.[59] Shortly thereafter, one of the last hindrances to construction was removed when the Sierra and San Francisco Power Company, the owner of the mining ditch, La Grange powerplant and property formerly held by the Yosemite Power Company, agreed to sell its plant, ditch, land and water rights to the districts for 10,000,000 kilowatt-hours of electricity per year from the Don Pedro powerhouse for twenty-five years.[60] After years of haggling, the power company was finally eliminated from the Don Pedro and La Grange area and the districts came into control of the only water right on the river older than their own. Preparations were now complete and full scale construction could begin.

During May and June of 1921 district crews were hard at work on the railroad under the general supervision of Sierra Railway engineer Newell, and a complete construction camp was built on the hills above the dam site. To house the hundreds of men who would come to build the dam, bunkhouses and tents were set up, with a "family street" of tents for the married men. A mess hall with seating for 400 to 500 men was built and placed in charge of Mrs. Ora McDermott. A ten-bed hospital, staffed by resident physician Dr. C. R. Fancher and a nurse, was erected, as was a recreational hall equipped with card and pool tables. Later, a school was opened at the camp and had twenty-two students the first year.[61] A modern water supply and septic tank system was installed, but garbage disposal was turned over to a large number of hogs. The camp was, by all reports, a well run and stable little community and a number of local boys, as well as engineering students looking for practical experience, joined the labor force. Dances and other social events were held from time to time in the recreation hall, but for some reason, the joint board consistently voted against showing movies at the camp. About the only excitement came on a couple of occasions when armed robbers appeared to relieve a few of the men of their paychecks.[62]

Ground was formally broken for the project on June 25, 1921, when TID director S. A. Hultman touched off a dynamite blast in the canyon. A month later, the railroad was completed and the first train arrived at the camp.[63] The engineers were eager to begin work so that the dam's foundation could be finished when the river was low during the late summer and fall. A wooden flume was built to carry water around the construction site, but work on the concrete walls went so rapidly that in October the flume was blown out to allow the river to flow through a temporary concrete channel at the base of the dam.[64] The initial work was done with small equipment working in the canyon itself, but by late October the first of two large cement mixers was installed at the top of the hill and on October 25, the first cement from the plant went down the chutes to the construction site. The big sluice gates in the base of the dam were installed in November, and by mid-December the work was ahead of schedule, with the dam up to sixty feet above the river bed.[65] Heavy rains that winter softened the railway roadbed to the point that the movement of the heavy gravel trains had to be suspended, and the highway to the camp was rendered virtually impassable. In all, the railroad could not operate for sixty-five days during the winter. Despite the inclement weather, excavation on the slopes where the dam would rise, and along the river where the powerhouse would be built, continued.[66] When work was resumed in the spring, it went so smoothly that the directors were forced to buy additional freight cars to keep up with the demand for enough sand and gravel to mix 1,000 to 1,200 cubic yards of concrete per day.[67]

With construction proceeding satisfactorily, attention turned once more to the question of what to do with the power the dam would soon be able to generate. In January 1922 the Turlock District sold the $1,028,000 worth of bonds it had earmarked for power development and ordered Chief Engineer Meikle to report on how much power could be sold and for how much.[68] Although distribution within the districts had long been discussed and was a popular idea, Pacific Gas and Electric, which had come to dominate the northern California electric industry, also wanted the power, or, more likely, wanted to keep the districts from becoming competitors to the company. In April, after a meeting with the company's general manager, John A. Britton, the two districts asked the company to make a proposal for the purchase of the entire output of the Don Pedro plant. In a week Britton was back with an offer that included the sale of drainage pumping power back to the districts essentially at cost.[69] Each district hired its own consulting electrical engineer to evaluate the company's offer and other options. Modesto's engineer, Louis F. Leurey, came out in favor of the PG&E wholesale offer; an attitude that seemed to reflect the opinion of some members of the Modesto board. Leurey's report showed that the districts could make a greater profit from wholesaling their power in accordance with the company's offer than by going into the distribution business themselves.[70]

R. W. Shoemaker, the engineer hired by the Turlock District, came to a different conclusion. He considered the terms of the power company's offer entirely inadequate. He agreed that the district would earn more money in the short run from wholesaling, but he also contended that, over time, distribution showed the greatest potential. Since Shoemaker's report became the cornerstone of the Turlock District's electrical policy, portions of it deserve quotation at length:

> . . . From the nature of the agricultural development it is not believed that there will be any requirements for pumping such as exist in sections of the country further south, and for the time being there will probably be very little requirements for power for industrial purposes. It is therefore believed that the greatest opening for power in the district is the development of the power requirements of the individual householder.
>
> As the district will be able to furnish power at a very small cost it is believed that a large load can be developed from the sale of power for lighting, cooking and water heating, in addition to small blocks of power that will be used around the farms for miscellaneous purposes.
>
> The use of electric power for cooking and water heating is largely a matter of rates, and it is believed that if an attractive rate can be arrranged the majority of the householders will eventually install electric ranges . . .
>
> As the cooking load is on a comparatively short part of the time it will be necessary to devise some method of developing a more constant load, or improving what is called the load factor, which is the ratio in percent. of the average load to the peak load. This can be done by giving a flat rate on water heaters such that the householder will allow the heater to run 24 hours a day . . .
>
> Considering all the sources of revenue in the district that might be developed along these lines it has been assumed that in 1935 there will be 8000 customers paying an average of $6 per month each for which the district could afford to give 500 KW hours per month.[71]

Shoemaker's proposed policy of low rates was meant to encourage the rapid growth of electrical consumption in the district, but it would still be years before the TID would need all of its Don Pedro power locally. In the meantime, Shoemaker recommended wholesaling the surplus to a power company. If the district put in an oil- or gas-fired steam plant to "firm up" its fluctuating supply of hydroelectric power, he thought it could sell a guaranteed block of power for twice what PG&E was offering.[72] Another possibility mentioned by Shoemaker was the use of Don Pedro power to make nitrogen fertilizer, a simple but energy intensive process.[73] With its consultant's report in hand, the Turlock board ordered a special advisory election for June 21, 1922, on the matter of selling the power wholesale or distributing it within the district, and the Modesto board scheduled a similar election for the same day.[74]

The head of the Turlock Chamber of Commerce asked the board, when it called the advisory election, if it would give some guidance in the matter, but the directors declined to do so.[75] Shoemaker, while also declining to give an outright endorsement of either distribution or wholesale, kept up a hectic speaking schedule the week before the election, and had nothing bad to say about distribution or good to say about the PG&E offer.[76] Meanwhile, without even waiting for the election they had just called, the Modesto board voted to wholesale their share of the power. The power company was also busy trying to drum up support for wholesaling, and a Taxpayers Association was quickly organized in Modesto to back the PG&E plan.[77]

What was happening in the Modesto District especially was symptomatic of a much larger struggle over public versus private power development. Californians in 1922 were being asked to vote on a constitutional amendment known as the Water and Power Act. It would have authorized state-owned reservoirs and hydroelectric plants, with no more than 20 percent of the power they produced going to private companies. The remainder would have been distributed by municipalities and other public agencies. The power companies spent large sums to attack and ultimately defeat the Water and Power Act, largely on charges that it represented a dangerous, socialistic expansion of state authority. Against the background of this controversy over public power, residents of the two irrigation districts were beign asked whether they wanted to have their own publicly-owned electric utilities. There was, however, little or no debate on the philosophical issues of public ownership in the Turlock District. A clear consensus for distribution was already present, and the proposition as presented by

Don Pedro Dam under construction, June 15, 1922.

Shoemaker only seemed to confirm its value. The campaign for wholesaling, which was tantamount to a campaign on behalf of private power, may have had some impact in the Modesto District, but clearly had none at all south of the Tuolumne. Electrical distribution in the Turlock District was approved by a vote of 2,433 to 116.[78] In Modesto, the ratio was a little less than three-to-one for distribution.

At Don Pedro, completion of the dam was expected by the end of 1922, and it was thought the powerhouse would be ready for operation a few months later. One of the most difficult problems associated with setting up the powerplant was getting the heavy generating equipment from the Sierra Railway to the powerhouse at the base of the dam on the opposite side of the canyon. A special and spectacular piece of railroad was built to accomplish the job. Electric hoists were used to lower a modified freight car down the steep slope beside the dam and across a track laid along the face of the dam.[79] The last piece of machinery was moved into the powerhouse on January 10, 1923, and by that time the dam itself was virtually finished, leaving only the spillway to be completed.[80] The original plan for the dam did not envision gates on top of the spillway crest, but in June 1922 it was decided that by installing them the maximum water level could be increased, resulting in more storage and a greater head on the powerplant. That became more important when the districts executed an operating agreement in February 1923 that in effect prohibited lowering the reservoir below the elevation from which water was drawn into the powerhouse.[81] Adding the spillway gates to increase the capacity of the reservoir compensated for the creation of "dead" or unusable storage below the powerhouse outlets.

By the summer of 1922 it was apparent that the reservoir would begin to rise that winter. The triangular opening fifteen feet wide and thirty feet high that, along with the three sluice gates, had been passing water through the dam, had to be closed in the fall, when the stream flow fell to its lowest point. Thereafter, if the river carried more water than the sluice gates could handle, the reservoir would begin to fill. Upstream, work was being rushed on a section of San Francisco's Hetch Hetchy aqueduct that would pass under the reservoir at Red Mountain Bar. Part of the city's construction railroad also had to be rerouted around the reservoir and a new bridge built for the line at Six Bit Gulch. The old Don Pedro cemetery had already been moved, and only two residents remained in

Loaded flatcar being lowered down the incline railway.
The tracks crossed the face of the dam and entered the powerhouse.

the canyon behind the dam; Mrs. Donohue, a rancher, and Lee Bung, an aged Chinese miner who had become too old to work. Lee Bung was said to feel "unmitigated resentment" against the districts for the loss of his home.[82] In November 1922 he reportedly decided that he would prefer to return to China, so Director Guyler of Modesto discussed the matter with the secretary of the Six Companies, a Chinese organization in San Francisco, and received a pledge of assistance in the matter. Earlier that month, the districts had informed Mrs. Donohue that storage had begun and water was expected to reach her buildings by December 15.[83] Lee Bung was offered a cabin downstream from the dam until other arrangements could be made, but in early January 1923 he told Superintendent Duncanson that he would not leave his longtime home. When he was not seen for a week, a search was begun that found only footsteps leading to the river. Old friends concluded that Lee Bung had drowned himself in the water rising over his old home.[84]

Except for a short time in April when the flood gates were opened to slow the rise of the reservoir so San Francisco could finish some of its work and remove its equipment, water rose steadily until, on June 11, 1923, the reservoir could hold no more and water went down the spillway for the first time. Overall, just under 300,000 cubic yards of concrete had gone into the dam and spillway, and to make that much concrete as many as forty of the 55-ton gravel cars had been moved each day from Oakdale to the dam site.[85] The 284-foot-high, arched dam was 177 feet thick at the base and 16 feet wide on top, where a roadway crossed its 1,000 foot length. In the face of the dam were two banks of six ports each for the release of irrigation water in excess of the amount that could flow through the five outlets that led to the water-wheels of the powerhouse. Passageways and galleries extended throughout the dam so that workers could operate and repair the valves and gates. To one side, a spillway with a crest 600 feet long and ten drum gates could bypass a flood flow of over 100,000 second feet safely around the dam. When it was full, the reservoir held 289,000 acre-feet and covered over 3,000 acres.[86] But Don Pedro was more than numbers. From the arched bridge over its spillway to the graceful curvature of its face, it was beautiful, if anything so utilitarian could warrant that description.

A formal dedication was held on June 25, 1923, and in honor of the occasion all the stores in Turlock closed for the day. A special

Excavating the Don Pedro spillway.

train brought 400 celebrants, and a count showed 375 cars parked near the ceremonies. The 2,500 or so visitors watched as the spillway gates were demonstrated and they listened to speeches by district officials, the Attorney General and even San Francisco city engineer M. M. O'Shaughnessy, whose dam at Hetch Hetchy had also just been finished.[87] The speeches made that day were not particularly memorable, but the occasion certainly was. The districts had the highest dam in the world, enough storage for a full irrigation season and a powerplant that would put them in the electric business, and they had financed and built the project themselves on schedule and within budget estimates.

How were these two districts able to do what most local water agencies could not; build their own large dam and powerhouse? There is no simple answer, but a number of fortuitous circumstances seem to have been at work. Both districts were relatively large and financially strong. They had passed through the settlement phase and had the kind of prosperous irrigated farms, and tax base, that could afford to pay for additional water. The Tuolumne was a large stream, with substantial hydroelectric potential, and no conflicts among its irrigation developers. Somewhat the same circumstances were to be found north and south of the Tuolumne, with similar results. On the Stanislaus River, the Oakdale and South San Joaquin irrigation districts built Melones Dam with financial aid from PG&E in exchange for power rights, and on the Merced River, the Merced Irrigation District erected Exchequer Dam and wholesaled its electrical output to the San Joaquin Light and Power Company. Both of these projects were built shortly after Don Pedro, and R. V. Meikle even served as chief engineer for the Merced Irrigation District during the construction of its dam. Only the Turlock and Modesto districts, however, entered the retail electric business, perhaps because the Tuolumne promised a greater supply of energy, or perhaps because only the older Tuolumne River districts could afford the risk and added expense of building an electrical distribution system and waiting for it to grow. If retailing Don Pedro power was, in fact, one sign of the districts' well-being even before Don Pedro, it was also a decision that would soon further enhance their strength. With Don Pedro, the era of pioneer irrigation was over and a time of electrical pioneering was at hand.

CHAPTER 9 – FOOTNOTES

[1] *Turlock Tribune,* Mar. 3, 1922; Don Pedro Recreation Agency, *Don Pedro Lake Guide,* (circa 1981).
[2] *Stanislaus County Weekly News,* Dec. 7, 1906.
[3] Board minutes, vol. 3, p. 63 (June 1, 1909).
[4] *Turlock Journal,* Dec. 3, 1909. [5] *Ibid.,* July 22, 1910.
[6] For example, see the TID and MID proposal in Board minutes, vol. 4, pp. 152-158 (Aug. 19, 1912).
[7] *Modesto Morning Herald,* Aug. 15, 1912; Board minutes, vol. 4, p. 379 (Apr. 28, 1913), p. 387 (May 12, 1913). [8] *Modesto Bee,* July 23, 1962.
[9] Board minutes, vol. 4. pp. 555-556 (Dec. 22, 1913); *Modesto Morning Herald,* Dec. 28, 1913. [10] *Modesto Morning Herald,* Dec. 3, 1914.
[11] *Ibid.,* Nov. 27, 1915; "Transcript of Meeting of the representatives of Yosemite Power Company, The City and County of San Francisco, the Modesto and Turlock Irrigation Districts, the Waterford Irrigation District, the Forestry Department of the United States, and the Department of Agriculture of the United States, in San Francisco, Nov. 23, 1915."
[12] *Modesto Morning Herald,* Dec. 16, 24, 25, 1915.
[13] Board minutes, vol. 5, p. 404 (Dec. 15, 1915).
[14] *Modesto News Herald,* June 11, 1924.
[15] "Transcript of Meeting, November 23, 1915," pp. 74-75, 79-80.
[16] Rhodes, "Thirsty Land," p. 148; Board minutes, vol. 5, p. 707 (May 14, 1917).
[17] R. V. Meikle to Arthur Powell Davis, Jan. 14, 1918, in Board minutes, vol. 6, p. 34 (Jan. 14, 1918).
[18] *Modesto Morning Herald,* Feb. 14, 1918. [19] *Ibid.,* Feb. 15, 1918.
[20] Board minutes, vol. 6, pp. 69-72 (Feb. 18, 1918). [21] *Idem.*
[22] *Modesto Morning Herald,* Feb. 22, 1918.
[23] The MID resolution of Feb. 27, 1918, is found in Board minutes, vol. 6, pp. 74-79 (Mar. 11 1918).
[24] The Second MID resolution of Feb. 27, 1918, is found in Board minutes, vol. 6. pp. 79-80 (Mar. 11,1918).
[25] *Modesto Morning Herald,* Mar. 1, 1918. [26] *Ibid.,* Mar. 3, 1918.
[27] R. V. Meikle to W. F. McClure, Mar. 6, 1918.
[28] *Modesto Morning Herald,* Mar. 12, 1918; Board minutes, 6, p. 82 (Mar. 11, 1918).
[29] Board minutes, vol. 6, pp. 96-97 (Mar. 18, 1918).
[30] *Ibid.,* pp. 111-112 (Apr. 12, 1918), p. 129 (June 3, 1918), p. 154 (July 9, 1918).
[31] *Modesto Morning Herald,* Oct. 29, 1918.
[32] A. J. Wiley, *Report on Don Pedro Reservoir for the Turlock and Modesto Irrigation Districts,* (Dec. 1918), p. 3. [33] Wiley, pp. 5-14, 22-28.
[34] Board minutes, vol. 3, p. 192 (Feb. 14, 1910), pp. 261-262 (June 7, 1910); *Turlock Journal,* July 8, 1910.
[35] R. V. Meikle to Allis-Chalmers Mfg. Co., Mar 3, 1916.
[36] *Modesto Morning Herald,* Nov. 16, 1912.
[37] See *Cal Stats.* 1911, pp. 813-821.
[38] *Modesto Morning Herald,* Mar. 27, Apr. 1, 12, 1919; *Cal Stats.* 1919, p. 778.

[39] Board minutes, vol. 6, p. 321 (Apr. 29, 1919).
[40] *Ibid.*, pp. 354-355 (June 23, 1919); *Modesto Morning Herald,* June 24, 1919.
[41] *Turlock Tribune,* July 18, 1919.
[42] Board minutes, vol. 6, p. 371 (July 24, 1919).
[43] *Modesto Morning Herald,* October 11, 1918.
[44] Board minutes, vol. 6, pp. 445-453 (Oct. 22, 1919).
[45] *Turlock Tribune,* Dec. 26, 1919.
[46] Board minutes, vol. 6, pp. 532-545 (Feb. 16, 1920).
[47] *Ibid.*, p. 551 (March 1, 1920); *Turlock Tribune*, Mar. 1, 1920.
[48] *Turlock Tribune,* Mar. 19, 1920; *Modesto Morning Herald,* Mar. 26, 1920.
[49] *Modesto Morning Herald,* March 31, 1920.
[50] Board minutes, vol. 6, p. 678 (June 21, 1920).
[51] *Ibid.*, vol. 7 pp. 58-61 (Dec. 14, 1920).
[52] *Ibid.*, pp. 142-145 (Feb. 26, 1921). [53] *Ibid.*, p. 151 (Mar. 8, 1921).
[54] *Ibid.*, pp. 165-166 (Mar. 11, 1921); *Turlock Tribune,* Mar. 14, 1921.
[55] *Turlock Tribune,* Mar. 16, 1921.
[56] Board minutes, vol. 7, pp. 185-188 (Mar. 17, 1921); *Turlock Tribune,* Mar. 18, 1921.
[57] Board minutes, vol. 7, p. 192 (Mar. 29, 1921); *Turlock Tribune,* Apr. 6, 1921.
[58] *Turlock Tribune,* Mar. 25, 1921, Apr. 22, 1921.
[59] Board minutes, vol. 7, pp. 228-231 (Apr. 26, 1921).
[60] *Ibid.*, pp. 240-245 (May 4, 1921).
[61] *Turlock Tribune,* June 8, 24, 1921, Nov. 21, 1921.
[62] *Ibid.*, Oct. 5, 1921, Mar. 10, 1922. [63] *Ibid.*, July 27, 1921.
[64] *Ibid.*, Oct. 17, 1921. [65] *Ibid.*, Nov. 21, Dec. 14, 1921.
[66] *Ibid.*, Jan. 25, Feb. 20, 1922.
[67] Board minutes, vol. 7, p. 749 (Mar. 21, 1922).
[68] *Ibid.*, pp. 662-663 (Jan. 11, 1922).
[69] *Ibid.*, pp. 790-791 (Apr. 25, 1922), pp. 793-797 (May 2, 1922).
[70] *Turlock Tribune,* May 19, 1922.
[71] R. W. Shoemaker, *The Don Pedro Project: Proposed Power Development Covering the Generation, Distribution and Sale of Electrical Energy by the Turlock Irrigation District,* (May, 1922), p. 25.
[72] *Turlock Tribune,* May 24, 1922. [73] Shoemaker, pp. 56-57.
[74] Board minutes, vol. 8, p. 16 (May 31, 1922).
[75] *Turlock Tribune,* June 2, 1922. [76] *Ibid.*, June 14, 16, 19, 1922.
[77] *Ibid.*, June 19, 1922; *Modesto Morning Herald,* June 18, 1922.
[78] Board minutes, vol. 8, pp. 35-36 (June 22, 1922).
[79] *Turlock Tribune,* Nov. 1, 1922.
[80] Board minutes, vol. 8, p. 169 (Jan. 10, 1923).
[81] *Ibid.*, p. 25 (June 6, 1922), pp. 193-194 (Feb. 20, 1923).
[82] *Turlock Tribune,* Mar. 3, 1922.
[83] Board minutes, vol. 8, p. 142 (Nov. 28, 1922), p. 145 (Nov. 21, 1922).
[84] *Turlock Tribune,* Jan. 17, 24, 1923. [85] *Ibid.*, Mar. 19, 1923.
[86] R. V. Meikle, "The Turlock Irrigation District," (Nov. 1971), pp. 6-7.
[87] *Turlock Daily Journal,* June 26, 1923; *Turlock Tribune,* June 27, 1923.

10

The Electric Age

Following the advisory election in June 1922, the Turlock Irrigation District wasted no time in beginning work on its electrical distribution system. Some thought had been given to building a single main transmission line to serve both districts, and the TID proposed running a jointly-owned line from Don Pedro to Keyes, where the power would be divided. The two districts, however, could not come to an agreement, and on July 7, 1922, the joint board decided that each district would be responsible for its own transmission facilities.[1] Just three days later, the Turlock District called for bids on the construction of its own transmission line. When the bids exceeded the engineer's cost estimates, the board, which by this time was accustomed to doing without contractors, ordered district crews to do the job.[2] Meanwhile, the Modesto District was in the midst of a controversy over its electrical policy. Dissatisfaction with directors who had originally voted to wholesale their power and whose dedication to the concept of distribution remained suspect resulted in the recall of three board members in late 1922. The reconstituted MID board soon reopened negotiations on the sharing of transmission. The Turlock board agreed on January 30, 1923, to sell 31.54 percent of its nearly completed main line to the Modesto District, but when a contract to that effect came before the joint board on February 9, it was rejected on the negative vote of all five Turlock directors.[3] There was no explanation for the sudden turnabout in TID sentiment, but when the proposed agreement was redrafted, the purchase of part of the line had been replaced by a lease.[4] In addition, the TID agreed to run a line to the Tuolumne River near Empire to connect the transmission line to the Modesto District's distribution system. The MID's share of Don Pedro power flowed through the TID system until at least 1928, when the Modesto District put the first circuit of its own transmission line into operation.

The Turlock transmission line was built during the winter of 1922-1923 southwestward from Don Pedro, passing to the south of Owens Reservoir and then along Keyes Road until it turned to reach a substation near the corner of Geer and Monte Vista roads, north of Turlock. The two-circuit, 66,000-volt line consisted of aluminum wires hung from steel poles. It was said to be the first high tension line to use aluminum wire. Aluminum proved to have several drawbacks. Its softness required extra care in installation, including the use of boards laid over rocky spots on the ground to protect the wire as it was uncoiled. Once the line was in place, it was found to be subject to vibration, which in turn led to cracks at the insulators where the wire was hung. The problem was so serious that Alcoa, the wire's manufacturer, was forced to develop a solution and pay for repairs. In 1924 the line was rebuilt with steel wire spliced in at critical points. Reconstruction and the use of simple vibration dampers or shock absorbers on the wire solved the problems associated with the aluminum high tension line.[5]

The transmission line was completed to the Geer substation on February 19, 1923. At the same time, the district's first distribution line was being built south to the Delhi area to serve drainage pumps and nearby customers.[6] By March the generators at the dam were running, producing just enough power to dry the transformers there and no more, because transformers for the substation were late in arriving. Finally, on Easter Sunday, April 1, 1923, the substation was ready and the first power came through the transmission line, but it was only enough to maintain the equipment for another two or three weeks until distribution was ready to begin.[7] Don Pedro power was first used to run a drainage pump at Delhi and the first retail customer was presumably in the same area. With the line to Delhi in operation, construction crews began stringing wire into Turlock itself, and by early August, after being in the power business only three months, the TID had over 400 customers. Two months later, its ninety miles of distribution lines were serving 600 homes and businesses.[8] At the end of 1924, electrical engineer R. W. Shoemaker's first formal report showed that the district had 3,220 customers and about 250 miles of distribution lines.[9] The district's career as an electric utility was off to a flying start.

Shoemaker's initial report in 1922 had emphasized that to make the retail distribution of Don Pedro power a paying proposition, local demand would have to be built up as rapidly as possible. To that end, the district put several Hotpoint electric ranges on display in the corrider of its Turlock office in early 1923 and had a Mrs. Finnerty demonstrate them to potential customers. The ranges were also for sale since there were no general appliance stores in the district at that time. In only a few weeks over one hundred ranges had been sold, and every member of the new electrical department had helped sell a stove or two.[10] By the time the main transmission line was energized in April, sales of stoves totalled $20,000, and in mid-September it was reported that the district had sold two freight car loads of electric ranges since the program began, as well as water heaters, space heaters and other appliances.[11]

The volume of appliance sales surprised the engineers, but it was an accurate measure of the enthusiasm that greeted the district's electrical enterprise. For rural residents, electricity had generally been unavailable or unaffordable. The La Grange Water and Power Company and its successors, the Yosemite Power Company and the Sierra and San Francisco Power Company, which was leased to PG&E by the time the district was ready to enter the power business, had run lines to towns in Stanislaus County. The San Joaquin Light and Power Company, which held the franchise in Merced County, extended its lines to Hilmar and Delhi. The private companies were reluctant to venture into the rural areas, where a limited number of customers per mile of line made the service expensive to install and difficult to maintain. If a farmer wanted electricity, he had to pay for the power lines needed to bring it in as well as the usual charges for the energy he used. For that reason, there were few rural lines before the district came on the scene. Nevertheless, there was a tremendous desire for electricity. As early as 1909 some houses were being constructed with built-in wiring even though there were no power lines in the neighborhood and no real expectation that there would be any in the near future.[12] Electricity was eagerly awaited because it meant an end to kerosene lighting, running a hot cookstove on a summer day or relying on windmills or gasoline engines to pump domestic water. It was, like the automobile or the radio, something that could break down the isolation and drudgery that had marked rural life.

The district sold its power on terms that were much more attractive than those offered by the private companies. Line extensions were free so long as monthly revenues were at least $24 per mile of line. When a line would not generate that much business, the district did require users to pay for construction, but with provision for a thirty-six month installment contract to lighten the burden.

Since the district had a relatively inexpensive source of power and was committed to developing a large local market for its energy, it rates were also lower than those set by the power companies. For example, before Don Pedro was built a Denair rancher signed a three-year contract with PG&E for $6.00 a month, which later rose to $7.50 a month when the company was awarded a rate increase by the Railroad Commission. The TID planned to charge only $2.60 monthly for the same service, prompting the company to reduce its rate to $3.00 in early 1923 to remain competitive.[13] Initially, the district followed Shoemaker's advice and instituted a service charge on all accounts and flat rates for water heating. There was even a so-called prepaid meter service for coin-operated meters. In 1930, the rate schedules were revised to eliminate the service charge, and were slightly altered again the following year.[14] The 1931 rate for domestic service priced the first fifteen kilowatt-hours per month at $1.00 and progressively lowered the rate per kilowatt-hour on succeeding blocks of energy. After 200 kilowatt hours a month, each additional kilowatt-hour cost only a penny.[15]

These rates made it impossible for the power companies to compete effectively against the district, although in some places, the district's lines ran on one side of the alley or street and the company's were on the other. In 1928 PG&E had only twenty-one customers in the City of Turlock, several of whom were company stockholders or were doing business with the power company.[16] In 1929 the San Joaquin Light and Power Company sold all of its facilities in TID territory north of the Merced River to the district, which paid just over $40,000, half in cash and the rest in surplus power, for the lines that still served sixty customers.[17] At about the same time, the San Joaquin company agreed to provide standby service in case the district's transmission line failed or Don Pedro was unable to generate. Two years later, in 1931, Pacific Gas and Electric sold its distribution system to the TID, giving the district full control of the electrical business within its boundaries.[18] As part of the purchase of PG&E facilities, the district found itself in possession of a substation in Hickman and the distribution lines in Waterford, on the other side of the Tuolumne River, serving sixty-nine customers. The Waterford system was sold to the Modesto District in 1933 and the Hickman substation was dismantled.[19]

The rapid spread of the district's distribution system created a large and steadily growing number of consumers in need of appliances and kept the district in the business of supplying them. In 1924, the

TID opened a retail store that soon carried five hundred items, not counting lamps, and did over $96,000 worth of business in its first year. It was located in the Broadway theater building at the corner of Broadway and Olive streets before it moved to the district's Broadway yard in 1929. Five men were kept busy with sales and service at the store, in addition to a four-man service crew that handled trouble calls and in 1924 was installing ranges at the rate of more than one a day.[20] For Christmas, 1926, the store advertised electrical gifts ranging from curling irons for $2.65, waffle irons for $8.00 and a chafing dish for $12.50, up to General Electric vacuum cleaners for $37.00, Thor washing machines for $145, electric ranges from $75 to $185 and General Electric refrigerators as low as $250 or as high as $440. In that year, 1926, the store sold 219 ranges, 161 water heaters, 142 air heaters and 43 motors, as well as wiring supplies, light fixtures and smaller appliances.[21]

Electric ranges and water heaters, along with lighting, were probably among the most common major uses of electricity in rural homes, followed closely by electric water pumps. Before electricity was available, domestic water was pumped into overhead tanks by windmills or by stationary gasoline engines. The first change was usually the substitution for an electric motor for the windmill or engine used to fill the tank, followed later by an electric pressure pump, which eliminated the need for the tank. However despite the emphasis on ranges and water heaters, a survey of the district's 4,782 domestic customers in 1931 revealed that almost 600 of them used power only for lighting, and another group of nearly 700 had only small appliances, including radios, toasters and washing machines, in addition to lights. Another 1,398 consumers used up to 70 kilowatt-hours monthly for lighting, small appliances and perhaps a hot plate or a refrigerator, but not a stove. According to the survey, well over half of the district's domestic customers did not have the electric cooking and water heating the district was trying to encourage. Only 484 customers used over 400 kilowatt-hours a month.[22] People added appliances as soon as they could afford them, but in many cases the desire for household conveniences outstripped the capacity of retro-fitted home wiring. The trouble and expense of installing wiring in older homes, especially two-story ones, made many homeowners opt only for lights, often with pull chains rather than wall switches, and a few outlets. Farmers occasionally tried to save a little money by doing their own wiring, though not always successfully. That was one of the reasons

The TID store stocked electrical appliances of all kinds.

why, in the days before the advent of county building inspectors, district employees routinely checked home wiring before hooking up the power.[23]

The district's status as an electric utility meant the creation of a whole new department and a larger work force. In 1924, as many as five line construction crews were busy setting poles and stringing wire all over the district. Shoemaker wrote that, "It has been the policy to use local men, preferably those owning property within the district, on all work, training them in their respective duties. As a result, the labor turnover has been far less than that experienced by other companies, and a more efficient organization obtained"[24] Much of the work done by the electrical crews was hard, physical labor, especially when it came to putting poles in the ground and stringing the wire. A line construction crew could set a mile of poles a day at twenty-six poles to the mile.[25] Holes were dug with shovels, then the poles were raised and pushed into place with pikes and the dirt tamped in around them. Hardpan could turn the simple job of digging a hole into an all-day ordeal. Dynamite could break the hardpan, but it was seldom used because it blew out such large pockets that it was difficult to pack dirt back in around the pole. Dynamite was also used on rare occasions when the ground was too soft rather than too hard. In waterlogged soil, the wet earth could sometimes ooze back into a hole almost as fast as dirt was shovelled out. A lineman who had worked in Florida told how dynamite was used to set poles in boggy conditions, but foreman A. C. Johnson was skeptical of the idea. One day when he was absent, a crew decided to try dynamite to set a pole in wet ground. A one-and-a-quarter-inch pipe was pushed into the ground about six feet deep where the pole was to go. A couple of sticks of dynamite were then shoved to the bottom of the pipe and the pipe was removed, leaving the dynamite in place. The pole was then put on top, held in position by four guy wires. When the dynamite was touched off, the pole went into the hole and the mud filled in around it. When he found out what had been done, Johnson was worried about the condition of the pole, but later, when it had to be removed because of fire damage on top, it was found to have been uninjured by the blast beneath it.[26] Once the poles were in place, the wires were either reeled out by hand or from trucks, insulators were installed on the poles and the wire was lifted into place. The wire was tightened by hand using a block and tackle, and then tied to the insulators. All this was the usual practice of power line construction, but it was an unusual business for an irrigation district to be in. Reflecting the season of farm work, orders for new services and the pace of construction slackened in the summer months, when the potential customers were preoccupied with their harvests.[27]

Power from Don Pedro was carried to the Geer substation by the 66,000 volt transmission line and stepped down at the substation to 11,000 volts for the network of distribution lines. The 66,000 volt line was later extended to the Hilmar substation. Another substation at the district's Broadway corporation yard dropped the voltage to 4,000 volts for the city of Turlock because it was then thought to be a safer voltage for lines going down alleys and over roofs. The substations at Geer and Broadway had to be run by operators, but the Hilmar substation was the district's first automated installation, controlled by the Broadway operator. Maintaining the system posed no unusual problems. The rebuilt transmission line suffered primarily from hawks, hunters' bullets and dust, especially from the red soil near Hickman that could short out insulators in the first light rain of the season. The hawks, like most birds, liked to perch on the wires, but their tendency to come up from below the line and a wingspan wide enough to touch both wires, occasionally led to the quick demise of the hawk, an outage and sometimes a grass fire touched off by the flaming bird.[28] The same kind of mishaps, along with falling trees and errant automobiles, also plagued the distribution lines. One particularly wild windstorm blew so many trees onto the lines around Hilmar that some sections were without power for up to two weeks.[29] When the power did go out, servicemen, known as troubleshooters, were dispatched to find the problem and correct it, unless damaged poles or wires required the use of a line crew. The first serviceman was said to be Clifford Plummer, who later became superintendent of the electrical department, and in 1936 was named chief electrical engineer of the Modesto Irrigation District. By 1937 the service department's foreman and nine troubleshooters were handling 700 calls a month.[30] Communication was a constant problem. The district had its own phone lines connecting Don Pedro, La Grange, the substations and even the ditchtender's houses, but most of the time, servicemen out on a call had to knock on doors in order to borrow farmer's telephones to get in touch with their headquarters. In the middle of the night, when such intrusions would be less welcome, they sometimes drove all the way back to Turlock to get their next orders.[31]

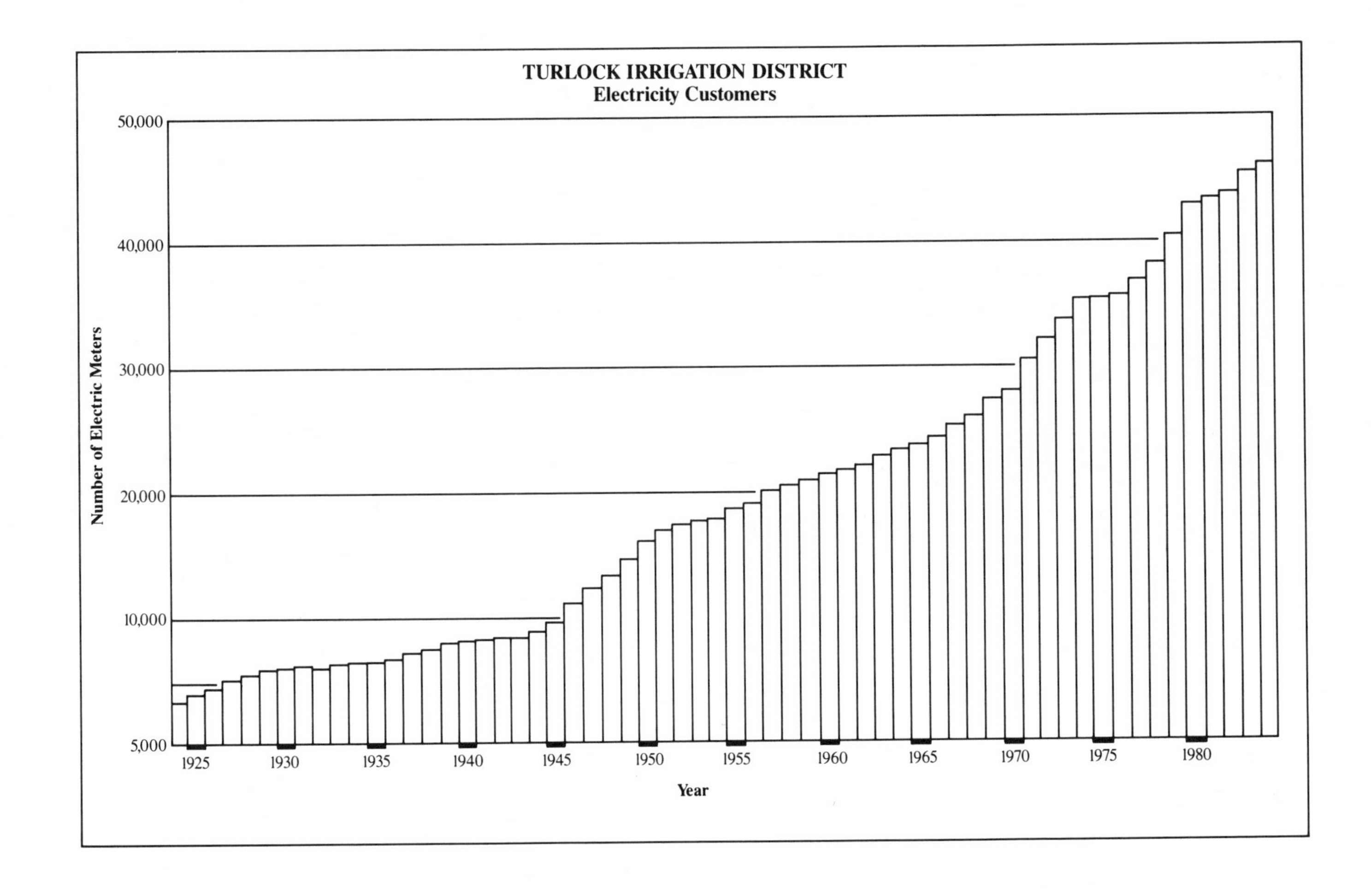
TURLOCK IRRIGATION DISTRICT
Electricity Customers
Number of Electric Meters
50,000
40,000
30,000
20,000
10,000
5,000
1925
1930
1935
1940
1945
1950
1955
1960
1965
1970
1975
1980
Year

The network of distribution lines, the retail store and other services were all designed to encourage the use of as much electricity as possible. With those inducements, the number of consumers increased from 3,220 in 1924 to over 6,000 by 1930. At that point, the increase virtually stopped as the Great Depression tightened its grip. Not until 1934 did the upward trend resume, slowly at first, until 7,000 consumers were on district lines by 1938 and 8,000 in 1941. Like the number of consumers, the amount of energy distributed by the district increased steadily, except during the early 1930s. The district was its own best customer, using as much as a third of the system's annual consumption to operate its drainage pumps, but larger numbers of consumers and the increasing use of electrical machinery and appliances reduced the proportion of energy going to the pumps to 25 percent in 1930 and 20 percent in 1940.[32] The largest commercial customers served by the TID before World War II were probably three gold dredgers in the La Grange area. The growth of the electrical distribution system was gratifying, but it would still be years before the largely rural district would be able to absorb its full 68.46 percent share of Don Pedro's output. In the meantime, it was highly important to find some way to dispose of the excess power.

Before the dam was completed, several irrigation districts west of the San Joaquin River were actively interested in buying power from the Turlock District to operate the pumps they depended upon for irrigation. There is evidence that the district gave serious consideration to the possibility of such sales, but for some reason, the idea was soon dropped. In his 1922 report, R. W. Shoemaker had suggested that surplus power could be wholesaled to one of the private utilities, and he emphasized that a firm block of power would be more valuable than a constantly fluctuating and uncertain surplus. Negotiations for the sale of power to the San Joaquin Light and Power Company began in early 1924. One of the worst droughts of the twentieth century was beginning, and the company was undoubtedly interested in obtaining additional power to make up for deficiencies in its own hydroelectric supply. The resulting contract, dated March 11, 1924, sold a guaranteed block of power from Don Pedro amounting to 6,500 kilowatts from June through December of each year, and 2,500 kilowatts during the remaining months. The price was 4½ mills ($0.0045) per kilowatt-hour. If the Tuolumne produced a runoff of under 1,900,000 acre-feet during the first seven months of any year, the company was guaranteed 65 percent of the district's generation rather than the 6,500 kilowatt block. Additional surplus power that might be available from time to time was also sold to the company, but at a lower rate. The district agreed to pay for a line connecting its lines with the San Joaquin system at Livingston, just south of the Merced River. The contract covered fifteen years, renewable at the district's option for an additional fifteen years after that.[33] For the Turlock District, it was a very favorable agreement, for it promised a long-term source of revenue from power that would otherwise be wasted until the district's own demand caught up with its supply.

In agreeing to wholesale a definite block of power without having the auxiliary steam plant originally suggested in Shoemaker's report, the district was compounding the risks it took in operating its share of the Don Pedro project as an electric utility in addition to its irrigation function. If the reservoir were operated strictly for irrigation, it could, if necessary, be drained by the end of the irrigation season and refilled beginning with the first winter rains to insure against below normal runoff later in the year. Sale of whatever incidental power the dam could produce directly to a private utility, as proposed by PG&E, would not necessarily have interfered with irrigation, but the decision to distribute Don Pedro power complicated operation of the dam. For example, the powerplant had to be able to operate all year, so the reservoir could not be lowered below the level of the powerhouse outlets, which meant a small portion of the storage capacity was, for all practical purposes, unusable. Also, water would have to be held in storage beyond the irrigation season to make power during the fall and winter, and instead of storing the runoff from winter storms, a certain amount of the inflow had to be run through the powerplant. The Turlock District's wholesale contract added to the requirement for power releases outside of the irrigation season, and consequently increased the risk that, in a dry year, water that could have been used for irrigation might have to be used before or after the irrigation season to produce electricity.

Had Don Pedro been the only reservoir on the river, the problem would have been much more difficult than it was. The completion of San Francisco's Hetch Hetchy reservoir in 1923 provided 206,000 acre-feet of storage capacity above Don Pedro, and evened out the flow of the river. There was no powerplant at Hetch Hetchy itself, but a large plant was put into service on the city's aqueduct at Moccasin Creek in 1925. Until 1934, when the first small deliveries of water to the Bay Area began, the Hetch Hetchy system was operated solely to maximize power production at Moccasin, which meant that low season inflows to Don Pedro were augumented by releases from the city's reservoir. As a result, the low flow entering Don Pedro rose from about 100 second-feet to 700 second-feet. Hetch Hetchy was a tremendous asset to the district's hydroelectric capability, providing water to run the generators that, had it not been stored by the city, would have been spilled by Don Pedro during the spring snowmelt. There were times when San Francisco's operation of Hetch Hetchy conflicted with the water releases guaranteed by the Raker Act and also impaired Don Pedro's electric output. In those instances, the city reimbursed the districts for any loss of power revenues.[34] Aside from such relatively minor inconveniences, Hetch Hetchy and the Raker Act were as much responsible for the success of the district's electrical program as anything done by the district itself.

Although Don Pedro could provide all the power the district needed, it might, like any powerplant, have to be shut down if the machinery failed or if its supply of fuel, in this case, water, ran short. The district needed some way to cope with such emergencies. Standby service from another utility would have been one way of solving the problem, but for a number of years no other utility was willing to take over the district's load in case of a failure at Don Pedro. The largest power company, PG&E, was still competing with the irrigation districts for customers and was in no mood to bail them out if anything happened to their systems. Since the Turlock District was on its own, it needed to build a backup plant for insurance. A steam plant that would burn oil or gas was one possibility, and that was the course followed initially by the Modesto District. The TID, however, decided on another hydroelectric plant. On November 16, 1922, the district applied for water rights for a powerplant just below La Grange Dam.[35] When the districts sold the equipment from the power company's old La Grange plant in 1923, the Turlock District was the successful bidder on machinery that included the two 1907-era generators.[36] The old generators and a larger Allis-Chalmers unit were installed in the new powerhouse the district built in 1924. The plant was fed by penstocks that tapped the main canal just after it left the diversion tunnel. The difference in height between the canal and the river was enough to generate 4,300 kilowatts, using about 700 second-feet of water. Since any water used by the powerplant was

La Grange powerhouse was built in 1924 near the outlet
of the TID's diversion tunnel from La Grange Dam.

lost so far as irrigation was concerned, it could not be operated all the time, but it could be started quickly to take over the distribution load in an emergency. It could also be run after the irrigation season to make more efficient use of water released for power generation at Don Pedro. In that case, less water would have to be released from storage because the same water was producing power at both plants.[37]

In the early years, the TID distribution system was isolated from other utility systems and had to rely entirely on its own resources. One sign of that isolation was the master clock at Don Pedro. Without interconnections to other utilities, the frequency of power delivered to consumers depended entirely on the careful operation of the Don Pedro generators. To monitor that operation, the master clock had both a pendulum and an electric motor, and if the system were running at exactly at 60 cycles per second, as it was supposed to, the single hand on the clock would remain stationary. The governors controlling the generators were not always able to maintain the precise frequency, and if the generators spun too fast or too slow, the master clock, and all the electric clocks in the district, would speed up or slow down. The error was usually small, but it could accumulate into an annoyance for customers relying on electric timepieces. To correct the problem, the operators at Don Pedro would check the clock each day, usually in the wee hours of the morning, and if it showed that the clocks had gained or lost that day, they would adjust the speed of the generators until the master clock's hand reached the neutral position and all the electric clocks in the distribution system were correct.[38]

Independent or not, the district's electrical system was considered reliable by its customers. In the days before automation, it took a lot of people to keep the service functioning that way. The electrical engineer was in overall charge, and until 1929 that post was held by R.W. Shoemaker. He was responsible for much of the strategy behind the distribution system and the wholesale contract, and was far-sighted enough to design the system to accomodate more rapid than expected growth. Drawing on their experience in setting up distribution in the Turlock District, Shoemaker and R.V. Meikle also acted as consultants for the Imperial Irrigation District in southern California when it entered the power business. After Shoemaker left to take a job in South America, Hale Parker became electrical engineer, remaining in the post until 1967.

Because it owned over two-thirds of the project, the TID actually

The interior of the Don Pedro powerhouse.

operated Don Pedro dam and powerhouse and maintained the equipment.[39] It took about eighteen men to run the powerhouse, making up three four-man crews plus relief workers and roustabouts. A chief operator watched the 200 or so dials on the switchboard balcony between the third and fourth floors of the powerhouse, while an oiler kept the machinary lubricated and two other men watched over the main generator floor.[40] These employees lived in a village run by the districts near the dam. There was a school for the little community and a store. Nearby, the districts built a clubhouse or recreation center with tables, chairs and an electric stove for the use of visitors. Until World War II, anyone was welcome to go through the dam and powerhouse, and as many as 200 people came on some weekends and holidays to see what was an important source of community pride.[41]

Within its first two years, the TID's electrical department established itself as a successful business. It had a complete generation, transmission and distribution system and a growing retail load, the Hetch Hetchy system firmed up the water supply to the generators and it had a valuable wholesale power contract. However, the district had no sooner achieved that enviable status than plans were begun to enlarge its power supply. The obvious place to start was at Don Pedro, where two additional powerhouse outlets had been built into the dam. In a letter to the districts' boards dated July 31, 1926, the Turlock and Modesto engineers pointed out that all three of the original 5,000 kilowatt generating units were running almost all the time, which would make repairs needed in the near future difficult to schedule. They recommended the installation of two additional 7,500 kilowatt units to double the capacity of the plant and allow greater flexibility in operation and maintenance. The directors soon approved plans for the new generators and once again R.V. Meikle took the post of chief engineer for the project.[42] Construction, which included adding over sixty feet to the powerhouse building, was without incident and both new units were in operation by June 1928.[43] Although the total capacity of the enlarged Don Pedro plant was given as 30,000 kilowatts, the generators were able to put out as much as 37,000 kilowatts. The plant's efficiency was due in part to the use of cool air drawn from passageways inside the dam to cool the generators.[44]

At the same time that the expansion of the Don Pedro powerplant was being proposed, the Turlock Irrigation District took the first steps toward a major new hydroelectric project upstream from Don Pedro. On July 12, 1926, Chief Engineer Meikle filed application number 5094 with the state Division of Water Rights for 1,200 second-feet from the Tuolumne River and Moccasin Creek for the purpose of power generation. The filing was a "complete surprise" to the Modesto District, though Meikle acknowledged that they were entitled to join in the project.[45] The project Meikle had in mind was described in the application as a dam and 20,000 acre-foot reservoir on the Tuolumne River near Wards Ferry and a low diversion dam on Moccasin Creek below San Francisco's powerhouse. Both dams would divert water into canals that would join and continue along the south side of the river to a powerhouse opposite the town of Jacksonville.[46] In November Meikle was back with another water rights application, this one ostensibly for irrigation. Application 5266 described diversions from the South and Middle forks of the river into a canal leading to a section of Big Creek near Groveland, where an 80,000 acre-foot reservoir would be built. From the reservoir, water would be conducted to the edge of the river canyon, where it would drop 1,700 feet through penstocks to a powerplant that could develop 30,000 kilowatts.[47] Also in November 1926, the Turlock District applied to the Federal Power Commission for a preliminary permit for its Groveland and Wards Ferry projects.[48] All of these actions were taken independently of the Modesto District, although the various filings stated that Modesto had the right to its proportional share of any project above La Grange Dam. Meikle expected the two districts would work together, and on February 7, 1927, the Modesto board officially joined in the projects.

During the approximately eight years that the upper river projects remained under consideration, their physical features underwent various changes. A proposal for a powerhouse at the inlet to Groveland reservoir was added in 1927, and dropped the following year. The development at Wards Ferry, which in the original scheme would have captured water released from the Groveland powerhouse as well as any water passing through San Francisco's Moccasin powerhouse that was not immediately rediverted into the aqueduct, was changed to eliminate any diversion from Moccasin Creek following advice from San Francisco that its future construction program would virtually eliminate spillage into the creek. By 1930 Wards Ferry was no longer a part of the Federal Power Commission application, and correspondence with the Division of Water Resources indicated it had been changed to a higher dam at Wards Ferry and a short conduit to a powerplant above Moccasin Creek.[49] Since a high dam at Wards Ferry would have backed water over the Groveland powerhouse, the two projects were no longer compatible, and in fact, Wards Ferry became an almost forgotton alternative to the Groveland reservoir and powerhouse.[50]

Since Don Pedro had only been in operation three years when the initial applications for the Groveland project were filed, it might be reasonable to wonder why, after having just completed their largest, most expensive project, the districts were ready to embark on another, nearly as large and probably more costly. An amended permit application to the Federal Power Commission offered this explanation.

> The Groveland Project is the next logical development on the Tuolumne River. It presents a reasonable, workable plan of water and power utilization which can be operated by the Districts in conjunction with the Don Pedro and La Grange Plants.
>
> The Don Pedro Plant is to a large extent of its capacity a seasonal plant and it is largely due to the regulation of the river by Hetch Hetchy that the Districts have been able to operate their project successfully. Diversion of Hetch Hetchy water from the Tuolumne water shed above Don Pedro in dry years will reduce the late irrigation supply of the Districts and limit the winter power output at Don Pedro. The Groveland Project was conceived to overcome this condition by storage of the South and Middle Forks water and releasing the entire storage after about the first of July in dry years. This plan of operation also brings into use spring power from the present installations which now has no value . . .
>
> It is the purpose of the Turlock and Modesto Irrigation Districts to insure an adequate supply of irrigation water to carry the Districts through the driest years . . . Water released from the Groveland Reservoir will be applied to the irrigation shortage, while the power generated by the released water, transmitted to pumps operating in wells in the Districts, will lift 20 times the amount of water required at the Groveland Power Plant to generate the power.[51]

The power benefits of the project were probably more persuasive than those relating to irrigation and they underscore the importance of electricity in district planning. The 30,000 kilowatt plant would equal the capacity of the enlarged Don Pedro powerhouse, and the Turlock District looked forward to wholesaling its share of Groveland's output to the San Joaquin Light and Power Company. The Modesto Irrigation District, on the other hand, was projected to need its share of the Groveland and Don Pedro plants for its distribution system by 1936.[52] The 80,000 acre feet of stored water, however, would probably have amounted to only one more irrigation on less than half the acreage of the two districts. Like Don Pedro, Groveland provided no significant carry-over capacity to store water from wet years for use in dry ones. The districts loudly proclaimed the value of the project's irrigation storage, but in fact from an irrigation viewpoint, it would have been a very expensive way of getting a little additional dry year water. The Groveland project was viable primarily because of its power features and its irrigation benefits were of secondary importance.

There was another reason why the districts so actively pursued the project when they did; they wanted to forestall any competing water developments between Don Pedro and San Francisco's portion of the watershed. Meikle made those intentions clear at the outset in a letter to San Francisco engineer M. M. O'Shaughnessy, written the same day he filed the first application for the Wards Ferry development.

> The Turlock Irrigation District has filed on 1200 second feet at Wards Ferry Bridge and Moccasin Creek. The Wards Ferry Bridge, as you know, is the lower power house site of the Yosemite Power Company's Wards Ferry Project. We also plan to file on the river between Early Intake and Wards Ferry, as soon as we find out the status of the Yosemite Power Company's Project. Our object is to control the river below your project and prevent any development by outsiders between the district's and San Francisco's projects.[53]

O'Shaughnessy wrote back, "I am very glad to see the District, rather than anyone else, doing this."[54] In fact, the Yosemite Power Company had already been busy on the same ground. The company, which had no powerplants or customers, was trying to develop its properties, including the old Golden Rock ditch, on the South and Middle forks of the Tuolumne. Although it had various alternative schemes, the centerpiece of its plans was a system of two reservoirs and three powerplants on the South Fork. The company's plan was incompatible with the district's Groveland project, and the argument over which must fall by the wayside delayed the issuance of any federal permit for years. The company contended that its proposal made better use of the water available in the South and Middle forks in terms of power production, while the districts disputed the claim and countered that their plan provided coordinated irrigation benefits that the company's could not offer.[55] Public agencies had a preference in the granting of power permits and licenses, but questions regarding the relative merits of the two projects delayed action until March 1931, when a preliminary permit was finally granted to the two irrigation districts for the Groveland project.[56]

The purpose of a preliminary permit for a powerplant, then as now, is to allow an applicant a certain amount of time to conduct complete surveys and prepare detailed plans that are then made part of an application for a Federal Power Commission license for construction and operation. As soon as they received word that the long awaited preliminary permit would be issued, the districts' boards directed Meikle to begin the necessary investigations.[57] In February 1932 a formal application for license was submitted, but by then, time had almost run out for the Groveland project.[58] A report that summer by Forest Service engineer John Beebe questioned the wisdom of the rock-and-gravel-fill dam design and suggested further consideration of dam sites on the main stem of the river.[59] These criticisms were of less concern than economic conditions that made any project impossible. In a letter to the Federal Power Commission in November 1932, the districts asked for a two year extension of their permit to May 1934 because of the impact of the national Depression. Although the Turlock and Modesto districts were still solvent, many other California irrigation districts were already in default on their obligations and others might fail, which reduced the market value of all irrigation district bonds. In any case, taxpayers struggling to keep their lands and livelihoods were hardly likely to approve more indebtedness. Farm prices had fallen so low that farmers could not afford to pay for more water to grow unprofitable crops. Electrical demand in the districts had stopped growing, but in the rest of the Central Valley, power consumption had dropped 21 percent and additional power could not have been sold at any price.[60] The commission agreed to keep the license process open until at least 1934. However, when conditions failed to improve by late 1933, the Turlock District informed the Federal Power Commission that there was no market for wholesale power and it was therefore withdrawing its application for the Groveland project.[61] The Modesto board soon took the same action, and the Groveland project officially died.

Although the Depression eliminated for the time being any major expansion of electrical generating capacity, it proved, as nothing else could, the value of the Turlock Irrigation District's electric system. The amount of power sold to retail customers fell slightly in the years 1932 through 1934, but while the Depression had an impact, the district did not suffer any major loss of customers or sales.[62] The basic policy of the district was to make retail power sales pay all electrical expenses, including the operation of the drainage pumps, and to keep the wholesale revenue as profit. In 1930, for example, wholesale power earned $245,919, "and the net profit from all electrical operations were $272,075."[63] The profits from the power business were not applied to a reduction in electric rates; in fact, rates were adjusted upward slightly in 1931 to reflect the true cost of providing service to various classes of users. Instead, it was the district's philosophy to use its profits to benefit the taxpayers who had bonded themselves to construct the system in the first place.[64] As a result, the district was able to reduce its tax rate to help farmers through the Depression. Power revenues, and the need to keep reserve funds available to meet any emergencies in the electrical system, also meant that instead of borrowing money for its operations, as it had in the past, the district had a working surplus to fall back on. The electrical system was a by-product of irrigation, but with a lucrative wholesale contract and a profitable distribution system, it brought the Turlock District a financial standing that helped it, and its taxpayers, weather the economic catastrophe of the Great Depression, while at the same time bringing the convenience of electricity within the reach of every rural home.

CHAPTER 10 – FOOTNOTES

[1] Board minutes, vol. 8, p. 41 (July 7, 1922); *Turlock Tribune,* July 10 1922.
[2] Board minutes, vol. 8, p. 42 (July 18, 1922), p. 66 (Aug. 10, 1922).
[3] *Ibid.,* p. 176 (Jan. 30, 1923), p. 179 (Feb. 9, 1923).
[4] *Ibid.,* pp. 191-193 (Feb. 20, 1923).
[5] Author's interview with Charles Rose, June 27, 1983; author's interview with Henry Schendel, Jan. 11, 1983.
[6] *Turlock Tribune,* February 19, 1923.
[7] *Ibid.,* Apr. 4, 1923; *Modesto Morning Herald,* Apr.9, 1923.
[8] *Turlock Tribune,* Aug. 6, 1923, Oct. 1, 1923.
[9] Electrical Department, TID, "First Annual Report, Year 1924," pp. 3, 6.
[10] *Turlock Tribune,* Jan. 24, Feb. 7, 1923.
[11] *Ibid.,* Apr. 4, Sept. 14, 1923.
[12] Author's interview with Christine Chance, Feb. 10, 1983; Arthur Starn interview.
[13] *Turlock Tribune,* Jan. 19, 1923.
[14] Board minutes, vol. 11, pp. 192-198 (Nov. 11, 1930).
[15] Memorandum on "New Power Rate: Schedule A." May 9, 1931, in Meikle files, vol. 5, item 37.
[16] R. V. Meikle, "Imperial Irrigation District, Proposed Power Project," (1931), p. 3.
[17] *Turlock Tribune,* Feb. 8, 1924.
[18] Board minutes, vol. 12, pp. 1-17 (Oct. 3, 1931); memorandum by R. V. Meikle Sept. 30, 1931, in Meikle files, vol. 5, item 27.
[19] Memorandum on sale of Waterford lines, in Meikle files, vol. 6, item 7.
[20] Electrical Department, "Report," 1924, pp. 1-2.
[21] *Turlock Tribune,* Dec. 17, 1926.
[22] "Rate Investigation Data," 1931, in Meikle files, vol. 5, item 36.
[23] Schendel interview.
[24] Electrical Department, "Report," 1924, pp. 1-2.

[25] Schendel interview. [26] Rose Interview. [27] Schendel interview.
[28] *Idem.* [29] Arvid E. Lundell interview.
[30] *Turlock Daily Journal,* June 9, 1937.
[31] Author's interview with Norman C. Boberg, Jan. 6, 1983; telephone interview with William Trent, Apr. 26 1983.
[32] See Electrical Department annual reports.
[33] Contract between TID and San Joaquin Light and Power Company, in Board minutes, vol. 8, pp. 355-359 (Mar. 10, 1924).
[34] Memorandum, Feb. 13, 1935, in Meikle files, vol. 7, item 2.
[35] Application to appropriate water, no. 3139, Nov. 18, 1922.
[36] Board minutes, vol. 8, p. 275 (Sept. 4, 1923), p. 365 (Mar. 24, 1924).
[37] Memorandum, ca. 1934, in Meikle files, vol. 6, item 4, pp. 3-4.
[38] Boberg interview.
[39] See Board minutes, vol. 10, p. 266 (Jan. 20, 1930).
[40] *Turlock Daily Journal,* June 9, 1937 (Golden Jubilee edition).
[41] Author's interview with Jim Arnold, June 22, 1983.
[42] Board minutes, vol. 9, p. 107 (Sept. 7, 1926), pp. 164-165 (Jan. 4, 1927).
[43] *Modesto News-Herald,* June 23, 1928.
[44] Electrical Dpartment, "Report," 1924, p. 2.
[45] *Modesto News-Herald,* July 13, 1926.
[46] Application to Appropriate Water, No. 5094, July 12, 1926.
[47] Application to Appropriate Water, No. 5266, Nov. 12, 1926. A good description of the project can be found in the *Amended Application for Preliminary Permit for Groveland Project Submitted to Federal Power Commission by Turlock Irrigation District and Modesto Irrigation District,* (Nov. 1928).
[48] "Application for Preliminary Permit from Federal Power Commission by Turlock Irrigation District: Groveland Project," (Nov. 1926).
[49] R. V. Meikle to Edward Hyatt, Nov. 10, 1930.
[50] R. V. Meikle to Everett Bryan, Jan. 7, 1932.
[51] *Amended Application for Preliminary Permit,* pp. 35, 37-38. [52] *Ibid.,* p. 54.
[53] R. V. Meikle to M. M. O'Shaughnessy, July 12, 1926.
[54] M. M. O'Shaughnessy, to R. V. Meikle, July 13, 1926.
[55] See *Amended Application for Preliminary Permit* for comparisons between the company's proposals and those of the districts.
[56] Federal Power Commission, "Preliminary Permit, Project 761," May 13, 1931.
[57] Board minutes, vol. 11, p. 322 (Mar. 26, 1931).
[58] TID and MID, *Application for License: Project No. 761-California,* (Feb. 1932).
[59] John C. Beebe, *Report to Federal Power Commission on Application for License, Project No. 761,* (San Francisco, Dec. 1932).
[60] TID and MID to Federal Power Commission, Nov. 14, 1932.
[61] Board minutes, vol. 13, p. 47 (Oct. 9, 1933).
[62] See the *Annual Reports* of the Turlock Irrigation District.
[63] Memorandum on irrigation tax, 1931, in Meikle files, vol. 5, item 29.
[64] Memorandum by R. V. Meikle, Apr. 6, 1933, in Meikle files, vol. 5, item 38.

11

The Irrigation Empire at Maturity

The decades following the completion of Don Pedro were in some ways unremarkable ones for the Turlock Irrigation District. The irrigation system of reservoirs and canals was in place, the area was well populated, and even though an electrical system was an unusual adjunct for an irrigation district, the routine of running wires, fixing outages and selling power quickly became commonplace. There were no great construction projects, no political turmoil and few dramatic incidents. These years, from 1923 through the 1950s, were, however, by no means unimportant. The challenges of the Great Depression and the war years had to be met, as well as the need for continued improvements in irrigation, drainage and power, as the irrigation empire founded in 1900 reached its full maturity.

The post-Don Pedro era got off to an unsettling start. The trouble began in September 1923 when TID assessor-collector-treasurer J. H. Edwards suddenly resigned, with the announcement that he had somehow lost about $18,000 of the district's money. Auditors were immediately brought in for a thorough examination of the district's books, but few irregularities were uncovered.[1] In February 1924 Edwards was indicted for embezzlement, and even though he steadfastly claimed that he had gotten none of the missing money, he served a brief term in San Quentin prison.[2] Meanwhile, news of the financial problems at the Turlock District office, and an argument over the handling of certain salvage sales of equipment used at Don Pedro, prompted the Modesto District to hire its own auditors to go over the joint Don Pedro books. Even before their report had been issued, affairs took a more ominous turn when the Modesto board unilaterally rescinded the 1920 resolution concerning the operation of the joint board; the resolution that had settled the last inter-district feud.[3] At the beginning of March, the report by the Wisler Audit Bureau was submitted to the MID board. It alleged a long string of either irregular or illegal practices on the part of the Turlock District during the Don Pedro project. Everything from the prices paid for land at the reservoir site to the sale of bonds, the purchase of cement and the handling of warrants was criticized. Since February 1923, when the Modesto District ran out of money to fund its share of Don Pedro construction, the Turlock District had been paying all project costs, and was owed, at that time, over $30,000 by Modesto. The Wisler audit, however, charged that because of poor management or actual malfeasance, the Turlock District really owed Modesto a balance of $113,000.[4]

Turlock officials reacted to the charges by requesting an investigation by the county grand jury, and on March 7, 1924, the board passed a further resolution asking that the State Board of Accountancy appoint an impartial auditor to examine the facts.[5] At first, it was thought the districts might agree to just such a plan, but a few days later the MID rejected it because, they claimed, the State Board of Accountancy was prejudiced against the Wisler firm.[6] The Modesto District thereupon sued the TID to recover the money it claimed it was owed. The trial lasted nineteen days and featured forty-two witnesses on every aspect of the Don Pedro project. Former TID director John Orr, who had resigned in 1921 because of his own concerns over the handling of joint funds, got into a shouting match with the TID attorney P.H. Griffin and spent five days in jail for contempt of court.[7] The charges were detailed and the atmosphere tense, as every animosity or complaint that had developed between the two districts came spilling out. Despite the complexity of the case, visiting Judge H. D. Gregory deliberated only about five minutes before finding in favor of the Turlock District on every count.[8] The audit dispute was the last of a series of arguments between the Don Pedro partners that had begun in 1918. After that, harmony reigned, and while the two districts did not always agree on every issue, they never again fought so bitterly or so publicly.

The disagreement over the financial accounts was the only unhappy footnote to the Don Pedro project. The extra water stretched the normal irrigation season from about the end of July or the beginning of August to the middle of October. The second year of operation, 1924, was one of the driest on record, but between water held at Don Pedro and 60,000 acre feet purchased from Hetch Hetchy for $1.50 per acre foot, the Turlock District kept water in its canals throughout most of the season.[9] In response to the availability of more water, the TID's irrigated acreage began to increase again after having remained around 100,000 acres since 1917. By 1928, over 130,000 acres were irrigated and in 1931 the total passed 140,000 acres. Only modest increases in irrigated land occurred during the Depression, but in the late 1930s, the numbers began to rise again, reaching 150,000 acres in 1942 and 165,000 acres in 1952.[10] The last figure amounted to nearly 90 percent of the land in the district, and represented, for all practical purposes, the full development of irrigated agriculture within its boundaries.

From its inception, Don Pedro had been viewed not only as a source of additional water but also, through the use of pumps powered by its electricity, as a solution to the drainage problem. Although the TID had been actively digging and maintaining open drains since about 1916, the best they had been able to do was to keep matters from getting worse. In 1918 the groundwater level finally stabilized, but it still averaged only four feet below the surface during the irrigation season and seven feet in the winter, when not as much water was being poured onto the soil. Since the open drainage ditches could not reach the subsurface strata where drainage would be most effective, pumps were the only real solution to the problem. As the electrical distribution system began to take shape, the first priority went to lines serving the waterlogged sections, so pumps could be installed. Drainage wells were generally located in low spots where the groundwater was closer to the surface. Since pumping had its greatest effect in the area surrounding the well, placing pumps in low areas would not only lower the water table but would make it conform more closely to the contours of the land. Tests run at a typical installation showed that the depth to groundwater a thousand feet away from the well stood at only 3.8 feet before pumping, and 7.1 feet afterward.[11] It was also discovered that pumping sharply reduced the seasonal fluctuation of the water table. By 1939 the district had ninety drainage pumps and more were added from time to time, until by the mid-1950s the number had risen to 170.[12] These pumps solved the drainage problem in general, although there remained small areas where the water table still rose too close to the surface for some crops. In 1958 there were still thirty-seven applications for drainage pumps on file at the TID office, though the district could hardly afford to drain every isolated low spot.[13] Water from the drainage wells went into the canal system to supplement the supply available from the Tuolumne, which made them especially valuable in dry years. In

1929, for example, 75,000 acre-feet were pumped into the canals, and the number rose as more pumps were added.[14] Drainage water could even be sold. For four years, from 1927 through 1930, a portion of it was run into the San Joaquin River so that an equivalent amount could be pumped out of the Delta into the East Bay Municipal Utility District's pipeline without infringing on the rights of Delta water users.[15]

Ironically, the first pumps had no sooner been started in 1923 than requests began coming in to shut them off. A.G. Crowell and others asked in June 1923 that the drainage pump on Monte Vista Avenue be stopped, "as it is drying their land too fast."[16] The pump was turned off, and similar pleas to shut down various pumps continued for some years. The reason was that melons, the Turlock region's most famous crop, were thought to do best when subirrigated by a high water table, and it was commonly held that surface irrigation would result in a lower quality crop. Petitions to turn off the pumps would seem to lend credence to claims that the lower water table was an important reason for the decline of the melon industry around Turlock, but the district's annual crop reports show otherwise.[17] Taken as a whole, the acreage in cantaloupes, casabas, honeydews and watermelons peaked in 1919 at over 13,000 acres. By 1923 it had fallen to about 5,000 acres, but then recovered to over 11,000 acres in 1925. The total remained near 10,000 acres annually until 1933, when it plummeted suddenly to less half that acreage and dropped still further by World War II.[18] Control of the water table may have affected melon growing, but if acreage figures are any guide, more melons were probably grown after Don Pedro than before the drainage system was constructed.

Don Pedro and the growing number of drainage pumps established a water supply that, with few changes, would serve the Turlock Irrigation District for almost half a century. The principal problem remaining so far as irrigation was concerned was the refinement of the distribution system that carried water from La Grange to each farm. Two decades of work on the main canal above Hickman had already increased its capacity and reliability. By 1923 some 55 out of about 250 miles of main and lateral canals had been concrete lined to reduce seepage and the chance of washouts. Ten years later, an additional fifty miles of lining was in place. Lining and other improvements contributed to a 30 percent increase in overall canal capacity between 1919 and 1929. As a result of increases in canal capacity and storage, the average interval between irrigations was

Cleaning dirt ditches: Cross-Ditch No. 2. About 1944.

reduced from thirty or thirty-five days in 1914 to only ten to fifteen days in 1934.[19] Despite these substantial improvements in water service, one major problem remained – the community ditches. There were at least 800 miles of them and most were poorly maintained. In many parts of the West, cooperative lateral companies were formed to build and operate such ditches, but apparently Turlock District irrigators rarely established formal organizations. It was often harder to get all the irrigators on a ditch to cooperate in its maintenance than it was to guide a team of horses and a plow along the slope of a weedy canal bank. Tenants and absentee landlords in particular often refused to join in the labor or expense of ditch maintenance. Then, too, more frequent irrigations and a longer season kept water in the ditches so much of the time that they did not dry out enough to be plowed during the season.[20] In a few places, where the irrigators were willing to work together and could afford it, some ditches were lined at least as early as 1915, but they were a rarity.[21] If the district was to have an efficient distribution system, something had to be done to maintain and improve the connection between the district's laterals and the farms.

One answer would have been for the district itself to assume control of the community ditches, and from time to time, the directors were asked to take over particular ones. They declined to do so because to take responsibility for one ditch would logically have meant taking in all of them. Some irrigation districts did operate a complete system running to each farm, but the Tuolumne River districts had never intended to offer that kind of service. In any case, if the district had taken over the community ditches it would have had to raise taxes to cover the cost of maintenance and repairs that otherwise could have been done at little or no cost by the irrigators themselves. In its search for an alternative, the Turlock board took the problem to the executive committee of the Irrigation Districts Association in 1927. The association and TID attorney P. H. Griffin worked out legislation allowing the establishment of local ditch improvement districts within irrigation districts. Sponsored by Modesto Senator J. C. Garrison, the measure became law in 1927. Chief Engineer Meikle wrote that, "This law . . . is, in my opinion, the most important irrigation district legislation since the original Wright Act."[22]

The new law had a great deal in common with the Wright Act. Like the Wright Act, it allowed a majority, in this case two-thirds of the irrigators on a ditch, to organize a district and levy taxes on the land benefitted even in the face of opposition from other owners. Improvements, like concrete lining, were to be paid for over a ten year period using interest-bearing warrants. To take advantage of the law, irrigators first signed a petition to the irrigation district Board of Directors asking for the formation of an improvement district to accomplish a particular purpose. The board then had to pass a resolution authorizing the district's engineers to make surveys, feasibility studies and cost estimates showing the assessment each landowner would have to pay. Once the engineering report had been accepted by the board, a formal hearing was called, after which a final order establishing the improvement district could be issued. The irrigation district's board served as trustees for each of the improvement districts, and the parent district sold the warrants needed to pay for the work. Assesments to repay the principal and interest, as well as any maintenance costs, were collected by the irrigation district along with its regular tax levy. The improvement districts were, in effect, subdivisions of an irrigation district. They used the technical and financial expertise of the district, while

Preparing a canal for concrete lining, 1951-52.

leaving the basic decision of whether or not to make the improvements in the hands of the farmers using each community ditch.[23]

In December 1927 the first improvement district petitions were presented to the board and on January 17, 1928, Improvement District No. 2 (numbers were assigned when petitions were taken out, not when districts were actually formed) became the first one organized. It was formed to line four miles of the McPherson ditch and install stop gates, side gates and bridges. Improvement District No. 1, the Maze and Wren ditch, was organized the next day.[24] As in the case of the Wright Act, there was an attempt to have the law declared unconstitutional, but this time the "anti's" had virtually no support and no success whatsoever.[25] Better ditches, extended payments and the fact that the TID handled all the legal, financial and engineering details made the improvement district concept popular with irrigators. The minute books of the district were soon filled with improvement district petitions, reports and organization documents. The number of improvement districts rose from three in the 1927-1928 season to sixty-one in 1932. The number remained nearly the same for three years in the depths of the Depression, but by 1936-1937 there were 120 improvement districts, and ten years later, 338 active districts covered 100,000 acres. By the late 1960s there were over 800 improvement districts of various kinds, including some representing second or third generation improvements on the same ditch.[26]

At first the most common purpose of an improvement district was ditch lining. Concrete-lined ditches virtually eliminated washouts and costly maintenance, and with their smaller profile and steeper slopes, they took up much less land than the wide, shallow dirt ditches they replaced. Some improvement districts went a step further and put in underground concrete pipelines, which wasted almost no land and could, to a certain extent, disregard grades, operating as a kind of pressure system. The pipelines built in TID improvement districts were monolithic, or poured in place, as opposed to factory- built sections put together at the job site. Lloyd Terrell, a Turlock contractor, was frequently credited with being the first to install pipe of this kind, beginning in the early 1920s when a few farmers used it on their own ranches. Pipelines were naturally more costly than lining. The trench had to be hand-dug to the correct shape and width and, until machinary was developed to do the digging, pipelines were not buried too deeply. The earliest lines were built with wooden forms, but these were cumbersome and wore out quickly. Semi-circular metal forms were soon substituted, and using them a four-man crew could lay 100 feet of pipeline a day, mixing the concrete on the site and using wheelbarrows to carry it to the trench.[27] When improvement district lines were installed, TID inspectors were on hand to insure that the concrete was mixed in the right proportions and the job was properly done.[28] Until the late 1930s concrete lining predominated in improvement district work and even in 1939-1940 less than 20 miles of the 132 miles of improved community ditches had pipelines. In the 1944-1945 season, however, a short stretch of lining was torn out to make way for pipelining and the trend continued. By 1951 the improvement districts had more miles of pipeline than lining.[29] In time the ditches that had once been such a prominent part of the local landscape disappeared from large sec-

Poured-in-place concrete pipeline being installed, 1947.

Only a row of irrigation gates marks the location of a pipeline.

tions of the district, their former course marked only by the pressure relief standpipes and gate structures of the underground lines.

The improvement district concept was flexible enough to include districts for maintenance only or for drainage ditches, and later for irrigation pumps. That flexibility was apparent in June 1930 when the TID's most complex improvement district was formed. Improvement District No. 52 covered over 6,000 acres in more than 200 parcels, but it was not just its size that distinguished it. It was the result of an ill-starred effort to establish a state land settlement colony at Delhi; a colony that was in itself an interesting episode in the history of the Turlock area.

The California land settlement concept was fathered by Elwood Mead, one of America's best known irrigation experts. After serving as head of the Department of Agriculture's Office of Irrigation Investigations, Mead went to Australia in 1907 to supervise the government development of irrigated communities there. He was so enthused with the program that on a visit to California in 1914 he suggested that the same sort of thing could be tried here. When he returned from Australia permanently the following year, Mead was appointed to head a Commission on Colonization and Rural Credits, which led to legislation in 1917 setting up a State Land Settlement Board with Mead as its chairman. The driving ideal behind the land settlement plan was essentially the same as that of the irrigation crusade in the Turlock region or of the national reclamation movement. It was Jeffersonian in its celebration of the virtues of family farms and the dignity of work on one's own land. Irrigation was essential to the creation of small farms in California, but by 1915 it was apparent that irrigation by itself was not enough to insure small, family-owned farms. It had happened in the Turlock Irrigation District, but all across California there were places where large irrigated farms were worked by tenants or laborers. Mead hoped the land settlements could do what water alone could not – create affordable opportunities for land ownership, restore dignity to agricultural labor and provide a model for a higher type of rural society. There were also darker overtones of racism in Mead's plans, based on a fear of Japanese domination of California agriculture.[30]

The first settlement was established at Durham, in the northern Sacramento Valley. The state bought the land, laid out the farms and ditches and sold them to selected settlers on easy terms. The settlers received paternalistic assistance and advice from state experts on their crops, their homes and, in fact, on every aspect of community development. With the Durham settlement off to what appeared to be a successful start, efforts were begun to locate land for a second settlement. One of the tracts considered was part of the old Mitchell estate near Delhi, then owned by Edgar Wilson.

Wilson's land was hilly and very sandy, and had never been considered especially fertile. Since it was outside the original Turlock Irrigation District boundaries, it was still being dry-farmed in grain. Frank Adams reported in 1918 that if the state decided to purchase the property, the TID board had no objection to including it in the district, if the state paid an amount equal to what the land would have contributed to the district had it been in from the start.[31] In 1919 the state bought the Wilson property, and in 1920 and 1922,a total of almost 6,000 acres of the land settlement was annexed to the Turlock District.

The terrain made open ditch irrigation difficult, and an early state report said that, "Last season an irrigator who tried it had to employ three men to watch the ditch banks and prevent breaks while one man distributed the water."[32] To overcome the problem, the state's irrigation specialists designed a distribution system that relied heavily on underground pipelines made of pipe sections manufactured at a pipe factory built in Delhi. The lines were sized to deliver what, in theory at least, was an ample irrigation supply to each parcel, ranging from one second-foot on the smallest holdings up to five second-feet; amounts which were far below the common practice on the adjacent TID system.[33] Original estimates called for 26 miles of pipeline but by 1923 it had been found necessary to lay 143 miles, which, with higher than anticipated land levelling expenses, raised the state's outlay to about $120 per acre, exclusive of the $92.50 per acre paid for most of the land.[34] These higher expenses were reflected in the price placed on the land and became a deterrent to potential settlers and a heavy burden for those who did come.

In the ideal rural community envisioned by the Land Settlement Board, farms would be small, growing primarily orchards or vineyards, and even laborers would have their own plot of land at least large enough for a garden, all in an atmosphere of cooperation and harmony. Except for the size of the farms, Delhi did not live up to the hoped-for image. Instead, it became the irrigation ideal gone haywire. The experts from the University of California, who laid out oddly shaped parcels with boundaries on the contour lines and a too-small irrigation system, failed to profit from the experience of those who had been farming the land near the settlement for years and had learned how to deal with the problems the sandy soil presented.[35] As a result, they seemed helpless to prevent a repeat of the hardships that had plagued the rest of the Turlock District when irrigation was young. Rabbits and windstorms played havoc with newly planted crops and many of the orchards died. The agricultural depression that followed World War I would have made it hard to succeed even under favorable conditions, and those at Delhi were far from favorable. When not enough prospective settlers applied for land, the state resorted to advertising and a reduction in the standards that had been set to insure that only people experienced in agriculture came to the colony. The arrival of novice farmers and a group of disabled veterans did nothing to help make Delhi successful or harmonious. By 1923 the gap between the dream and the reality was painfully obvious and the colony was in serious trouble. Most settlers were behind in their payments and generally discontent with their situation. Elwood Mead was hanged in effigy and a fortunately defective bomb was set for the superintendent. In 1925 the state wrote off the delinquencies and reduced the interest rate on the remaining debt. These steps were still not enough to satisfy many of the settlers, and negotiations were begun to extricate the state from the disaster that Delhi had become. Finding a plan that would satisfy all the disgruntled settlers was difficult, but in 1931 agreement was reached on the terms under which their indebtedness to the state would be adjusted and transferred to the Federal Land Bank. In all, the state lost over $2,000,000 before it got out of the abortive colonization scheme.[36]

One of the problems facing the state in its withdrawl from the Delhi debacle was the operation of the irrigation system. It was a complex combination of gravity water from the Highline Canal, booster pumps, drainage wells and pipelines, many of them far too small to give adequate service. The state was obligated to operate the system until other arrangements could be made. The solution finally agreed upon was to make the entire colony a single improvement district so that all the landowners could be assessed for the maintenance and operation of the whole integrated system. Because of the unique circumstances present at Delhi, the improvement district act had to be amended before anything could be done. That was accomplished in 1929. However, before the district would take responsibility for the system, it required the state to make $25,000 worth of changes to correct obvious deficiencies, especially the use of small diameter pipe.[37] In June 1930, Improvement District No. 52 was formed and the state relinquished responsibilty for irrigation to the TID, which in turn worked with an advisory committee elected annually by the settlers. Even with the changes mandated by the district in 1929, the network of pipes and pumps proved difficult to maintain and still provided unsatisfactory service to many of the irrigators.[38] It became an enduring and unhappy legacy of an unhappy experiment.

By 1930, of course, Delhi was not alone in its problems. Propelled by the stock market crash in 1929, the American, and world, economy collapsed like a house of cards. The prices paid for farm products plummetted, land values declined and farmers everywhere fell behind in payments due on their land. Here and there in the Turlock area, farms were simply abandoned, and foreclosures would have been more widespread than they were but for the forebearance of many of the banks. It was not charity that caused them to temporarily ignore unpaid loans as much as a realization that they were no more likely to make a profit or sell the land for a reasonable figure than the farmers themselves.[39] With crop prices at or below the cost of production, cash for any purpose, including irrigation taxes, was scarce. In 1930 about 11 percent of the district's tax levy went unpaid, and the deeds for property taken by the district for nonpayment of that year's taxes cover forty-nine typewritten pages in the minute books.[40] In the following year, 1931, delinquencies totalled 15 percent, and by 1933 the number was over 20 percent.[41] Not only farmers but some owners of town lots or business property were unable to make their tax payments. There was no pattern to the failures and delinquencies – no one crop or part of the district that was hit harder.[42] Small farms were no more likely to fail than their larger neighbors. It all seemed to depend on the circumstances of each landowner. Some could survive but others, who were saddled with greater debts or who might have had lower yields or planted the wrong crop at a critical time, were unable to stay in business or keep their land.

Despite the beginning of hard times in 1930, there were few outward indications of a departure from business as usual for the TID itself, apart from the difficulty of selling improvement distict warrants in the stricken financial markets.[43] When the prosperity that was supposed to lie just around the corner failed to materialize, increasingly desperate taxpayers began to appeal to the district for whatever relief it could give. Salaries were being slashed across the country as deflation increased the real value of money, and by July 1931 many farmers were urging the directors to reduce the pay of district employees. Some months later, the employees themselves suggested a wage cut that was adopted.[44] Any reduction in expenditures was expected to produce a lower tax rate, and with cash incomes reduced to almost nothing, that was critically important to many farmers. District assessments, however, had already begun to fall. In 1928, the rate had been $4.50 per hundred dollars of assessed valuation on land only, since improvements were not taxed by the district. It had been cut to $4.25 in 1929 and to $4.00 in 1930, reflecting the generally healthy financial condition of the district since Don Pedro. The year 1931, however, presented a dilemma. Tax revenues the previous year had been reduced by delinquencies, and since 1931 was a very dry year, the district had little excess power to sell and, beginning in September, it even had to buy outside power for the first time. Despite increased costs and falling revenues that might normally have justified a tax increase, taxpayers continued to clamor for a further reduction in the rate. The board agreed, and cut the rate by fifty cents, using the surplus accumulated from electrical operations and a reduction in planned construction to cover the shortfall.[45] In a further effort to trim expenditures at the end of 1931, the district purchased $30,000 worth of its own immature bonds at below par prices, which reduced the interest it owed, and the practice was continued in later years.[46] The district continued to curtail its expenses and use a portion of its electrical revenues to reduce taxes. In 1932, when all canal lining and construction work not necessary to maintain water deliveries was eliminated and the drainage program was slowed down, the rate dropped to $3.00[47] It remained at that level until 1935, when profits from the electrical system once again allowed a gradual reduction to a rate of $2.00 in 1941 and $1.00 in 1942.

The history of the Turlock Irrigation District during the Great Depression stood in stark contrast to that of other irrigation districts. Many, hard hit by the agricultural depression of the 1920s, were already in trouble when the general Depression began. By the end of 1930, sixteen California irrigation districts were in default on their obligations and more followed. The Reconstruction Finance Corporation provided over fifty of the state's irrigation districts with loans to enable them to purchase and retire their old bonds at an average rate of only fifty-four cents on the dollar.[48] The Turlock District remained financially healthy throughout the period. Obviously, electrical revenues played a vital role in maintaining its solvency, but merely having a powerplant was not, by itself, enough to avoid default. The Merced, South San Joaquin and Oakdale irrigation districts all required RFC loans to solve serious financial difficulties even though they either sold wholesale power or had sold power rights to pay for dam construction.[49] It would be conjecture to claim that it was the Turlock District's retail system that made the difference, although the surplus accumulated as a result of that enterprise undoubtedly helped. It may also have been that as an older district, the TID had fewer outstanding debts in relation to its size and tax base. Whatever the reason the Turlock District was one of only a few districts to survive the Depression unscathed.

Retrenchment was the first and most obvious response to the economic crisis, and indeed the tax reductions made possible by reducing wages, construction and other expenses were about all that could be done to help property owners weather the storm. A noticeable change in district policies occurred in 1933 in response to national programs designed to put men back to work. Among the earliest of President Franklin D. Roosevelt's New Deal agencies was the National Recovery Administration – symbolized by the blue eagle – which hoped to increase employment through the creation of industry-wide codes covering wages and hours. The TID was covered by the electrical industry code and the plan it adopted in the summer of 1933 to comply with the code called for the hiring of twelve additional men, a stenographer and three men at Don Pedro.[50] In a related move the board decided to resume the canal lining program during the winter of 1933-1934 to provide a source of employment during the off-season.[51]

The New Deal soon produced a spate of other agencies and programs to extend credit where it was needed and, through the use of federal dollars for so called "pump priming," to create jobs that would take people off the unemployment rolls. When money became available, the TID wasted no time in applying for its share. As early as 1932 the district considered applying to the Reconstruction Finance Corporation for aid in refinancing its bonds to reduce the tax rate to $1.00, and in 1933 the district actually made an application for $6.5 million for that purpose, but apparently to no avail.[52]

Also in 1933 an application was made for a federal loan of $200,000 and an outright grant of $100,000 from the Public Works Administration (PWA) for a canal lining program to put men to work.[53] It was not until May 1934 that the $311,000 project was approved. Under terms of the agreement, a PWA grant would cover 30 percent of the cost of labor and materials and the government would take revenue bonds from the district at par to cover the 70 percent that constituted a low interest loan. The revenue bonds, earning four percent interest, would be paid off over six years at the rate of about $40,000 annually from the proceeds of electric sales. The district had originally requested that delinquent taxes be used, when paid, to reimburse the government for the district's share of the project, but the PWA wanted securities it could sell for cash. The grant and loan would pay for lining about 25 miles of the district's canals and would employ up to 400 men.[54]

Following a brief delay while the exact amount of the grant was confirmed, the board endorsed the PWA proposal and scheduled an election on the revenue bonds for August 16, 1934. Board members, except E.O.McCombs of Hughson, considered the project an excellent opportunity for the district. Normally, the annual lining program cost about $60,000, but with the grant that figure would drop to $40,000 a year plus interest. Assurances were given that the loan would not involve any increase in taxes or electric rates.[55] Many taxpayers failed to share the directors' enthusiasm for the PWA project and, with delinquencies still running at 18 percent, there was opposition to incurring any more debts, no matter how painless they were promised to be. As a result farmers in the Gratton, Keyes and Ceres Granges opposed the bonds.[56] When the ballots were counted on election day, the revenue bond issue had been defeated on a vote of 1,472 "no" to 1,318 "yes." Seven of the eleven precincts had voted in favor of the bonds, but heavy opposition in Hughson, Keyes and Denair denied the measure the majority it needed.[57] Without the bonds, the district had no choice but to abandon the entire project.

It was not long before the district was able to participate in other programs that had fewer controversial strings attached. At the end of 1934 an application was made to the State Emergency Relief Administration (SERA) for men to work on canal maintenance and weed eradication along the canal banks.[58] The application was apparently well received for in the following months the SERA was called upon to provide the labor, or more precisely, pay for it, on a host of district projects. A total of 6,000 man-hours were used in cleaning out a drain and hauling dirt to widen canal banks.[59] Other projects included removing rock slides in the main canal, grubbing trees and stumps out of the canal, concrete lining of a local ditch and preparing for drain improvement work.[60]

State funding did put men to work but the projects were mostly small, short-term maintenance jobs rather than major construction. In July 1935 the district made another application for a grant from the PWA for canal improvements and lining on various improvement district ditches. This time it was successful in getting a simple grant of $78,750, which was 45 percent of the $175,000 worth of improvements. The district was to pay the remainder in any way it chose, eliminating the loan and grant provisions that had scuttled the earlier PWA project. Because the project was not approved by all parties until mid-December, 1935, construction could not begin until the winter repair season was drawing to a close. The district received an extension until the spring of 1937 for the completion of the project. Work was done on various parts of the main canal and Lateral 6 as well as on improvement district ditches, providing employment at a season when jobs were the hardest to find in the Turlock area.[61]

Shortly after it applied to the PWA for funds, the district also applied for aid from a rival New Deal agency, the newly formed Works Progress Administration (WPA), for a variety of projects including more concrete lining of improvement district ditches. It was reported that WPA administrator Harry Hopkins turned down the application because it would get too little labor off the relief rolls.[62] Although several applications to the WPA for canal work appear in the minutes, there is no record of their acceptance or the use of any WPA funds for that purpose. The district had more luck with the PWA. In 1938 it won another canal lining grant from that agency.[63] All of these federal projects, actual and proposed, required some form of cost sharing. Since the voters had vetoed borrowing money for that purpose, the district would not have been able to participate in the federal largess had it not been for the money it received from its electrical operations. As it was, taxes were reduced and, with federal assistance, major improvements were made in the canal system. The availability of PWA funds for community ditch improvements lowered the cost of such work to the farmers and encouraged the formation of new improvement districts, which continued to proliferate following the lull from 1931 to 1934.

Although the TID apparently failed to get WPA money for its canal work, it did get the agency to contribute to a project begun with SERA support in 1935 – the construction of a park along the Tuolumne River at Owens Reservoir. The forested bottomland below the bluff from the reservoir was owned by the district, and in 1934 or 1935 the Turlock Sportsmans Club got the idea of establishing a park on the property. They collected 637 signatures on a petition asking that land be set aside for that purpose.[64] Recreation on the reservoir itself had been provided for as early as 1932 when the district leased land to B.E. Handy, and then Leonard Weeks when Handy quickly defaulted, for use as a "public resort and amusement park."[65] In the spring of 1935 SERA funds provided men to clear brush at the proposed park site and build a steep road to the river bottom. Later that year, the WPA allocated $5,594 to continue the project with more road work, well drilling and the construction of fish rearing ponds.[66] In applying for more WPA labor in March 1936 the district noted that two artesian wells and the road, as well as picnic tables and benches, had been installed and the park was in use by various community organizations. More roads and trails were planned, along with more work on the fish hatchery. By November 1936 the Turlock Sportsmans Club was beginning to stock their hatchery ponds with fish rescued from TID canals when the water was pulled out at the end of the season.[67] Federal and state relief organizations, along with the district and community groups, had built a pleasant riverside park that would become part of Turlock Lake State Park when it was established in 1950.

With electrical revenues and federal and state relief funds, the district itself fared better during the Depression than many of the farmers it served. The reductions in the tax rate came too late to prevent many delinquencies and were too small to forestall others. Some owners eventually managed to pay their overdue taxes but hundreds of parcels were not redeemed and became the property of the district. The first sales of property taken by the district to new owners seem to have been made near the end of 1936.[68] The following year the district embarked on a program designed to return all delinquent property to the tax rolls. In cases where parcels had also been taken by the county or city for failure to pay their taxes, arrangements had to be made to eliminate conflicting claims. The district entered into agreements with Stanislaus and Merced counties to buy out the counties' interest in delinquent

Lateral 6 was one of the canals improved with WPA funds.

property, while similar property in the city of Turlock was sold to the city for the nominal fee of one dollar per parcel.[69] Once it had full title to property, the district put it on the market. Some land was sold for as little as $10 per acre, while in another case, Chief Engineer Meikle ordered a large parcel west of Turlock subdivided into smaller farms and sold at the price of $25 per acre.[70] On March 9, 1942, the TID completed its program of returning land to the tax rolls when it sold the last eleven parcels. In five years the district had sold 6,000 acres of land that had come into its hands during the depths of the Depression.[71]

The 1930s were hardly the happiest of times but nevertheless a grand celebration was held to commemorate the fiftieth birthday of the Turlock Irrigation District. There was, of course, really a good deal to celebrate. The district had so far successfully weathered a Depression that left most California irrigation districts bankrupt or close to it, and its stability had in some measure helped the community as a whole through the crisis. The district's success was a source of community pride and the residents of the predominantly rural area then less than four decades removed from the grain era were fully aware of the significance of irrigation and the agency that had made it possible. For all of these reasons the district's golden anniversary was a party to be remembered. The festivities began on Friday, June 11, 1937, with a meeting in Turlock of the Irrigation Districts Association of California at the high school auditorium. That evening, the Association held a banquet addressed by Governor Frank Merriam, while in the business district a special vaudeville show was presented three times and a street dance lasting until 1:00 a.m. attracted thousands of people. The next morning, the delegates to the irrigation district convention went on a tour of the district and Don Pedro that ended with a barbecue lunch at the new park at Owens Reservoir.

They were back in time to join the crowd estimated at 20,000 people who watched a long parade wind its way through Turlock on Saturday afternoon. The TID Board of Directors, dressed, as were many of the celebrants, in old time garb, rode in a four-horse stagecoach at the head of the parade, while special cars carried Chief Engineer Roy Meikle and Secretary James McCoy. All kinds of district equipment, some of it antiquated and not used for years and even an electric drainage pump, joined the parade, along with floats entered by various departments of the district. On the float entered by the TID shop crew, welding techniques were demonstrated, and other entries came from the service department, the store, the drainage department, the troubleshooters and others. Following the TID portion of the parade came horses, bands and floats entered by other local organizations. The parade lasted more than an hour. A picnic supper was held at Crane Park before the Jubilee finale was presented at the Turlock High School football field.

A stage at the north end of the field had three levels and stretched 150 feet, with a 60-foot-high backdrop of Don Pedro Dam painted by Turlock High art instructor Viola Siebe. The pageant itself was directed by Lura Flint and was based on a history of the Turlock District written a few years earlier by Helen Hohenthal. A dozen horsemen carrying American flags galloped in front of the stage, and Frank Mancini of Modesto led the massed bands in the National Anthem. In a prologue taken from Genesis, 200 girls took part in a choreographed interpretation of the heavens and the earth, the waters and the dry land. Episodes of the region's history followed, each featuring the students of one of the local high schools. Hilmar students portrayed the Indians; Denair, the early Spanish explorers; Hughson, the pioneers and railroad builders; and Ceres, the canal builders and attorneys who struggled through the years of litigation. A hundred girls represented the water from La Grange Dam, and Turlock students in ethnic costumes symbolized the many nationalities that had come to the area. At the end, a line of silver-colored fireworks was set off along the portion of the massive backdrop depicting the Don Pedro spillway to give the impression of a sheet of water pouring down it. More fireworks completed the program. It was a spectacular and memorable conclusion to the celebration of California's oldest irrigation district's first half century.[72]

As the 1930s drew to a close, signs of a return to prosperity were evident in the Turlock District. The final PWA grant was used in the winter of 1938-1939 to line more canals and community ditches, and the program of disposing of delinquent property was well underway. The number of electrical customers was rising again and the addition of two gold dredges, one near La Grange in 1938 and the other at Roberts Ferry in 1941, boosted retail power sales substantially.[73] With healthy electric revenues, the tax rate continued to fall, and in 1941 the district even had enough surplus money to buy the warrants of Improvement District No. 260, the west branch of the McMullen ditch.[74] That was the first time the TID itself had purchased the warrants of one of its improvement districts rather than selling them on the open market, and it quietly introduced a policy of financing all new improvement districts with TID funds. The district's ability to do so was another indication that the Depression decade was finally over.

An inflationary trend in the national economy became noticeable in 1941 and led to increases in the wages paid district employees of $10.00 a month or 50¢ a day for daily employees in September 1941.[75] Further adjustments were made that in all amounted to a 50 percent raise for common labor down to a 17 percent increase for the ditchtenders. These increases, however, were not enough to prevent the district's first serious labor dispute. Disgruntled employees joined the International Brotherhood of Electrical Workers (IBEW) and tried to conduct negotiations with the district, but talks collapsed. With perhaps 90 percent of the employees in both the electrical and irrigation departments enrolled in the union, a strike vote was taken on May 18, 1942, over wages, working conditions and alleged discrimination against employees who had been active in the union. A strike was called for midnight, May 25, and plans were made to throw the switches at the Don Pedro powerhouse to shut down the electrical system. George

Mulkey, an IBEW representative, hoped that the strike threat itself would be enough to win intervention from the War Labor Board and make an actual walkout unnecessary. A few days later Mulkey announced the receipt of a telegram from the War Labor Board asking that the strike be delayed until the board had an opportunity to look into the problem.[76] The district's board, meanwhile, released a statement of its position. It argued that under the National Labor Relations Act the district was a political subdivision and was therefore not subject to collective bargaining requirements and, in fact, could not formally recognize the union or negotiate a contract with it. The board insisted that it had raised wages, had not fired anyone because of union activities and had met with its employees whenever asked.[77] The strike was deferred, and Paul A. Dodd of the War Labor Board announced that he would come to Turlock. When he did not appear and nothing more was done about their grievances, the employees voted on June 11 to strike at midnight the following day.[78]

As the deadline approached, there were real fears that an attempt would be made to shut down the electrical and canal systems, but during the single day between the strike vote and the midnight walkout, a representative of the California Industrial Relations Bureau arrived in Turlock, the governor intervened and IBEW headquarters cancelled the strike.[79] Both the state and the federal War Labor Board were now involved. Secretary of Labor Frances Perkins contacted Governor Culbert Olson in early July to ask him to look into the problem at the TID. The governor did meet with both sides in Los Angeles and promised to ask the War Labor Board to formally enter the case.[80] It did so immediately, appointing Paul Dodd to head a Fact Finding Commission that would also have one representative for the district and one for the union. Hearings held in early August succeeded in reestablishing direct negotiations between the two sides and by the time a final meeting was held on September 27, 1942, the controversy was settled. A pay raise of better than 15 percent had been agreed to, the overtime pay issue had been dropped and the reinstatement of one employee, Homer Alquist, had been settled by giving him first consideration in hiring when he returned from military service. The union acknowledged that the district could not at that time embody these terms in a written contract or deal formally with a union.[81] A few employees subsequently left the district, prompted, it was thought, by the labor dispute. The whole episode was over less than half a year after it began, and though it was not the last labor dispute the district would have, it was by far the most bitter.

The threatened strike was only one of the problems facing the TID during World War II and was the shortest lived. The most immediate concern after Pearl Harbor was for the security of the dams and powerplants. The districts closed Don Pedro and La Grange dams without delay, and a few weeks later the Turlock District closed Owens Reservoir and put it, too, under guard. The district paid, at least in part, for the guards stationed at its facilities, and it asked, without apparent success, for additional help in preventing acts of sabotage.[82] The threat, of course, failed to materialize and by April 1944 the situation had eased enough to allow the reopening of Owens Reservoir except for the area around Dam C and the outlet gate.[83]

Dealing with the shortages imposed by the war turned out to be a far greater concern than security. Tires, for example, were hard to come by. Ditchtenders, who furnished their own vehicles, were in an especially difficult position. Early in the war, the state rationing administrator informed the district that the ditchtenders would not be able to get tires or retreads and said, "In view of the extreme rubber shortage, we feel the irrigation district can work out the problem themselves, and where possible use bicycles, horse and wagons, steelrimmed wheels, etc, if they can not buy used tires."[84] To save rubber and gasoline and reduce wear and tear on their irreplacable cars, the district read electric meters only every other month. Bills were still mailed out monthly but the first month's used an estimate based on the previous month's consumption and the next bill covered the remainder of the energy actually used.[85] The curtailment of the construction program also helped reduce the use of district cars and trucks.

Electrical materials, such as copper wire, were also in critically short supply. Retail sales of electricity dropped slightly in 1943 when the gold dredges were forced to shut down, but by the end of the war the system was selling more power to more customers than ever.[86] The lack of wire and other supplies made it difficult to build the lines that were needed and some of the services that were run were patched together and barely usable. At one point the only wire the district could get was salvaged from a wrecked ship and it had to be stripped of its stainless steel coating before it could be used. The fairgrounds in Turlock were converted into an Army stockade, and from time to time the district was called upon to install additional electrical lines. The Army was supposed to furnish the wire, but all it had was a pile of short pieces, most of them unusable without tedious splicing. After having to use their own scarce wire instead, district crews loaded up the whole pile to salvage what they could, and above all, force the Army to order a new and hopefully more serviceable supply.[87]

Shortages also extended to finding enough employees. By the end of 1942 a quarter of the district's work force had left to enter military service or do other war-related work. The electrical department was forced to hire so-called "boomer linemen," who were highly skilled in their trade but were unpredictable drifters and sometimes drunks who might walk off the job at any time. They were hardly desirable workers and in other circumstances would not have been tolerated by the district, but during the war there was little alternative but to take whatever they could get.[88] The war also brought the district its first and only female ditchtender. When Norman Moore, a second generation TID ditchtender in the Ceres area, was preparing to enter the military, his wife was faced with having to leave the ditchtender's house furnished by the district. To keep it, she asked if she could take over her husband's job. The district agreed and she worked as a ditchtender for the three years he was absent.[89]

Before the war was over, one of the strangest events in the district's history took place. The board minutes of February 5, 1945, recall it in unusually graphic detail:

> At the hour of 2:00 o'clock P.M. Improvement District No. 342, known as the Hodges Ditch, came on for hearing, pursuant to notice, and there were present at this hearing Tony B. Silveira and Tony Rozario, owners of property within this improvement district.
>
> At approximately ten (10) minutes past 2:00 o'clock Arie Wyngarden, one of the owners of property within said Improvement District No. 342, came to the East door of the Board Room and without provocation produced an automatic revolver, took deliberate aim at the Chief Engineer, R. V. Meikle, and fired, the bullet entering his right hand between the second and third finger, passing through his hand and grazing his abdomen.
>
> ADJOURNMENT
>
> Moved by Director Commons, seconded by Director Thornberg, and duly carried, the meeting adjourned.[90]

Fortunately, when Wyngarden pulled the trigger a second time, the .32 Savage jammed. The board members, who ironically had once decided to retreat into the vault in case of just such an occurrence, sat stunned, while one observer left the building through the glass door at the back without bothering to open it.[91] Wyngarden himself walked out the front door and down Main Street toward Highway 99, followed at a safe distance by Director C. N. Ahlem, who pointed him out to the police.[92] Wyngarden was commonly thought of as a strange old man who lived alone and never hired any help.[93] He gave the deputy district attorneys who took his statement a rambling half-hour discourse that, among other things, accused Meikle or his hirelings of poisoning his food and depriving him of his water rights. Wyngarden was quoted as saying, "I didn't intend to kill him, I only wanted to hurt him. That's why I shot low down. Meikle is a robber. This has been going on for years."[94] Apparently the formation of the improvement district, which Wyngarden opposed, triggered the shooting. Meikle's wounds were superficial and he was back at work in less than two weeks. Wyngarden was even tually committed to the Napa State Hospital, where he died.[95]

With the end of the war, the TID reached a prosperous maturity. Its electrical business was booming. Although the number of meters rose only from 8,567 in 1941 to 9,950 in 1945, energy consumption increased by 35 percent over the same period. Postwar growth was even more dramatic with 60 percent more customers in 1950 than in 1945, and sales of 43 percent more electricity.[96] Such growth meant a demand for more lines and more substations to handle the load. Besides the old Geer, Broadway and Hilmar substations, one had been built near Hughson in 1941 and soon after the war another was erected at Ceres, followed by one on F Street in Turlock in 1949. From the end of the war until about 1947 the demand for new lines exceeded what district crews could build, so two outside contractors were brought in for line construction work. When the contractors were no longer needed, the district purchased some of their equipment, including the first mechanical digger for planting power poles.[97]

Change was perhaps less noticeable in the irrigation system, but the canal lining and drainage programs continued and there was considerable growth in the number of improvement districts; so much so that the district once again sold some of the warrants on the market rather than buying all of them itself.[98] In 1948 a new type of improvement district appeared; one whose purpose was the installation of deep-well pumps to provide supplemental irrigation water. With a good supply of gravity water from the TID, most farmers were reluctant to spend the money to put in an expensive irrigation pump, but a sharp drought in 1948 convinced many growers that an auxillary or emergency supply was well worth the price. In February 1948 as many as 140 wells, not all of them for improvement districts, were being drilled, mostly in the Hughson area. The shortage of electrical transformers and the lack of hydroelectric power in a dry year meant that many of the pumps had to be installed with gasoline or diesel engines rather than electric motors, at least at first.[99]

Even in normal years these private wells could be useful in pro viding more convenient irrigation for small acreages or other special services, but their greatest value was in dry years. There were several very dry years between 1957 and 1961. In those years, the owners of pumps were allowed to sell water to irrigators who needed more than they could get from the district, using the canal system to transfer the water. In other words, a farmer with a pump was permitted to run his pumped water into the district's canals and anyone purchasing water from that farmer would be entitled to take that amount out of the canal system. In 1959 for example, the district was able to supply only thirty-three acre-inches to each irrigated acre from Don Pedro and its drainage pumps. That was enough for many farmers, but those who needed more could buy it for from three to six dollars per acre-foot from the owners of private wells.[100] The next year was even drier, but 67 private pumps put almost 39,000 acre-feet into the canal system, and once again, crop losses were avoided.[101]

Apart from the development of private wells, the major innovations in irrigation after World War II were new policies rather than new construction. They demonstrated that water distribution methods can, and do, change; an important point to consider in the study of any irrigation system. In the case of the Turlock District, the changes were not part of any long-range plan, but were simply practical adjustments in the way water was handled. The most important of these came about gradually, almost imperceptibly, in response to physical improvements in the irrigation system and changes in crop patterns. Lining and enlarging the district's main canal and lateral system increased the amount of water that could be delivered at any one time and that, in combination with the increased supply available from Don Pedro and the drainage pumps, meant that the rotation interval between irrigations could be shortened. In about 1936, Chief Engineer Meikle noted that rotations varied from a week to thirty days apart, depending on the crop and soil conditions.[102] The basic premise of rotation, that irrigators would receive water only at regular intervals, was still adhered to in general, but the intervals could be set with so much flexibility that in some instances, especially where ditchtenders were lenient, the system was not far removed from providing water whenever it was wanted.

Changes in crop patterns called for still more flexibility. In the late 1930s large acreages were planted to ladino clover, a shallow-rooted, permanent pasture crop. Clover was frequently grown on somewhat marginal land, where the soil layer might be too thin to support other crops. Heavy tractors and scrapers made it easy to grade the land into the long, narrow strip checks common in irrigated pastures. Because its mat of roots was so shallow, clover required frequent irrigation to keep it alive, in some cases as often as once a week in hot weather. Clover was grown on so many acres, up to 35,000 acres at its peak in 1952, that it dictated the rotation schedules in many parts of the district. With clover on a ten-day rotation, alfalfa growers were practically forced to water their crops every twenty days, or every other clover run, even if it was sooner than they would have preferred. Other crops were in a similar situation, prompting the solution that Meikle outlined in late 1949. Community ditches would be allowed to vote on the length of rotation they needed. "The other crops on the ditch," he wrote, "will be irrigated under a modified rotation system where the irrigator will notify the Ditchtender when he will be ready to take the water and the Ditchtender will make the delivery as close to that date as possible."[103] The "modified rotation system" Meikle referred to was not really a rotation system at all but a type of demand system, with the irrigator deciding when he wanted to get water for his crops and then asking the ditchtender to schedule the water for him. This became known as the call or call-in system, since it required farmers to call their ditchtenders when they needed water rather than relying on a fixed rotation schedule.

The call system would not have been possible without the lining and pipelining of the community ditches. Under the rotation system, water was put into a ditch for a run and not pulled out until the end of the run. That made maintenance somewhat easier and reduced the losses that were inevitable when moving water from one dirt ditch to another. Improvements in the ditch system meant that water could easily be shifted around within a ditchtender's division. The increased capacity of the canals and the work of the improvement districts made the system flexible enough to meet the fluctuations in demand that were bound to result from letting farmers themselves determine when they should irrigate, or at least request the water. That last distinction was an important one to Meikle, who always refused to think of the call system as a true demand system. Irrigators could not demand water exactly when they wanted it, but could request it when they were ready and the ditchtender would then schedule it when he could, which might mean it would reach the farmers in a few hours or a few days from the time of the request.

The call system's flexibility was a boon to farmers but an added burden to the ditchtenders, who now had to plan the movement of water around the community and private ditches in their division and arrange the transfer of water from one farmer to another. Some crops had more difficult scheduling requirements than others but probably few could try the ditchtender's patience as much as peaches. They had always been one of the district's most important tree crops, but in the 1930s the acreage began to expand at a pace that increased dramatically in the late 1940s and 1950s. In the heyday of the canning peach industry, large areas around Ceres, Hughson and Denair were devoted almost entirely to that crop. Each farm would typically have five to ten varieties, ripening from July into September, and each variety would need irrigation on a different schedule as it approached harvest. Not only was timing critical but since a farmer might irrigate only a few acres at a time, water had to be moved frequently between farmers. No matter what the crop, however, planning the delivery of water more or less when it was wanted by sometimes impatient irrigators could test a ditchtender's tact as well as his ingenuity.

Despite all the improvements in the irrigation system, one problem stubbornly remained – the excessive use of water by some farmers. Applying more water than the crop needed had been the most common sin of early irrigators. Experience in handling water, regrading of unlevel checks and time limits imposed by the district to give everyone a fair share of the supply helped reduce, but did not eliminate, waste. As in the case of the rotation system, irrigated pastures brought matters to a head. By the 1940s much of the excessive use was occurring on ladino clover, where frequent irrigation was a necessity. On sandy soil, great quantities of water could go beyond the shallow root zone and be lost. There was also land that was still not adequately levelled, and in a few cases, irrigators would turn the water into several checks at once to slow down the irrigation so they could do something else for a while. In the latter case, the slowly flowing water saturated the soil near the check gate rather than quickly covering the entire check the way it should have.[104] As early as 1926 Meikle had insisted that if irrigators had to pay for the water they wasted they would be more careful and efficient in its use.[105] His suggestion that irrigators be charged $1.50 per acre-foot for anything over an annual total use of five acre-feet per acre was not accepted at that time. Increasing pasture acreage in the 1940s, and a consequent increase in the demand for water, threatened to overtax the canal and drainage systems. The alternative to enlarging the canals was to enforce more efficient use of water. In normal years Don Pedro and the drainage pumps could supply four acre-feet to each irrigated acre in the district, but in 1949 a quarter of the land in the district received more than that amount. For the 1950 season, it was decided to bill irrigators a dollar for each acre-foot over four acre-feet per acre that they used. The results were impressive. Only 5,000 out of 160,000 acres exceeded the four acre-foot standard, the peak flow in the main canal was reduced, the season lasted two weeks longer and the groundwater level dropped three and a half inches.[106] From that time forward, the four acre-foot limit was a basic policy of the Turlock Irrigation District.

The four-acre-foot limit, like so much else, could be credited to the district's chief engineer, Roy V. Meikle. In 1950 he turned sixty-five years old, but he was still over twenty years away from retirement. His name appears frequently in this or any other telling of the TID's history, and at times has an almost institutional ring, so closely was the man identified with the district. His influence was so pervasive that it would be appropriate to briefly consider R. V. Meikle in greater depth. He was almost universally respected as a brilliant engineer, and in addition to his work for the TID he served as chief engineer for the construction of the Merced Irrigation District's Exchequer Dam, was a consultant on the design of the Central Valley Project and was later appointed to the State Water Commission. However, these activities and his other business ventures, which included the ownership of ranch property near Hickman, were never more than peripheral to his main interest and preoccupation – the Turlock Irrigation District.

The chief engineer was often the de-facto general manager of an irrigation district and that was certainly true in the case of the Turlock District and Roy Meikle. His management style could be described as authoritarian and his attitude was conservative, especially when it came to spending money. Although the district was actually under the control of the Board of Directors, it was often said that Meikle dominated the boards he served under. That was probably not far from the truth. Once Don Pedro had been built under Meikle's guidance and the electrical business, which he obviously favored, was a success, he had earned the confidence of board members and their constituents. He did not, however, bypass the directors in his management of the district. He reported to them regularly and took care to offer advice rather than directives. He was so respected by the board that his advice was in time accepted almost without question, especially on major issues of policy. Negotiations with San Francisco or other agencies on the development of the Tuolumne River were left entirely in Meikle's hands, and he kept the details to himself. The authoritarian style continued in his control of operations within the district's organization. His word was widely regarded as law and his decision was final. While overseeing the entire system, including irrigation, electrical, shop and office functions, Meikle remained most involved with irrigation, as befitted his training and background. He was out almost every day checking on the course of work on the canals, and on many mornings he was up at dawn to drive the main canal as far as the dam to keep an eye on every detail of the system. Although he set overall policy for the electrical department, he readily delegated authority for day-to-day operations and did not meddle in the way his subordinates went about their work.

Personally, Meikle was a shy man who did not mingle easily with others. He was always accessible to anyone who wanted to see him, but he was known to be business-like to the point of bluntness. He simply had no interest in small talk. If a farmer or an employee needed to see him, he wanted to know exactly what the problem was and he then gave his opinion, advice or solution, as the case might be, and made it clear that the interview was over. Such behavior, and his obvious authority, could certainly be intimidating, but his door was said to be open to anyone with legitimate business. Farmers especially found it was easy to present their problems to Meikle, who firmly believed that, although it had also become an electric utility, the TID was first and foremost an irrigation district. At times he might even leave a board meeting to see a farmer who

Chief Engineer Roy V. Meikle.

had come to the office. No one remembered when Meikle ever took a vacation, and even when he had to be absent due to illness he kept up with his mail and other business. He had little patience with delays of any kind and in some cases of possible conflicts told those under him to do the job, "and we'll stand the suit later."[107] He was also modest by nature, usually unwilling to talk at any length about himself. Although suggestions were made, most notably during the construction of New Don Pedro Dam, of naming things after him, he always quickly and firmly put a stop to such talk.

By the 1950s, the Turlock region had become the world the district's founders had dreamed of. The small farms, the variety of crops, and the prosperous small towns were the benefits that irrigation had promised. The district, too, had become what they had hoped it would; the smoothly functioning supplier of all the life-giving water the irrigators needed. The work started with so much difficulty in the nineteenth century had matured; its prophecy was fulfilled. But the tranquility and simplicity of the irrigation empire at its height was short lived. Changes were already taking place that would alter and complicate the affairs of the Turlock Irrigation District in the following decades.

CHAPTER 11 – FOOTNOTES

[1] *Modesto Morning Herald,* Sept. 28, 1923.

[2] *Ibid.,* Feb. 17, 1924; *Turlock Tribune,* July 1, 1925.

[3] Board minutes, vol. 8, p. 349 (Feb. 21, 1924); *Modesto Morning Herald,* Feb. 22, 1924.

[4] *Modesto Herald,* Mar. 2, 1924; *Turlock Tribune,* Mar. 26, 1924.

[5] Board minutes, vol. 8, pp. 352-353 (Mar. 3, 1924), p. 354 (Mar. 7, 1924).

[6] *Turlock Tribune,* Mar. 12, 1924.

[7] *Modesto Morning Herald,* June 15, 17, 1924; *Turlock Tribune,* June 20, 1924.

[8] *Modesto Herald,* Aug. 15, 1924.

[9] Board minutes, vol. 8, p. 399 (July 10, 1924).

[10] TID, "Comparison of Crops" (annual). The compliation in Meikle files, vol. 19, item 39 covers 1915-1952.

[11] Memorandum on drainage, Mar. 12, 1929, in Meikle files, vol. 5, item 1; memorandum on a typical drainage well, Feb. 11, 1929, in Meikle files, vol. 5, item 2.

[12] "Questions raised at Improvement District No. 52 meeting," Mar. 27, 1957, in Meikle files, vol. 28, item 44.

[13] Drainage memorandum, July 16, 1958, in Meikle files, vol. 29, item 22.

[14] Memorandum on drainage, Mar. 12, 1929, in Meikle files, vol. 5, item 1.

[15] Board minutes, vol. 9, pp. 218-219 (Mar. 28, 1927).

[16] *Ibid.,* vol. 8, p. 246 (June 26, 1923).

[17] See *Streams in a Thirsty Land,* p. 212.
[18] "Comparison of Crops," in Meikle files, vol. 19, item 39.
[19] Summary of TID development, by R. V. Meikle, ca. 1933, in Meikle files, vol. 1, item 13; undated memorandum in Meikle files, vol. 6, item 4; memorandum on drainage, Mar. 12, 1929, in Meikle files, vol. 5, item 1.
[20] Memorandum on irrigation system by R. V. Meikle, ca. 1927, in Meikle files, vol. 8, item 8. [21] Henry Schendel interview.
[22] Memorandum on irrigation system by R. V. Meikle, in Meikle files, vol. 8, item 8. Additional information can be found in Meikle files, vol. 6, item 4.
[23] *Cal Stats.* 1927, pp. 1415-1418.
[24] Board minutes, vol. 9, pp. 305-315 (Dec. 5, 1927), pp. 340-346 (Jan. 17, 1928), pp. 347-349 (Jan. 18, 1928). [25] *Turlock Tribune,* Jan. 17, 1930.
[26] "Yearly Summary, Turlock Irrigation District Improvement District Data," ca. 1968, in Meikle files, vol. 41, item 44.
[27] Telephone interview with Paul Terrell, Feb. 28, 1983; Grant Paterson interview.
[28] Author's interview with Ronald Hawkins, Feb. 17, 1983.
[29] "Improvement District Data," in Meikle files, vol. 41, item 44.
[30] On Mead's ideas see Paul Conkin, "The Vision of Elwood Mead," *Agricultural Hist.,* 34 (Apr. 1960), pp. 88-97; and Kevin Starr, *Inventing the Dream: California Through the Progressive Era,* (New York, 1985), pp. 170-172. The land settlement concept and the establishment of the Durham colony are discussed in the *Report of the State Land Settlement Board* (June 30, 1918).
[31] Frank Adams, "Report on Water Supply for the Wilson Tract, Merced County," prepared for the State Land Settlement Board, 1918, pp. 1-3.
[32] *Report of the State Land Settlement Board,* Sept. 30, 1920 in *Appendix to the Journals of the Senate and Assembly,* 44th Sess. (1921), vol. 1, p. 26.
[33] *Report of the State Land Settlement Board,* Sept. 30, 1920, p. 30.
[34] Roy J. Smith, "The California State Land Settlements at Durham and Delhi," *Hilgardia,* 15 (Oct. 1943), pp. 411-413.
[35] Schendel interview.
[36] The story of the Delhi settlement is told most succinctly in the article by Smith cited above and in Tony Kocolas, "California's Experiment at Delhi: Evaluation of the Delhi Land Settlement Project of 1920 to 1931," (student paper, Stanislaus State College). The interview of Delhi superintendant J. Winter Smith by J. Carlyle Parker, Feb. 25, 1971, in the Stanislaus College archives is also useful. [37] Smith, p. 414.
[38] "Delhi Colony, Improvement District No. 52," May 20, 1952, in Meikle files, vol. 19, item 40; memorandum on State Land Settlement, Feb. 4, 1959, in Meikle files, vol. 30, item 15; memorandum on exclusion of land in Improvement District No. 52, by R. V. Meikle Apr. 3, 1959, in Meikle files, vol. 30, item 31. [39] Christine Chance interview.
[40] Board minutes, vol. 15, pp. 204-253 (July 20, 1936).
[41] *Turlock Tribune,* Aug. 5, 1932; *Modesto Bee,* Aug. 7, 1934.
[42] Author's interview with Richard Vollrath, Mar. 1, 1983.
[43] *Turlock Tribune,* Oct. 10, 1930. [44] *Ibid.,* July 17, 1931, May 13, 1932.
[45] Board minutes, vol. 11, p. 463 (Sept. 14, 1931).
[46] *Modesto News-Herald,* Dec. 30, 1931; *Modesto Bee,* Feb. 22, 1933.

[47] *Turlock Tribune,* Sept. 23, 1932.

[48] "Conclusions and Recommendations of the Report of California Irrigation and Reclamation Financing and Refinancing Commission," Dec. 1, 1930, in *Appendix to the Journals of the Senate and Assembly,* 49th Sess. (1931), vol. 5, p. 13; S. T. Harding, *Water in California,* (Palo Alto, 1960), p. 85.

[49] See Kenneth R. McSwain, *History of the Merced Irrigation District,* (Merced, Calif., 1978), pp. 108-128; and W. Turrentine Jackson and Stephen D. Mikesell, *The Stanislaus River Drainage Basin and the New Melones Dam: Historical Evolution of Water Use Priorities,* (Davis, 1979), pp. 14-15.

[50] *Modesto Bee,* Aug. 15, 17, 1933.

[51] *Ibid.,* Aug. 4, 1933.

[52] *Turlock Tribune,* Sept. 2, 1932, Sept. 23, 1930; Board minutes, vol. 13, p. 42 (Oct. 2, 1933).

[53] *Turlock Tribune,* Oct. 13, 1933.

[54] *Modesto Bee,* May 19, 1934; Board minutes, vol. 13, pp. 208-209 (June 18, 1934); memorandum on "Grant and Loan from the Federal Government," July 5, 1934, in Meikle files, vol. 6, item 18.

[55] Memorandum in Meikle files, vol. 6, item 18.

[56] *Modesto Bee,* June 28, 1934, Aug. 7, 1934.

[57] Board minutes, vol. 13, p. 242 (Aug. 20, 1934); *Modesto Bee,* Aug. 17, 1934.

[58] Board minutes, vol. 13, p. 355 (Dec. 3, 1934).

[59] *Modesto Bee,* April 9, 1935.

[60] Board minutes, vol. 13, p. 429 (Jan 21, 1935), p. 451 (Feb. 4, 1935).

[61] *Ibid.,* vol. 14, p. 420 (Dec. 21, 1935), pp. 439-440 (Dec. 30, 1935), vol. 15, p. 201 (July 13, 1936), vol. 16, p. 159 (Apr. 12, 1937); *Turlock Tribune,* Mar. 27, 1936, Oct. 2, 1936.

[62] Board minutes, vol. 14, pp. 157-159 (Sept. 5, 1935); *Turlock Tribune,* Sept. 13, 1935.

[63] Board minutes, vol. 17, pp. 374-375 (Aug. 8, 1938).

[64] *Turlock Tribune,* Apr. 26, 1935.

[65] Board minutes, vol. 12, p. 282 (Apr. 11, 1932).

[66] *Turlock Tribune,* Oct. 18, 1935.

[67] Board minutes, vol. 15, pp. 88-89 (Mar. 30, 1936); *Modesto Bee,* Nov. 4, 1936.

[68] Board minutes, vol. 15, pp. 471-472 (Dec. 14, 1936).

[69] For example see Board minutes, vol. 16, p. 70 (Feb. 23, 1937), vol. 17, pp. 376-378 (Aug. 15, 1938).

[70] Vollrath interview

[71] *Modesto Bee,* Mar. 10, 1942.

[72] *Turlock Tribune,* Souvenir Edition, June 11, 1937; *Turlock Daily Journal,* June 11, 12, 14, 1937; Paterson interview.

[73] *Turlock Tribune,* May 16, 1941.

[74] Board minutes, vol. 20, p. 399 (Apr. 28, 1941).

[75] *Ibid.,* vol. 21, p. 151 (Sept. 29, 1941).

[76] *Modesto Bee,* May 19, 21, 1942.

[77] Board minutes, vol. 22, pp. 190-191 (May 21, 1942).

[78] *Modesto Bee,* June 12, 1942.

[79] *Ibid.,* June 13, 1942; Author's interview with Frank Clark, June 29, 1983; Charles O. Read interview; Norman Moore interview.

[80] *Modesto Bee,* July 11, 1942.

[81] *Turlock Tribune,* Oct. 9, 1942.

[82] *Modesto Bee,* Feb. 14, 1942; Board minutes, vol. 22, p. 153 (May 4, 1942); author's interview with Reynold Tillner, Jan. 12, 1983.

[83] Board minutes, vol. 24, p. 244, (Apr. 24, 1944).
[84] *Turlock Tribune,* Mar. 20, 1942. [85] *Ibid.,* Feb. 19, 1943.
[86] *Ibid.,* June 4, 1943. [87] Read interview. [88] *Idem.* [89] Moore interview.
[90] Board minutes, vol. 25, p. 53 (Feb. 5, 1945).
[91] Author's interview with Arlene Abel, Edna Anderson, Clarice Espinola, Isabelle Martin, Velma Openshaw, June 23, 1983; Trent interview; Tillner interview.
[92] *Turlock Tribune,* Feb. 9, 1945. [93] Trent interview.
[94] *Turlock Daily Journal,* Feb. 6, 1945. [95] Trent interview.
[96] The number of electric meters can be found in the district's Annual Reports. Energy sales are from a compilation made in 1984 and hereafter referred to as "Historical Electrical Load Data."
[97] Clark interview. [98] Board minutes, vol. 27, p. 179 (Apr. 28, 1947).
[99] *Turlock Tribune,* Feb. 27, Mar. 12, 1948; Hawkins interview.
[100] Memorandum on the 1959 irrigation season, in Meikle files, vol. 30, item 91.
[101] Memorandum on 1960 irrigation season, by R. V. Meikle, Oct. 20, 1960, in Meikle files, vol. 31, item 18.
[102] Report on TID irrigation, drainage and weed control, 1936, in Meikle files, vol. 7, item 53.
[103] Memorandum on four-acre-foot limit, Dec. 1949, in Meikle files, vol. 18, item 48.
[104] Memorandum on irrigation and the four-acre-foot limit, by R. V. Meikle, Dec. 29, 1950, in Meikle files, vol. 18, item 49.
[105] Board minutes, vol. 9, p. 116 (Sept. 27, 1926).
[106] Memorandum on irrigation and the four-acre-foot limit, Meikle files, vol. 18, item 49.
[107] Hawkins interview. General information on Roy V. Meikle was gathered from district employees and others interviewed by the author.

12

Cooperation Comes to the Tuolumne

Since 1913 the Turlock and Modesto irrigation districts had known they would have to share the Tuolumne river with San Francisco, and it was commonly assumed that at some point they would come into conflict with the city over how much of the river each would get. The Raker Act did not settle the question. It did guarantee the districts 2,350 second-feet of the natural flow, and 4,000 second-feet for sixty days each spring, but the water rights were a matter of state, not federal, jurisdiction. Although Congress was entitled to put whatever condition it wished on the grant of federal lands for reservoir sites and rights-of-way, there was nothing to prevent the districts from pressing claims in the state courts for more than the Raker Act assured them. The numbers included in the Raker Act reflected only the status of the districts' canal systems and irrigation development in the early years of the twentieth century. Their original filings were prior to, and therefore superior to, those of San Francisco. These filings covered more water than they could use at first and were meant to protect their ultimate needs for full development, even if it took decades to reach that point. On the other hand, the claims of the two districts – 4,500 second-feet for Turlock and 5,000 second-feet for Modesto – were far in excess of the natural flow that they could ever beneficially use for irrigation. The actual extent of the districts' rights, then, could be found in neither their filings nor in the Raker Act. The only way to determine those rights appeared to be through adjudication. Both sides anticipated a legal contest.

The Modesto Irrigation District filed the first suit seeking a settlement of Tuolumne River water rights in 1914, soon after the passage of the Raker Act. The Turlock District did not join in the action, just as it had not joined the Modesto District in the frantic last minute attempt to defeat the Raker bill in Congress. San Francisco promply hired two leading irrigation experts, B. A. Etcheverry and Thomas H. Means, to gather information on how much water the districts needed for irrigation and how much waste existed, just as it had hired J. H. Dockweiler to make a similar investigation in connection with Freeman's work. Etcheverry and Means set up test plots on farms in the districts where they measured water use and even experimented with using different amounts of water to see what effect that might have on crop yields. The field research program continued until 1920, and in the meantime the suit was not brought to trial.[1] San Francisco was in no hurry to test its water rights and was content to see the case lapse while it went on building its project. While Modesto was apparently ready to resume its legal attack at any time, the Turlock board remained unwilling to cooperate. Finally, in December 1932 the TID authorized its attorney to take steps to protect its water rights against San Francisco, and the following month, the two irrigation districts filed the long awaited suit to determine rights in the Tuolumne River.[2]

The timing of the Turlock District's acquiesence in bringing the matter of water rights to trial may have been due to two things. One was the approaching completion of the city's tunnel and pipeline system that would enable water to reach the Bay Area. The other was a lawsuit begun in 1932 against San Francisco and the irrigation districts by a riparian water user. Meridian, Ltd., owned the 4,000 acre El Solyo Ranch on the western bank of the San Joaquin River just below its junction with the Tuolumne. El Solyo Ranch had been developed since about 1919 as a model farm, with extensive orchards and vineyards as well as vegetable and forage crops and a dairy. Its irrigation water was pumped from the San Joaquin River near the mouth of the Tuolumne, but even though it was located on the San Joaquin, the greater part of its supply actually came from the Tuolumne. Besides its right as a riparian owner to use water in preference to appropriators, it also held a small appropriative right dating from 1919. Although both the legal and engineering issues were complicated, what the Meridian, Ltd. suit hoped to accomplish was an adjudication of the Tuolumne that would define and limit rights the districts were acknowledged to have and prevent San Francisco from exporting water or even storing additional water that would otherwise flow down to El Solyo. The actual amount of water the ranch needed was negligible, only about sixty-six second-feet, but the first unit of San Francisco's pipeline could carry nearly that much, and to insure good water quality at the El Solyo pumps, more water would have to flow down the river and past the ranch than could be used. The Meridian case made it seem clear that the time to adjudicate the Tuolumne River once and for all was at hand, and the districts must now press their own case against San Francisco while they all fought Meridian, Ltd.

Even though they had filed suit against the city, the Turlock District was not inclined to try to stall or overturn the Hetch Hetchy project. The city planned eventually to raise the height of O'Shaughnessy Dam at Hetch Hetchy by over eighty-five feet, and with the possibility of federal aid from New Deal agencies, it pushed the project. Considering their suit against the city, it might have seemed that the last thing the districts would have wanted to see was an expansion of the city's facilities. On October 5, 1933, however, the TID board endorsed the proposed addition to the dam.[3] Some observers were concerned that the district's action, so soon after finally agreeing to sue the city, meant that it was ready to abandon the litigation.[4] Not so, said Meikle, and he pointed out that San Francisco would need to increase its storage capacity if it was going to bring its project up to its ultimate size without invading the rights of the districts. Meikle noted that there was plenty of water in the river in an average year for both the city and the districts. What was needed was enough reservoir space to fully manage the river and carry over water from wet years into drier ones.[5]

Despite disclaimers that it would proceed with the litigation, the Turlock District's policy reflected Meikle's opinion that reservoirs rather than verdicts were the best way to reduce friction over dividing the river. San Francisco city attorney Dion R. Holm hoped an out-of-court settlement would be possible, and he and TID attorney Thomas Boone asked for and received a delay in starting the trial from July to December 1934, over the objections of attorneys for the Modesto District.[6] On November 26, 1934, Holm and other San Francisco representatives appeared before a joint meeting of the Turlock and Modesto boards to propose that a commission of engineers, two from each of the three agencies, be established to devise an out-of-court settlement. What the city was looking for was some mutually agreeable plan to manage the river. The Turlock board, presumably guided by its chief engineer, had already been moving in that direction, and at the meeting with Holm, the Modesto board, too, agreed to the city's request.[7]

San Francisco made two specific proposals for dividing the river in 1935, known as Plan A and Plan A-1, which described operations under a 400 million gallon per day diversion to the city and under a 180 million gallon per day diversion, respectively. The presentation of the plans illustrated their operation over a period similar to 1921 through 1934, which included two critically dry years. The numbers were less significant than the concept the city was proposing. In exchange for being allowed to make the diversions it wanted, the city would manage its present and future reservoirs to guarantee certain amounts of water to the districts. Those guarantees would, in all but the wettest years, exceed the minimum levels specified in the Raker Act but to get the extra water, the districts would have to give the city a first priority to the river and accept a type of secondary right protected only by the city's promises. This was, of course, the reverse of the actual situation in which the districts held the first rights and the city was entitled only to the remaining water. The districts had always been extremely jealous of their water rights in the past, but the prospect of getting more water by a cooperative agreement rather than by gambling on the outcome of a lawsuit was attractive enough to merit serious discussions. However, Paul Bailey, a former State Engineer hired by the districts to advise them in the Meridian and San Francisco cases, warned in November 1935 that until the city had proven its own rights were valid, it was offering benefits to the districts it might not be able to deliver.[8]

Bailey had good reason to sound a note of caution. In July 1935 the city lost the first round in the Meridian, Ltd. case. Judge Raglan Tuttle of Nevada County, sitting in the Stanislaus County Superior Court in Modesto, essentially upheld the irrigation districts' rights, subject to minor restrictions, but severely limited San Francisco's right to take water from the Tuolumne. He ruled that the city was entitled to store no more than the original capacities of its reservoirs at Hetch Hetchy and Lake Eleanor and to divert only the flood waters stored in those reservoirs.[9] If allowed to stand, the decision would have crippled the Hetch Hetchy project in order to protect the small amount of water withdrawn at the El Solyo Ranch pumps. While the city's appeal to the state Supreme Court was pending, there was little point in going forward with talks on dividing the Tuolumne. A future settlement was still considered likely, however, and the city and the districts cooperated from year to year to maximize water conservation. One particular problem was that the city finished raising O'Shaughnessy Dam by early 1938, but an injunction based on Judge Tuttle's decision prohibited San Francisco from using the extra 142,000 acre-feet of storage capacity. The more that could be stored on the watershed the better, so the TID gave the city permission to store the district's share of the 4,000 second-foot Raker Act spring release for 1938 in Hetch Hetchy's extra space.[10] The following year, 1939, was drier than normal and the maximum use of all available storage was made even more urgent. El Solyo's attorneys, however, refused to permit the city to use its extra storage capacity. Finally, on May 5, 1939, the Turlock District approved a temporary arrangement to remove the impasse by allowing San Francisco to fill Hetch Hetchy in exchange for the release by the TID of 30,000 acre-feet through the La Grange powerhouse for El Solyo, the exact timing of releases to be scheduled by Meikle and the El Solyo Ranch engineer.[11] The very next day, the Supreme Court overturned Judge Tuttle's decision and freed the city to operate and expand the Hetch Hetchy system without further interference from El Solyo. Nevertheless, the arrangement entered into for 1939 was still carried out by the TID.[12]

With the riparian rights issue finally disposed of, the districts directed their engineers to resume negotiations with San Francisco.[13] They did so, and on February 20, 1940, the engineers' committee made its report. No specific plan was recommended largely because it was impractical to, in effect, amend the Raker Act by adopting a different release schedule. In fact, Meikle believed that the people of the two districts had come to accept the Raker Act so much that any suggestion of changes would bring a cry to "protect the sanctity of the Raker Act."[14] There was enough water in the river, if properly developed, to meet the ultimate needs of all parties, so the report urged continued cooperation within the basic framework of the Raker Act. Neither side would waive any of its rights, but the litigation pending since 1933 would be postponed for fifteen years. These simple recommendations were approved by the districts at a joint meeting on February 29, 1940, and by the San Francisco Public Utilities Commission on March 4, 1940.[15] As adopted, the recommendations became known as the First Agreement between the city and the districts. The agreement to live and let live in friendship was more than just a fifteen-year truce, it was the basis for a whole new era of development on the Tuolumne River.

Actually, plans for more dams were being made even before the First Agreement was signed. San Francisco had long planned to expand Lake Eleanor and develop a small reservoir on Cherry Creek, which would ultimately be linked by tunnel to Lake Eleanor, from which another tunnel would lead to Hetch Hetchy reservoir. By 1937 revised plans called for a 300-foot-high dam on Cherry Creek impounding a quarter-million acre-feet of water. Instead of a tunnel running from Cherry to Eleanor to Hetch Hetchy, a short tunnel would go in the opposite direction, from Eleanor to Cherry reservoir. Releases from Cherry would go down Cherry Creek and into the Tuolumne River. Since Cherry Creek joined the river a short distance downstream from the city's Early Intake diversion dam, water from the Cherry and Eleanor reservoirs would, without further construction, be out of reach of the aqueduct carrying water to San Francisco. That meant that the primary purpose of the Cherry project would be to provide releases to protect the districts' rights as required by the Raker Act. In its decision on the Meridian, Ltd. case, the Supreme Court confirmed that the city had water rights on Cherry Creek, "provided that said proposed reservoir shall hereafter be completed and put into use with due diligence."[16] The court's reminder that due diligence was required to hold water rights prompted the city to begin surveys at the dam site in July 1940.[17]

Even before city crews went to work in the summer of 1940, there was already talk of participation by the irrigation districts in the construction of the reservoir, although no detailed proposals had been exchanged.[18] Complicating the matter were new issues – flood control and an attempt to put a federal dam on the Tuolumne. During the 1930s, the Army Corps of Engineers was given a nationwide responsibility for flood control and it began planning flood control reservoirs. The TID board advanced money for a cooperative flood control study of the Tuolumne in 1939, and that same year, the districts and the city proposed a flood control plan for the existing Don Pedro and Hetch Hetchy reservoirs and the districts' foothill reservoirs.[19] The proposal did not offer the degree of flood protection the Army was seeking. Instead, local Corps of Engineers officers made plans to build a new dam just above Don Pedro reservoir primarily for the regulation of floods. The Jacksonville reservoir site had been surveyed by the state in the 1920s as part of a comprehensive investigation of potential water development in the Central Valley, leading to a State Water Plan. The dam site had a poor geologic foundation and an earthquake fault ran through it, causing the state to recommend instead the consideration of a greater Don Pedro plan, with a high dam located just downstream of the districts' existing structure.[20] Despite its well known drawbacks, the Army announced plans for a 317,000 acre-foot Jacksonville reservoir. The Turlock Irrigation District took a dim view of the proposal. Its physical configuration would result in large evaporation and seepage losses, and it had relatively little hydroelectric potential. Worse still, it would interpose a federal dam between the city's reservoirs and Don Pedro, which would complicate the management of the river. By the time the First Agreement was signed, the city and the districts believed that a better course would be for the Corps of Engineers to contribute to the construction of Cherry reservoir in exchange for flood control by Cherry and the other reservoirs on the river. The Cherry dam site was a favorable one and had the possibility of a large power development to take advantage of the 2,000 foot drop from the reservoir to the Tuolumne. Operated under the framework of the First Agreement, Cherry promised benefits to both the city and the districts that the Corps' Jacksonville plan did not. The problem then became one of convincing the Corps of Engineers that their flood control requirements could be met without putting a dam of their own on the Tuolumne.

Roy Meikle and Hetch Hetchy engineer Max Bartell took the lead in evolving a cooperative program for the full development of the Tuolumne watershed. Keeping other agencies off the river motivated much of the thought given to long-range policies in the early 1940s. Fighting federal reservoir construction rather than asking for it was somewhat uncommon. Almost all of the major dams and multipurpose developments in the West have been the work of the Bureau of Reclamation or the Corps of Engineers. Local agencies seeking to increase their water supply have routinely turned to Congress and to government dam builders to get the water they could not afford to develop themselves. On the Tuolumne, however, the city and the irrigation districts had already made substantial investments in water development and were on the verge of coordinating their operations in accordance with the Raker Act. Interjecting another agency, and with it another set of goals or purposes, could have jeopardized, or least complicated, their use of the river. What's more, the Turlock and Modesto districts, unlike most local irrigation agencies, could afford to undertake further development, thanks largely to their successful electrical systems, and so, of course, could San Francisco. The Tuolumne, then, presented a rare situation; a major stream with relatively uncluttered water rights, no interstate complications and local water users who had been able to afford substantial development and could do still more. The districts and the city had good reason to adopt a position of "hands off the Tuolumne".

The Corps of Engineers posed the most immediate threat of intervention, but as Bartell pointed out in an April 1941 memorandum to one of his superiors, the California Division of Water Resources had proposed a larger Don Pedro dam and reservoir as far back as 1931, during the studies that led to the Central Valley Project. The larger Don Pedro reservoir, which according to the state could be built to hold as much as 2,500,000 acre-feet of water, was not made an immediate part of the state's development plans.[21] Bartell was still concerned, however, because, "This study suggests that there is an excess of water from the Tuolumne River available for additional irrigation use."[22] Bartell thought it might be necessary for the districts and the city to build the larger Don Pedro themselves to forestall any future encroachment. Besides providing increased water and power benefits, enlarging Don Pedro to 1,000,000 acre-feet or more would provide as much flood control as the Jacksonville reservoir, which any future addition to Don Pedro would innundate. Bartell attached a draft agreement to his memo which called for the districts and the city to cooperate in planning Cherry and the larger Don Pedro.[23] Subsequent revisions of the proposed agreement were made by Meikle, and on November 15, 1943, it was approved by the districts and became known as the Second Agreement.[24] Like the First Agreement, it was a brief statement of principles. The time had come, it said, to make public the plan the districts and the city had been working on so that there would be no misunderstanding of their intentions. Cherry and New Don Pedro had both been determined to be necessary and all parties to the agreement would cooperate in their construction. It was a strong warning to anyone else who might covet the waters of the Tuolumne that the current proprietors of the river intended to complete its development themselves. The statement was a timely one, for only a few months earlier the Bureau of Reclamation had informed the House Flood Control Committee that Jacksonville reservoir should be expanded to 416,000 acre-feet and turned over to the Bureau for its Central Valley Project.[25] It was exactly that kind of thinking that the Second Agreement was designed to prevent.

A settlement with the Corps of Engineers was reached soon

after the Second Agreement was adopted. In a series of conferences in Sacramento in January 1944, the Army agreed to abandon its Jacksonville reservoir if the city and the districts would build their proposed facilities. New Don Pedro, with a flood control reservation equal to the size of the Jacksonville reservoir, would provide full control of Tuolumne River floods, and the Corps' approval of any plan was based on having at least that much reservoir space to hold and regulate floods at about that point on the river. However, the city and the districts wanted to build Cherry first because it would be a cheaper dam and an excellent site for a high-head powerplant. New Don Pedro would not be scheduled for construction until about 1955. The Corps' Jacksonville project would have cost over $8,600,000, so the local Corps of Engineers office in Sacramento recommended that the same amount be paid toward the construction of New Don Pedro. It was understood that since the operation of the existing reservoirs and Cherry would accomplish a significant degree of flood control until New Don Pedro was completed, a large portion of the overall payment would be advanced to help pay for Cherry.[26] The Chief of Engineers in Washington pared down the prospective payment but approved the cooperative arrangement, which became law in the Flood Control Act of 1944.

San Francisco and the Turlock and Modesto districts were now free to develop the river without interference. Just how the responsibility for doing the job would be shared still had to be negotiated, but before it was, the city and the districts found themselves linked in yet another way. In 1945 the districts agreed to buy power produced by the Hetch Hetchy system. How that agreement came about is an interesting and important aspect of the development of the Tuolumne's power resources. Like so much else, the story begins with the Raker Act. Congress required the city to develop the hydroelectric potential of its dam and aqueduct system, but placed specific restrictions on what could be done with the power it produced. True to the Progressive principles embodied in the act, Section 6 prohibited San Francisco from selling to any individual or private corporation the right to resell Hetch Hetchy water or power. In other words, power could be sold by the city directly to consumers or to some other public agency, which would in turn sell it to consumers. The penalty for violating this provision was simple – forfeiture of the entire grant to the federal government. It was generally assumed that San Francisco would acquire its own electric distribution system and sell Hetch Hetchy power directly to its own citizens. Even if it did not do so, Section 6 insured that the hydroelectric bounty of the Hetch Hetchy grant would not fall into the hands of a private power company.

The doctrine of public ownership was at the heart of the Raker Act, but there is some room for doubt about the dedication of Hetch Hetchy's builders to the concept, at least where the disposal of electric power was concerned. In 1918 City Engineer M. M. O'Shaughnessy promised to develop powerplants "which will more than take care of the municipal needs of San Francisco and enable the city to sell at a profit on the whole project the excess power which it does not need."[27] The municipal needs mentioned by O'Shaughnessy included electric street railways, lighting public buildings and similar functions. Four years later, with work on the dam at Hetch Hetchy well underway, former Congressman William Kent asked in an open letter how O'Shaughnessy planned to distribute power without violating the Raker Act.[28] The engineer's response that he would use private enterprise to do the job must have been a keen disappointment to Kent, who in spite of his deep respect for John Muir and Muir's wilderness ideals had nonetheless supported the Raker Act largely because it promised the municipal control of water and power. Gifford Pinchot, the former chief forester who had pled the city's case to Theodore Roosevelt and Congress, warned San Francisco that municipal distribution was mandated by the Raker Act and that, "all parties to the grant are under obligation to see to it that the consuming public receives the full benefit of all the water and power available, at cost, solvently calculated," Pinchot added that, "Congress definitely protected San Francisco against weak-kneed surrender to these interests by the clear and definite forfeiture clause of Section Six."[29]

The Hetch Hetchy dam was finished in 1923 and work was pressed to complete the Mountain Tunnel from Early Intake to the Priest regulating reservoir, from which huge penstocks would carry the water down to the powerhouse on Moccasin Creek. Power revenues from that plant were expected to help pay for building the aqueduct under the San Joaquin Valley and through the Coast Ranges to San Francisco. In 1923 a citizens advisory committee was set up to deal with the problem of handling Hetch Hetchy power. The two private utilities then serving San Francisco – Pacific Gas and Electric and the Great Western Power Company – were not interested in selling their distribution lines to the city, and O'Shaughnessy estimated that to build an independent municipal system capable of serving the entire city would cost $45 million.[30] Plans were begun to condemn all or part of the private distribution networks, and other public agencies were given an opportunity to take the power on an interim basis, but no offers were received. With the Moccasin plant nearly ready for operation, the city decided to market its power temporarily through the private utilities.[31]

Any direct sale of Moccasin's output to a private corporation for resale would have been an obvious violation of the Raker Act. To solve the problem, the city entered into a so-called agency contract with PG&E in 1925. The contract read: "The City hereby employs the Company and the Company accepts employment as temporary distributor for and on behalf of the City of the electric energy to be generated at Moccasin Power House and transmitted to Newark by the City over its own transmission lines."[32] The company would take all the power generated at Moccasin and sell it to consumers at its normal prices as the city's agent, keeping a fixed portion of the money collected to pay for the services it provided and turning the rest over to the city. The legitimacy of this arrangement was debated for years, but it was clear from the first that, semantics aside, the contract had the same result as an outright sale of Hetch Hetchy power to PG&E.

Before the agency contract was signed, the citizens advisory committee emphasized that "all possible speed be given to the completion of a transmission system from Newark to San Francisco, and a distributing system in San Francisco, which, indeed, is absolutely and legally necessary in order to carry out the provisions of the Raker Act."[33] In 1927, a bond issue to finance transmission lines and a step-down station failed. The Railroad Commission, a forerunner of the California Public Utilities Commission, completed its valuation of the two private power systems in the city in 1929, and the following year, four separate propositions for bonds to finish the transmission line, buy out the two existing systems and build another powerhouse at Red Mountain Bar, where the aqueduct crossed the Tuolumne River, were put before the voters and all failed. Secretary of the Interior Ray Lyman Wilbur had been investigating the city's compliance with the Raker Act and felt the agency contract was a violation of the act. When the bonds were voted down in 1930, he advised the city to try another election, amend the Raker Act or dump the agency contract. City officials appealed to him to delay taking any action until the aqueduct was completed, presumably since power revenues were helping finance construction. Wilbur agreed to the request.[34]

Harold L. Ickes became secretary of the interior in 1933, and it was not long before the self-styled "curmudgeon" was at odds with the city over the disposal of Hetch Hetchy power. Water from Hetch Hetchy reached the Bay Area in 1934, and Ickes was one of the speakers who celebrated its arrival. However, he lamented to his secret diary that the city was so tightly in the grip of the private power companies that it could not sell its power to its own citizens. He vowed to do what he could to remedy the situation.[35] Armed with an opinion by the Interior Department's solicitor that the agency contract was in violation of the Raker Act, Ickes summoned San Francisco officials to a meeting in his office on May 6, 1935. By that time two more elections had seen the defeat of measures that would have made municipal distribution possible. The city tried to defend its contract but Ickes was unconvinced, and in August 1935 he formally ruled the contract was in violation. Two plans were proposed by the city for distribution, but because they would have involved the sale of some off-peak power to PG&E, Ickes rejected them. He finally approved a plan, but it suffered the same fate as all the others at the hands of the voters in March 1937. Ickes demanded assurances that the city would stop violating the Raker Act or he would go to court to stop them. The city could give no such guarantees, and the government filed suit in April 1937. On June 28, 1938, Federal District Judge Michael Roche ruled the contract was a violation of the Raker Act. The city appealed to the Ninth Circuit Court of Appeals and won a reversal in its favor in 1939. The Department of the Interior then carried the case to the U.S. Supreme Court.[36]

The Supreme Court ruled on April 22, 1940, that the contract was indeed a violation of the Raker Act. In reaching that conclusion the Court relied not on the kind of technical contract analysis that the Court of Appeals had employed but on the clear intent of Congress. Justice Black wrote that, "Congress clearly intended to require – as a condition of its grant – sale and distribution of Hetch Hetchy power exclusively by San Francisco and municipal agencies directly to consumers in the belief that consumers would thus be afforded power at cheap rates in competition with private power companies, particularly Pacific Gas & Electric."[37] Enforcement of the Court's ruling might have meant that the Moccasin plant would have to be shut down if no satisfactory alternative to the illegal contract could be found. That possibility worried TID chief engineer Meikle because the regulated releases from Moccasin were needed to maintain operation of the Don Pedro powerhouse. Ickes learned of Meikle's concern and telegraphed his reassurances that the irrigation districts would be protected. The telegram read in part:

> . . . If ultimately the City is required to close the power plant to comply with the decree the City will not be prohibited by the Raker Act from regulating its reservoir or from allowing water to pass downstream. There is no requirement on the City under the Act to make you pay for water released from the reservoir and not required for city use. Furthermore, if the power plant is shut down I have authority to assume the operation of it.[38]

Instead of requiring immediate compliance by the city, Ickes granted them more time to come up with a workable and legal plan to use Hetch Hetchy power. After several plans were vetoed by Ickes, one finally received his approval, but it was rejected at the polls in November 1941. The city also tried to amend the Raker Act, but its bill died in committee in early 1942.[39]

The city seemed to be stuck between the proverbial rock and a hard place. Its contract with PG&E was illegal but every effort to find a replacement ended in rejection by the government or the voters. Ickes intimated that selling the power to a defense plant might be a way out. In February 1942 San Francisco officials went to Washington and succeeded in having an aluminum plant located along the transmission line at Riverbank, northeast of Modesto. The plant, which began operation in April 1943 used all the power Moccasin could produce plus supplemental power from PG&E. It ran for a little more than a year before the growing aluminum stockpile and complaints of crop-damaging emissions forced its closure in August 1944. The city then had to resume its search for a legal means to dispose of its power.[40] It was at that time that the irrigation districts began negotiating to buy Hetch Hetchy power. The Modesto District was already buying substantial quantities of PG&E power and was looking for a cheaper source of electricity. The Turlock District was still wholesaling power to PG&E under a contract that would run until 1954, but its distribution load was growing rapidly and it, too, would soon need to buy additional power. Early negotiations discussed the sale of the entire Hetch

Hetchy output to the districts, which would use what they needed and then resell the remainder to PG&E. That plan would have been little better than the agency contract.[41] By the end of 1944 the scheme had been revised to sell the districts only what they needed, with PG&E being paid to deliver power from Hetch Hetchy to the city for its municipal needs. Meikle and MID chief engineer Clifford Plummer went to Washington in January 1945 to finalize the terms of a power purchase contract.[42]

The contract between the Turlock and Modesto irrigation districts and San Francisco was signed on March 12, 1945. Although it contained a clause binding the districts to dispose of all power received from the Hetch Hetchy system in accordance with the Raker Act, Ickes demanded a firmer guarantee that the districts would not use Hetch Hetchy to meet their own needs and sell more of their Don Pedro generation to PG&E. Ickes noted that the Interior Department staff had suggested a clause prohibiting wholesale power sales by the districts to PG&E in excess of the levels reached in 1944, and he preferred that kind of guarantee to the "vague promise" found in the completed contract.[43] Meikle drafted a response to Ickes' criticism, which was sent to San Francisco over the signature of TID secretary James McCoy. It read:

> ... the Turlock Irrigation District will not agree to limit the disposal of the output of Don Pedro Power Plant as it is now constructed or in the future enlarged. The District has distributed power in this area and has sold surplus power to the Pacific Gas and Electric Company for the last twenty years and the contract for the sale of this surplus power extends to 1954.
>
> There would be no reason or purpose at this time for the District to submit to the arbitrary dictum of the Secretary of the Interior limiting operations of the Don Pedro Power Plant to the output of the subnormal year of 1944 or any other year for that matter.[44]

The districts were, however, willing to make a stronger statement of their honorable intentions and amended their contract to prohibit the sale of Hetch Hetchy energy directly or indirectly to any power company. Ickes was still not satisfied until a further amendment specifically prohibited the substitution of Hetch Hetchy energy for energy generated at Don Pedro and sold to PG&E.[45]

Negotiation of the power contract was a milestone in relations between the districts and San Francisco. It helped cement and expand the cooperative bond between them that was leading to the coordinated operation and development of the Tuolumne. For the districts the bargain provided a large and reasonably priced supply of energy to supplement the Don Pedro powerplant, which was no longer able to meet all the demands placed upon it. Because of the energy resources it made available, the contract was, for the Turlock district at least, the most important single event in the district's electrical history since the signing of the wholesale contract in 1924.

The Second Agreement of 1943 and the Corps of Engineers' agreement the following year to contribute to the construction of New Don Pedro and Cherry in lieu of building a separate flood control reservoir established the basic guidelines for development but left the details to be worked out later. Two agreements were needed. One involving the Corps, the city and the districts would define the amount of the federal payment for flood control and what was expected from the agencies in return for the money. The other agreement would have to be between the city on the one hand and the districts on the other to allocate the responsibilities and benefits from the anticipated flood control contract and the dams and powerplants that would be part of the cooperative development. It was conceded that the agreement between the city and the districts – the Third Agreement – would have to be completed before the federal flood control contract to insure that the promises made in the flood control agreement would be kept. Although statements were made during negotiations in 1944 that construction of Cherry dam would begin soon after the war, negotiations on a Third Agreement were not even begun until September 1947, when Max Bartell sent a short draft agreement to the districts for their consideration. Although substantial changes were made in the details during almost two years of subsequent negotiations, the basic concepts were present from the start.[46]

At the heart of the cooperative plan was the idea that the city would contribute to the construction of New Don Pedro Dam, which the districts would own and operate, just as they did the original Don Pedro. The Raker Act required the city to guarantee that the districts' rights to the natural flow of the river be protected. The city was entitled only to water available over and above the districts' rights and that water came only in times of high flows in the spring or during floods at other seasons. For that reason, the city's rights were sometimes referred to as flood flow rights. To take advantage of its rights to the Tuolumne's high water, the city had to build reservoirs to store the floods and then release the water later into its aqueduct. The Tuolumne River averaged a total runoff of over 1,800,000 acre-feet annually. The districts estimated their ultimate requirements at about 1,090,000 acre-feet, and the city needed about 453,000 acre-feet a year to meet its eventual 400 million gallon per day demand.[47] Their total claims amounted to about 85 percent of the river's annual average, but the average was, of course, deceiving and many years the river yielded less water than the two systems would need. Carry-over storage was needed to level out variations from year to year and it was especially necessary for the city, since its rights were junior to those of the districts. It was estimated that the city would have to have 1,400,000 acre-feet of storage above Jawbone Creek to insure that it could deliver 400 million gallons per day and still meet its obligations to the districts under the Raker Act. Since Hetch Hetchy and Lake Eleanor held only about 366,000 acre-feet, and even Cherry would raise the total to just over 600,000 acre-feet, it was clear that the city faced an expensive program of building more storage in the upper watershed.[48] The alternative was to build storage on the lower river, in the districts' section of the stream below Jawbone Creek. That was what the city proposed to do with New Don Pedro.

If built to a capacity of 1,400,000 acre-feet as suggested by Bartell in 1947, New Don Pedro reservoir would include the 290,000 acre-feet the districts held in their old reservoir plus 410,000 acre-feet of flood control storage and 700,000 acre-feet of so-called exchange storage space for San Francisco. Flood water in excess of the Raker Act's natural flow requirements would be collected in the reservoir for the districts so that the city could later divert an equal amount of Raker Act water at its upstream facilities. The plan allowed the city to build the storage it needed at much less cost than a number of high country reservoirs, assuming such sites were even available. Since it was the major beneficiary of New Don Pedro storage, it was expected that the city would pay for most, if not all, of the new dam.

The other major conceptual element in the Third Agreement involved flood control and Cherry reservoir. Although the city's reservoirs and Don Pedro would all be operated for flood control until New Don Pedro took over the task, the heaviest burden of interim flood control would fall on Don Pedro because it was the furtherest downstream and would control runoff from the large area below the city's dams. The agreement therefore had to pro-

San Francisco's Cherry Valley Dam was one part
of the long-range plan to develop the Tuolumne River.

vide for storage in the city's reservoirs to protect the districts from the loss of water and power that might result from keeping a portion of Don Pedro empty to hold and regulate floods. Although the districts would not be contributing directly to the cost of the city's Cherry project, their participation in the cooperative flood control program was expected to help make federal payments to Cherry possible, so they expected to receive some definite benefits from Cherry at least until New Don Pedro was finished. All this meant that for the first time the city and the districts had to agree to some specific contractual arrangements rather than the broad declarations of policy found in the First and Second agreements.

Negotiations on the Third Agreement took place in late 1947 and early 1948 and again in the spring of 1949. There were differences of opinion and some hard bargaining along the way, but the agreement signed on June 30, 1949, described the terms under which San Francisco and the irrigation districts would share in the development of the river. By 1949 the federal flood control contribution had been increased to an estimated $12 million, three-quarters of which would be paid toward the construction of Cherry dam because of the flood control it could provide, along with Don Pedro, until New Don Pedro was built. Once Cherry was in operation, the city would be allowed to intercept and store Raker Act water and then release it to the districts when they needed it. In fact, until Cherry powerhouse was completed the districts were allowed the free use of water stored in Cherry, except for storage needed to allow the city to maximize production at its Moccasin and Early Intake powerplants. After Cherry powerhouse was running, the districts were guaranteed a release of 25,000 acre-feet per month from April through October, and they received the right to purchase 35 percent of the power produced at the plant at cost. New Don Pedro was to be built on five years notice by either party, or when the city's diversions from the watershed reached 200 million gallons per day. The minimum size of the reservoir had been reduced to 1,200,000 acre-feet, which was broken down as 290,000 acre-feet to replace old Don Pedro, 340,000 acre-feet for flood control, and 570,000 acre-feet of exchange space for the city. The flood control reservation had to be kept empty during most of the year to hold incoming flood waters, but as the spring runoff ended it could be filled for storage until it had to be emptied in anticipation of winter storms. Any water stored in the flood control space was to be divided evenly between the districts and the city.

Don Pedro's irrigation outlets and spillway were opened to handle the 1950-51 floods. Operation of the dam for flood control reduced damage downstream.

San Francisco agreed to pay for the construction of the new Don Pedro Dam to a size capable of storing 1,200,000 acre-feet, with the districts supplying all lands and rights-of-way for the project. The districts were to be solely responsible for the powerplant at the new dam. The site for New Don Pedro, a short distance downstream from the old dam, could accomodate a larger dam and reservoir than the one the city promised to build, so the agreement gave the districts the right to increase the height of the dam and control the additional storage created, provided that they paid the difference between the cost of the high dam and the one the city would pay for. Finally, Article 19 stipulated that if either the city or the districts planned to build any powerplants on the river or its tributaries between Cherry powerhouse and New Don Pedro, the other party had the right to a 50 percent share if it desired.[49]

Once the Third Agreement was ratified, the flood control contract with the Corps of Engineers followed in less than two months.[50] The value of the arrangement was proven in late 1950, when what may have been the largest flood on record surged down the river. Controlled releases were supposed to be no more than 9,000 second-feet, but on November 21, 1950, Don Pedro was spilling and up to 38,000 second-feet of water was going over La Grange Dam. When the storms let up, the 9,000 second foot releases were resumed to lower the reservoir, but a fresh onslaught filled Don Pedro again by December 7. The flood peak passed at noon the following day, when 64,500 second-feet filled the river, and high flows continued until December 14. In all some 870,000 acre-feet had flowed down the Tuolumne in less than a month. Had there been no dams on the river and no effort to regulate the flood, the highest mean 24-hour flow would have been about 90,000 second-feet, and the peak would have been higher yet, but the actual mean daily figure did not exceed 49,000 second-feet.[51] Another major flood occured in 1955. Without more storage on the river, damage from these floods could not be eliminated, but the severity of the problem could at least be held in check.

The 1940s ended with an era of good feelings on the Tuolumne. Of course, compromise in sharing the river did not begin in the 1930s and 1940s; the Raker Act itself was a compromise in which Congress attempted to divide the water according to what were thought to be the districts' rights and to divide the watershed by implication at Jawbone Creek. Even while planning to put their rights to the test of litigation, the districts could not help but acknowledge that having San Francisco on the river had so far been a benefit. Regulation of the river by Hetch Hetchy and Moccasin helped make possible the successful operation of the Don Pedro powerhouse and the districts' electrical distribution systems. Meikle estimated that by 1933 the Turlock District had earned $700,000 in additional power revenue due to San Francisco's presence on the river.[52] For this reason, the Turlock District and its chief engineer were never really in favor of litigation except as a last resort to defend the district's rights. They were ready to cooperate with the city according to the dictum that more reservoirs rather than lawsuits were the only thing that could produce the water each side wanted. The city and the districts were also drawn together in defense of the river against others who wanted its water, from the El Solyo Ranch to the federal government. Cooperation was the only way to keep what they had been prepared to fight over among themselves. An entirely different chain of events made the districts customers of Hetch Hetchy power and gave both sides another reason to join forces in the development of the river. Cooperation had replaced conflict among the river's users. The next step was to implement the plans they had made.

CHAPTER 12 – FOOTNOTES

[1] B. A. Etcheverry and Thomas H. Means, "Report on Use of Water from Tuolumne River by Turlock Irrigation District," (1915); Etcheverry and Means, "Report on Use of Water from Tuolumne River by Modesto Irrigation District and Turlock Irrigation District," (1920).

[2] Board minutes, vol. 12, p. 410 (Dec. 5, 1932); *Modesto News-Herald,* Jan. 17, 1933.

[3] Board minutes, vol. 13, p. 48 (Oct. 9, 1933).

[4] *Modesto Bee,* Oct. 17, 1933.

[5] Memorandum on storage by San Francisco on Tuolumne watershed, Oct. 23, 1933, in Meikle files, vol. 6, item 9; *Modesto Bee,* Oct. 24, 1933.

[6] *Modesto Bee,* July 17, 1934.

[7] *Ibid.,* Nov. 27, 1934.

[8] M. J. Bartell, "Proposed Plan for Dividing the Flow of the Tuolumne River," June 10, 1935, in Meikle files, vol. 2, item 9; "A Preliminary Report dealing with A Settlement of Litigation with City and County of San Francisco," Sept. 30, 1935, in Meikle files, vol. 2, item 10; report by Paul Bailey, Nov. 20, 1935, in Meikle files, vol. 2, item 11.

[9] *Modesto Bee,* July 15, 1935; *Meridian Ltd. v. City and County of San Francisco,* Judgement and Decree by Judge Raglan Tuttle (Superior Court), June 15, 1936, pp. 6-7.

[10] Board minutes, vol. 17, p. 212 (Mar. 14, 1938).

[11] *Turlock Tribune,* May 5, 1939; Board minutes, vol. 18, pp. 309-310 (May 5, 1939).
[12] *Modesto Bee,* May 6, 1939; *Turlock Tribune,* May 12, 1939.
[13] Board minutes, vol. 18, p. 374 (June 28, 1939).
[14] Comments on proposed division of water, Meikle files, vol. 15, item 8.
[15] Board minutes, vol. 19, p. 133-134 (Feb. 26, 1940), p. 144 (Feb. 29, 1940).
[16] Quoted in San Francisco statement on Cherry Valley Project, in Meikle files, vol. 14, item 1.
[17] *Idem.*
[18] *Modesto Bee,* Mar. 6, 1940.
[19] Board minutes, vol. 18, p. 216 (Mar. 13, 1939), vol. 24, pp. 100-101 (Jan. 25, 1944).
[20] Calif., Div. of Water Resources, *San Joaquin River Basin,* Bulletin No. 29 (1931), pp. 224-225.
[21] *Ibid.,* pp. 227-232.
[22] Memorandum to L. J. McAfee, by M. J. Bartell, Apr. 9, 1941, in folder BART 21, Calif. Water Resources Center Archives, Berkeley.
[23] *Idem.*
[24] Board minutes, vol. 23, p. 498-A (Nov. 15, 1943).
[25] See especially paragraph 11 in "Recommendations of the Chief of Engineers to Secretary of War on San Joaquin River and Tributaries Flood Control," 1944, in Meikle files, vol. 15, item 29.
[26] "Statement of R. V. Meikle Before the Senate Flood Control Committee Relating to Flood Control on the Tuolumne," May 26, 1944; Board minutes, vol. 24, pp. 100-101 (Jan. 25, 1944).
[27] O'Shaughnessy, *Hetch Hetchy,* p. 92.
[28] *Turlock Tribune,* Mar. 27, 1922.
[29] *Ibid.,* May 24, 1922; Nash, *Wilderness and the American Mind,* pp. 172-175.
[30] O'Shaughnessy, p. 99.
[31] Public Utilities Commission of San Francisco, "History of Efforts to Dispose of Electricity Generated on Hetch Hetchy Project," Dec. 1944, pp. 2-3, in Meikle files, vol. 15, item 37; O'Shaughnessy, p. 103.
[32] Quoted in O'Shaughnessy, p. 106.
[33] *Ibid.,* p. 103.
[34] "History of Efforts to Dispose of Electricity," pp. 3-4.
[35] Harold L. Ickes, *The Secret Diary of Harold L. Ickes,* vol. 1 (New York 1953), pp. 214-215.
[36] Ickes, vol. 1, p. 357, 563; Ickes, vol. 2 (New York, 1954), pp. 124-125; "History of Efforts to Dispose of Electricity," pp. 4-6.
[37] Supreme Court decision, Apr. 22, 1940, in Meikle files, vol. 13, item 33.
[38] Harold L. Ickes to R. V. Meikle (telegram), May 25, 1940.
[39] "History of Efforts to Dispose of Electricity," pp. 6-7
[40] *Ibid.,* p. 7
[41] Memorandum on San Francisco power, by R. V. Meikle, Oct. 23, 1944, in Meikle files, vol. 15, item 39.
[42] "History of Efforts to Dispose of Electricity," pp. 7-8; Summary of Hetch Hetchy power hearings, Washington, Jan. 1945, in Meikle files, vol. 15, item 38.
[43] Harold Ickes to Mayor Roger Lapham, June 11, 1945.
[44] J. F. McCoy to E. G. Cahill, June 14, 1945, in "Report of Turlock Irrigation District and Modesto Irrigation District to Committee on Interior and Insular Affairs on Engle Bill, H. R. 2388, 84th Cong.," in Meikle files, vol. 21.
[45] *Turlock Daily Journal,* June 15, 30, July 10, 1945.

[46] Draft agreement by M. J. Bartell, Sept. 4, 1947, in Meikle files, vol. 17, item 44.
[47] Memorandum on storage by San Francisco on Tuolumne watershed, Oct. 23, 1933, in Meikle files, vol. 6, item 9.
[48] "Statement of R. V. Meikle Before the Senate Flood Control Committee," May 26, 1944. [49] Third Agreement, June 30, 1949.
[50] Board minutes, vol. 29, pp. 288-288-A (Sept. 6, 1949).
[51] "The November-December 1950 Flood: Tuolumne River," in Meikle files, vol. 18, item 44.
[52] Memorandum on Tuolumne River development, by R. V. Meikle, Apr. 6, 1933, in Meikle files, vol. 5, item 38.

New Don Pedro Dam in June 1969, as seen from upstream.
The dark strip in the center of the dam is the impervious clay core.
Source: Bechtel Corporation.

13

New Don Pedro

The cooperative development program contained in the Third Agreement got off to a slow start. Two-thirds of the money for Cherry dam was to come from federal flood control payments, but those depended on congressional appropriations. In late 1950 the districts told a congressional committee that inadequate appropriations had already set the timetable back three years.[1] Turlock and Modesto directors frequently appeared before commitees dealing with flood control funding to urge Congress to vote enough money to finish Cherry as rapidly as possible. The districts' interests were plain enough; without the additional storage in Cherry, they ran a greater risk of water and power shortages at Don Pedro, especially when they had to keep storage space open for flood control. Cherry dam was finally completed in 1956.

In the mid-1950s the city and the districts had to fight another threat to the viability of their plans, this time from Congressman Clair Engle and Tuolumne County Water District No. 2. The water district, which covered four of the five county supervisorial districts, was formed in 1947 to help develop water supplies for irrigation and other uses. Funding the dams and canals it wanted to build was a major problem, but the district came up with a daring and imaginative solution. Between Hetch Hetchy and Early Intake, releases from O'Shaughnessy Dam flowed down the Tuolumne River. John R. Freeman's plans for the Hetch Hetchy system called for the eventual construction of a tunnel to carry water from Hetch Hetchy to a point high above Early Intake, where it would then fall through penstocks over a thousand feet to a powerhouse adjacent to the Early Intake diversion works. Development of this Early Intake powerplant by the city had not yet begun when the Tuolumne County Water District No. 2 gave notice that it wanted the project for itself. Congressman Engle, representing Tuolumne County, introduced legislation in 1954 to give the district the right to use that power drop. Engle said,

> Tuolumne County has lost most of its water to outsiders. The

> remaining water is limited, difficult and expensive to develop. To keep the cost of new irrigation and domestic water within the limits the users can pay, it is necessary to have help from outside sources. The new revenues from the operation of a powerhouse on the Tuolumne can under our bill provide that vital assistance . . . We do not expect to take any of San Francisco's water or take anything from San Francisco which that city has put to use. In addition we expect to share the profits with San Francisco.[2]

Although the water district promised to treat San Francisco fairly, the city could see nothing fair about taking away a site it planned someday to develop.

The Turlock and Modesto districts were more than just interested bystanders in the dispute over the Early Intake site. The districts expected to need power from the Cherry and Early Intake plants to meet their future loads. More significantly, San Francisco planned to use revenues from its power sales, including power from Early Intake, to finance its share of New Don Pedro. If San Francisco lost the right to develop the Early Intake site, New Don Pedro could be delayed or even lost, and the districts' power supply reduced. The irrigation districts passed resolutions opposing Engle's bill and sent some of their directors and engineers to Washington for committee hearings on the measure.[3]

At the March 1955 hearings, Engle charged that San Francisco had forfeited its exclusive right to the Early Intake site by lack of diligence in developing it, and in addition, he asserted that the city was in violation of the Raker Act because Hetch Hetchy power had been illegally sold to PG&E through the irrigation districts, especially Turlock.[4] The accusations were vigorously denied, but Engle insisted on having more information from the districts. In supplying it, R. V. Meikle commented:

> By agreeing to buy Hetch Hetchy power, the Districts relieved a very difficult situation. Turlock District's position with respect to its obligation to deliver power to the Pacific Gas and Electric Company until 1954 was understood at the time by all parties including the Secretary of the Interior.
>
> Turlock District has never before been accused of violating the Raker Act. Now it has happened, a year after the completion of our 30 year contract with Pacific Gas and Electric Company.
>
> If we had been in violation of the Raker Act by any material departure from the regulations during the last 10 years, we would have expected such action to be taken by the Secretary of the Interior as would be necessary to enforce the regulations.[5]

Despite evidence that the TID had not resold any Hetch Hetchy energy to PG&E, Engle asked the comptroller general to investigate the city's Raker Act compliance. In a report the following year, it was found that the city had committed only a technical violation by selling excess, or dump, power to PG&E, but despite that fact, its power sales were still in reasonable compliance with the Raker Act.[6] Engle's bill did not pass, and in November 1955 San Francisco voters approved $54 million in bonds to pay for the Cherry and Early Intake powerplants, effectively ending the threat to the city's plans.[7] However, Tuolumne County's interest in sharing the waters that originated there did not end. The city and the districts increasingly had to take notice of the county's aspirations, and accomodate them in their plans.

The first step toward the construction of New Don Pedro Dam, the last and largest part of the cooperative plan, was taken in January 1951 when the districts applied for water rights for over 1,500,000 acre-feet of water to be stored behind a 485-foot-high dam.[8] Permits for that amount were issued by the Division of Water Rights in 1953, and in August of that year, Fairchild Aerial Surveys was hired to make topographic maps of the reservoir area. Using these maps, the estimated capacity of the reservoir was increased to just under 2,000,000 acre-feet with a 515-foot dam in 1954. Roger Rhodes, a consulting geologist hired in 1955, reported that dam site number four was the most promising location, and tunnels were run into the hills there to determine the subsurface geology.[9] In 1957 Rhodes reported that the site, about a mile and a half downstream from old Don Pedro, was suitable for a high dam. The decision on the type of dam to be built was placed in the hands of the Bechtel Corporation in 1958. Since San Francisco was committed under the Third Agreement to pay for a dam capable of impounding 1,200,000 acre-feet, estimates were made for a dam of that size as well as for one as high as the site would allow. In its 1959 report Bechtel estimated that the minimum reservoir would require a 455-foot concrete arch dam costing about $62 million. The high dam would be 90 feet higher and would hold 2,030,000 acre feet. It would be a rock-fill dam, expected to cost about $78 million.[10] Although the formal choice of a high or low dam was not made until 1961, the districts never expressed much interest in the smaller project. The added costs of the high dam that the districts would have to pay were low in comparison to the advantages of controlling an additional 830,000 acre-feet of storage space and increasing the head on the new powerplant.

New Don Pedro proved once again the truth of the adage that for almost any project the engineering problems are minor compared to the institutional, legal or financial ones. This project was by far the largest and most expensive the Turlock Irrigation District had ever been involved in, but there was no controversy over its size or design or even the need to build it. It did, however, produce a number of unexpected controversies that delayed and complicated its construction and operation.

One of the underlying problems was that the project had four partners; the Turlock and Modesto districts, the City and County of San Francisco and the U.S. Army Corps of Engineers, all of whom had to reach agreement on how to share the various costs. The Corps had to determine how much it would pay for the final stage of the Tuolumne River flood control plan, and under the 1949 contract, it could ask for more than the 340,000 acre-foot flood control reservation it was guaranteed if it payed a proportionately larger share of construction costs. By 1960 the Corps had made it clear that it wanted to increase its storage space to 700,000 acre-feet, over twice as much as originally planned. The districts had the right to approve or deny any increase in flood control, and they soon rejected the Corps request because any increase in federal payments would be more than offset by the loss of power revenues that would result from giving the government the right to control more of the reservoir.[11] The Corps of Engineers finally contributed about $5.5 million to New Don Pedro construction, in addition to the $9 million that had already been spent on Cherry reservoir.[12]

Before the city and the districts could complete their Fourth Agreement covering the allocation of costs and how the project would be operated, several preliminary steps had to be taken. One of these was arranging the necessary financing. Studies made in 1961 by Stone and Youngberg and Blythe & Co. for the districts showed that they could best pay for their share of the project with general obligation bonds. The financial analysts predicted that the TID would not have to raise either its taxes or its electrical rates to pay for New Don Pedro.[13] On November 7, 1961, voters in the Turlock District approved $34 million in bonds for the project by an overwhelming majority of 5,745 to 126. The result was the same in the Modesto District, and it was reported that the majorities were unprecedented in Stanislaus County history. San Francisco voters approved the bonds for their city's share of the project the same day.[14] With money assured, only the approval of the Federal Power Commission, in the form of a fifty year license, was still needed. The license application had been filed in May 1961, and no problems were anticipated. Even before the bonds went before the voters in November, however, it had become clear that getting the license was going to be a lot more complicated than anyone would have imagined a few months earlier.

The problem was a fish – the chinook (or king) salmon. They are an anadromous species; that is, they reproduce in fresh water but spend most of their lives in the ocean. After two to four years in salt water, they return to their native streams to spawn, and then die. Their annual spawning run occurs in the fall in the Central Valley rivers. Before Wheaton dam blocked the Tuolumne, salmon spawned above La Grange, perhaps as far upstream as Wards Ferry. In the right conditions of water temperature, depth and velocity, the salmon scooped out the gravel of the riverbed to make their nests, or redds, and deposited their eggs. The eggs hatched in late winter or early spring and the young salmon went down to the sea with the spring freshets. The effect of Wheaton's dam was described in 1877.

> Immense quantities of salmon have been prevented from reaching their breeding grounds further up the stream in consequence, and much indignation is expressed regarding the obstruction. The ranchers and others have been taking wagon loads of salmon from the river below the dam during several months past, killing the fish with clubs as they passed over the riffs. The Fishery Commissioners should compel the construction of a fish ladder to the dam, as the law requires.[15]

Although the salmon did spawn in the stretch of the river below La Grange, M. A. Wheaton was twice brought before the courts for failing to provide a fish ladder. The last time, in 1889, his attorneys included C. C. Wright and P. J. Hazen, and the jury delivered a rapid verdict of not guilty.[16] Illegal salmon "fences" in the San Joaquin River erected by poachers impeded the annual migration in some years, but around the turn of the century, more determined enforcement of the fish and game laws reduced the practice and there were reports of thousands of salmon at La Grange Dam. Salmon were commonly caught with spears, and some people were said to be gathering great numbers of fish to be salted down.[17]

Until 1940 there seem to have been no estimates of how many salmon spawned in the gravel riffles above Waterford, and the number varied considerably from year to year. Salmon numbers between 1940 and 1960 ranged from a high of 130,000 fish in 1944 to a low of 3,000 in 1951, although after 1944 there were only four years of 45,000 fish or more, and none above 61,000.[18] The salmon run on the San Joaquin River itself was eliminated by the construction of Friant Dam in the mid-1940s, and runs in San Joaquin tributaries like the Tuolumne may have suffered as well.[19] As early as 1946, the California Department of Fish and Game (DFG), in commenting on a federal water development report, had recognized that more dams and additional diversions from valley rivers could endanger the salmon population. To save the salmon the department recommended that controlled minimum flows be required to provide enough water for migration and spawning. On the Tuolumne River the 1946 report recommended flows below La Grange Dam ranging from at least 750 second-feet during the spawning season, down to 100 second-feet in the late spring and summer.[20] Despite the loss of the San Joaquin River run and a forthright recognition that the salmon fishery needed some kind of protection, the DFG inexplicably raised no objection to the districts' applications for New Don Pedro water rights, even though they were informed of them. As a result, the water rights permits issued in 1953 carried no conditions or restrictions relating to fish or the maintenance of flows in the river below La Grange Dam.[21]

There is no evidence that the districts were given any reason to consider the annual salmon run in designing their project until, without apparent warning, the DFG asked in June 1961 for annual releases totalling 219,000 acre-feet specifically for the benefit of salmon.[22] With the San Joaquin Valley fishery already in serious trouble, it must have seemed imperative to the DFG that any new dam be operated to protect and, if possible, enhance spawning conditions. There was a fear that without guaranteed releases, New Don Pedro could virtually dry up the river, especially when San Francisco increased its diversions. In fact, even without releases for fish, New Don Pedro would provide more water in most years than the smaller dam it would replace, but the long-term plan to increase the irrigation and domestic use of the Tuolumne boded ill for the fishery unless something was done.

The releases requested by the state posed a major problem for the districts. Fishery releases meant that water that could have been saved for irrigation or to maximize power revenues would have to be run into the river instead. If the dam had been a federal project, all or part of the costs chargeable to fish and wildlife enhancement or recreation could have been deemed non-reimbursable, which meant they would have been paid by the federal treasury rather than by the project's users. But New Don Pedro was not a federal dam. Its entire cost, apart from the small flood control allocation, had to be recouped by the water and power it produced for its owners. Anything that affected the output of water and power would certainly affect the economic justification for the project. The districts were naturally unwilling to voluntarily sacrifice valuable water, especially since they felt they had the paramount right to its use. The DFG, therefore, had to find some way to compel them to make the releases it sought. The opportunity to include such requirements in the project's water rights had long since passed, but the districts still needed a license from the Federal Power Commission, and that fact offered the last chance for the DFG to get what it wanted. In July 1961 the state petitioned for the right to intervene in the licensing proceedings and formally protested the granting of a license for New Don Pedro without first including in it a measure of protection for the fishery. Despite arguments to the contrary by the districts, the FPC granted the state the right to participate and ordered full hearings into the license.[23]

During the summer of 1961 the DFG refined and clarified the conditions it wanted incorporated in the New Don Pedro license. At a meeting at the Modesto District office on September 27, the DFG and the U.S. Fish and Wildlife Service presented a list of twelve demands or conditions for the project. Some, like maintaining a minimum pool in the reservoir or leaving the gates of the old dam open to allow the passage of fish through it were no problem. Others such as purchasing additional lands for fish and wildlife enhancement or building unspecified fish traps or other facilities, were unwelcome, but probably still negotiable. The real problem was with how much water would be released below La Grange Dam to benefit the salmon and how it would be done. The state's proposed minimum flow below La Grange was set at 550 second-feet from November through March of normal water years, and at lesser amounts during the remainder of the year. Releases were to be larger in wet years and smaller in dry ones. Meeting the DFG demands would require 199,926 acre-feet in a normal year , 243,909 acre-feet in a wet year, and 116,106 acre-feet in a dry one.[24] The districts objected that any use of their water for fish, except for occasional surpluses not needed for irrigation, would infringe on their basic water rights. In some years at least, it was likely that the proposed releases would mean that water needed for crops would instead be used for fish, defeating the purpose of the new dam and leaving the districts worse off than they were with the old one. Other conditions relating to the releases were scarcely less onerous. The powerplant would normally be expected to operate on a schedule that reflected electrical demands that peaked during the day and went down at night, which meant that the river, too, would rise during the day when more water was passing through the turbines and fall at night. The DFG objected to such fluctuations in water levels, and suggested allowing only the most limited changes in the river's elevation during the fall spawning season. Furthermore, since salmon preferred the kind of cold water found deep in the reservoir, one of the twelve demands would have required fishery releases to come through the dam's low level flood control outlets rather than through the powerhouse, further reducing energy production. Meikle later wrote that the demands as submitted would have killed the project.[25] MID chief engineer Plummer agreed that they could not be granted, but said he was "pretty sure" they were just for bargaining purposes.[26]

The Department of Fish and Game could, of course, accomplish nothing by killing New Don Pedro since without it there was little hope of saving the Tuolumne River salmon run, especially when San Francisco increased its diversions.[27] Negotiations on a compromise were therefore begun in January 1962. Meikle privately doubted they would have much value if the districts ultimately decided to defend their rights as a matter of principle, but he also admitted that until San Francisco took its full share of the river, there would be enough water to allow some releases for salmon spawning. In early 1962 a series of test releases were made through the La Grange powerhouse to enable biologists to observe the spawning gravels at various flow levels.[28] Discussions continued into the spring of that year, at times involving the governor, various consultants and local legislators. By May both sides were in general agreement on releases of about 100,000 acre-feet in normal years and half of that in dry years. There would be no releases at all in critically dry years, but the state insisted that a critically dry year could, by definition, only follow one classified as dry, although the districts argued that nature was not always so consistent. The last important stumbling block appeared to be the operation of the powerplant and the allowable limits of fluctuation in water levels during spawning. The DFG wanted no more than a two-inch variation in the level of the river, but was willing to settle for eleven inches, with up to an eighteen-inch variation for an hour and a half a day, which was still highly restrictive in terms of how the powerplant could be operated.[29] Despite the substantial progress that had been made, by May 21 the likelihood of agreement seemed so remote that the TID board seriously considered the possibility of abandoning the project.[30] Just over two weeks later, a tentative agreement covering the first twenty years of the project was announced, including a normal year release of 105,000 acre-feet to benefit salmon.

The agreement should have ended the almost year-long impasse over New Don Pedro and the Tuolumne River fishery, but it was soon apparent that it did nothing of the kind. Within two weeks of the June 5 announcement that agreement had been reached, pressure was reportedly building against it from sport and commercial fishing organizations.[31] Just how these groups, which had apparently played no part in the negotiations nor made any previous public comment on New Don Pedro and salmon, became so quickly informed and ready to take concerted action is uncertain, but by the time the State Fish and Game Commission met in San Francisco on June 29, 1962, to consider the agreement, the fishermen were prepared to attack the compromise. The Associated Sportsmen of California, the California Wildlife Federation, The Aquatic Resources Committee, Tyee Club, Marin Rod and Gun Club and the California Seafood Institute were among the groups condemning the agreement as insufficient to protect the salmon, especially since it covered only twenty years rather than the full fifty year license period. Cecil Phipps of the California Wildlife Federation even demanded that the commission reprimand DFG director Walter T. Shannon for having the "effrontry" to present such an inadequate agreement. TID board president Abner Crowell and Milo Bell, a fishery consultant hired by the districts, defended the agreement, and the DFG's Shannon argued that it was the best deal available, pointing out that, "We got a lot more under this agreement than we started with."[32] The Fish and Game Commission, however, was not convinced that it was the best that could be done. It ordered Shannon to renegotiate with the districts and report back in two months.

Until the compromise came under fire, the fishery controversy had not been treated publicly as a major crisis by the two districts, but officials now had to concede that the project so overwhelmingly approved by the voters was now in jeopardy. Meikle and others could see no point in further discussions with the DFG because they felt they had nothing more to give.[33] The commission had refused to ratify the agreement, but in ordering more negotiations, it had not formally rejected it either. Since they did not plan on making any changes in the agreement, the districts simply asked the commission to either approve or reject it at their July 20 meeting.[34] The Department of Fish and Game tried to do what it had been told to do by the commission and proposed a new agreement calling for 135,000 acre-feet of releases and other conditions not dramatically different from their original demands. District representatives attended one meeting, but made it clear they had no wish to start all over again.[35] When the Fish and Game Commission met in July, only one member had been persuaded to change his mind, and the agreement died on a three-to-two vote.[36]

The Federal Power Commission had originally scheduled hearing for the spring of 1962, but they were delayed until October at the request of the districts and the DFG when it appeared a negotiated settlement might be reached. When the agreement fell through, it became obvious the major public showdown over New Don Pedro would come at the hearings. They began on October 16 in the basement of the Federal Building in San Francisco, and lasted through two weeks of legal arguments and conflicting technical testimony. There were a great many issues raised, including questions about the role San Francisco would have in the project or ought to have in the licensing process. However, FPC examiner Francis Hall, who conduced the hearings, made it clear at the outset that in his mind the main issue was how much water the districts could afford to release for fish and still have a viable project.[37] By the time of the hearing, the state was asking for 123,210 acre-feet annually to maintain the fishery at historic levels, and they had biologists on hand to show that it took that much water to cover enough gravel deep enough to give the fish room to spawn. In addition the state asked the FPC to impose restrictive standards on fluctuations in water levels and on the temperature of releases.

The districts countered those assertions with their own corps of experts, who argued that less gravel, and consequently smaller flows, were sufficient. For example, Milo Bell stated that 27,000 acre-feet annually would be enough to protect the salmon population.[38] One of the reasons why the districts and their advisors thought the DFG had overestimated the release requirement was a failure to appreciate the nature of the river. The Tuolumne was what was called a "gaining stream," meaning that seepage from irrigated lands added to the river's flow. Testimony showed that in March of the critically dry year of 1961, the mean flow at La Grange bridge, about two miles below the dam, was only nine second-feet, but near Hickman, nineteen miles away, the mean flow was 109 second-feet, and at Tuolumne City, west of Modesto, it was 241 second-feet.[39] Since salmon spawned at various gravel beds between Waterford and La Grange, releases from La Grange Dam were not the sole determinant of streamflow. The districts also tried to prove that they could not guarantee as much water as the state wanted without undermining the economic value of the project. Bechtel studies based on the water years 1923-1961 showed that the DFG's proposed release schedule would seriously reduce the irrigation supply in dry years and would reduce power revenues by $23 million over the fifty-year life of the license.[40]

An interesting feature of the hearing was the eleventh hour attempt by Secretary of the Interior Stewart Udall to intervene in the proceedings just before the hearings began. The action was so tardy that in fairness the FPC could allow only limited participation by the Department of the Interior. In general, Interior supported the position taken by the state on behalf of the Department of Fish and Game, but they also made further, and more ominous, demands. One of these was the suggestion that the FPC should compel the districts to sell or donate land to the federal government for the construction of fish facilities, or force them "to improve streamflow by providing Central Valley Project water for the Licensees' service areas, such water exchange to be on a reimbursable basis. Water thus made available from the facilities under this license should be released . . . downstream from La Grange dam."[41] In other words, if the Bureau of Reclamation built the Central Valley Project East Side Canal it was then planning, the districts could be forced to buy water from it so they could, without compensation, release their own water for the fish. The Interior Department also drew the commission's attention to the responsibilities of the Central Valley Project and the State Water Project to protect water quality in the Delta, and asked that FPC retain the authority to increase releases from New Don Pedro, if necessary, to help them fulfill those obligations.[42] These proposals would have tied New

Don Pedro to the Central Valley Project entirely at the districts' expense. Since the Interior Department had the right to only limited participation in the licensing process, they were not able to argue forcibly for these provisions and they were eventually forgotten.

Ironically, while the often technical dispute over the flows needed to save the Tuolumne River salmon run was at its height in 1961 and 1962, salmon numbers were dropping precipitously. From 45,000 fish in 1960, the run fell to 500 in 1961, to 200 in 1962, and to only 100 in 1963. A severe drought from 1959 to 1961 undoubtedly made it harder for the fish to return to the rivers, and that reduced the potential size of the runs three or four years later, but other factors were at work as well. The Bureau of Reclamation's Tracy pumping plant, which fed the Delta-Mendota Canal, was responsible for reverse flows in some southern Delta channels, and for reduced flows in the San Joaquin River near Stockton, where cannery wastes decaying in the slow-moving stream used up the dissolved oxygen in the water. Migrating salmon were confused by the flow reversals and were unable to penetrate the oxygen-depleted stretch. Corrective action was taken beginning in 1963, but the crisis demonstrated that the fate of the Tuolumne River salmon population depended on more than conditions in the Tuolumne itself.[43]

In the months following the FPC hearings, briefs were filed by all parties and the commission's staff developed its own set of recommendations. The central feature of the staff proposal was the acceptance of a normal year release of 123,210 acre-feet as recommended by the DFG, but only for the first twenty years of project operation, followed by a review and modification if necessary. The staff also suggested allowing greater fluctuations in water levels than the DFG desired. Increases of up to ten inches in river height would be permitted by the staff proposal during the critical spawning period, and as much as eighteen inches for two hours a day to permit some peaking with the powerplant.[44] Although the recommendations offered a compromise, neither side was favorably disposed toward them. The state and federal governments considered them inadequate, especially because they did not cover the whole length of the license. On the other hand, the districts still had two important concerns. One was that any requirement to release water for fish could infringe on established water rights for irrigation and power, and, like any water agency, the districts were extremely protective of their fundamental rights. There was also a fear that reopening the whole issue after twenty years would make it harder to sell thirty-five year bonds to finance the project, since revenues needed to repay them could be affected by changes in release schedules.[45]

The Federal Power Commission did not give its "Opinion and Order Issuing License" for New Don Pedro until March 10, 1964. The commission adhered to its staff's recommendations and outlined the reasons for doing so in its opinion. In terms of water releases for salmon, it said:

> Our analysis of the record shows California's studies to be reasonable, although somewhat conservative. For example, California's estimates of the number of adult fish to be accomodated and the amount of gravel required for each female are higher than the comparable estimates of the applicant's experts. Also, its studies do not take into account accretions in the river due to the return flow of irrigation water. Under these circumstances, we find it appropriate to adopt the release schedules recommended by California, but with the modification in their use suggested by staff and discussed below. This modification, as will be seen, results in a greater incidence of California's dry-year schedule and a consequent over-all reduction in the releases to be required.[46]

The schedule of releases originally prepared by California and adopted by the Federal Power Commission are:

	Schedule A (Normal Year)		Schedule B (Dry Year)	
Period	*cfs*	*acre feet*	*cfs*	*acre feet*
Pre-Season Flushing Flow	2,500	4,960	——	——
October 1-15	200	5,950	50	1,490
October 16-31	250	7,930	200	6,350
November	385	22,900	200	11,900
December 1-15	385	11,450	200	5,950
December 16-31	280	8,880	135	4,280
January	280	17,210	135	8,300
February	280	15,550	135	7,500
March	350	21,520	200	12,300
April	100	5,950	85	5,060
May-September	3	910	3	910
Total Acre Feet		123,210		64,040[47]

The change instituted by the commission was to define years as normal or dry based on actual inflow into New Don Pedro rather than natural flow computed as though no dams existed on the river. In that way, San Francisco's diversions would be taken into account and the districts would not be required to release water that never reached them. The commission estimated that Schedule B might be used once every four years rather than once every ten years as the state had suggested.[48]

The license ordered fish releases for only the first twenty years, and again the commission discussed the reasoning behind that decision:

> In our opinion the 20-year limitations on the release of water for fish is justified by the circumstances in this case. It is a reasonable limitation which is both appropriate and necessary to serve the best interests and legitimate needs of all of the parties, and of the public as well. There are basically three reasons for this conclusion. First, it appears that some modification of the prescribed releases from New Don Pedro will be necessary at the end of 20 years if the economic feasibility of the project for power is to be preserved. Second, there are a number of factors bearing on the need for fish releases in the future which cannot now be properly evaluated and which should be considered before requiring the applicant to make releases from New Don Pedro beyond the 20-year period. Third, it appears that there will be sufficient water available in the Tuolumne River for 20 years to meet the essential requirements of all interests.[49]

By adopting these conditions, the commission endorsed its staff's attempt to forge a workable compromise between the needs of the districts and those of the Department of Fish and Game.

Meikle believed the commission's decision was basically a fair one, but he and other district officials were still worried over the impact the twenty year provision would have on financing, although the FPC's Opinion indicated that future release schedules would not be allowed to impair the project's economic feasibility.[50] The potential impact of the license conditions on water rights and financing were substantial enough for the districts to order their attorney, Robert McCarty of Washington, D.C., to request a rehearing by the FPC.[51] The Department of the Interior and the DFG still wanted definite conditions covering the full fifty year license, so they, too, asked for a rehearing. The commission denied all such requests. Even though they were not happy with all its terms, the districts decided to accept the license, but before they could do so, the State of California, representing its Department of Fish and Game, and the secretary of the interior began an action in the Ninth Circuit Court of Appeals to have the license revised. The districts then counter-sued to protect their rights.[52] The court's decision in May 1965 upheld the conditions imposed by the FPC. However, McCarty warned that while the result was what the districts had hoped for, the reasoning followed in the decision was far less favorable. By holding that the FPC could reallocate water held under state-granted water rights to other purposes, such as fish, the Court of Appeals had, in McCarty's view, contradicted part of the Federal Water Power Act and attempted to make new water rights law. He recommended that the districts request a rehearing to clarify the situtation.[53]

The districts accepted McCarty's advice, but their request for a rehearing was quickly denied. Their only recourse was to appeal to the U.S. Supreme Court, and they did so. McCarty convinced several state governments to join the districts in their petition, on the theory that the court's ruling, if followed elsewhere, could affect the state control of water rights. In December 1965, however, the Supreme Court refused to review the lower court's decision on the New Don Pedro license. Some consideration was given to asking the Supreme Court to reconsider its refusal to hear the case, but when the Modesto District unilaterally decided to drop the matter, the TID had no choice but to abandon the last attempt at solving the legal problems created by the Federal Power Commission and the Court of Appeals.[54]

The four year battle over fishery releases was finally at an end. Viewed in the larger perspective, the salmon controversy that enveloped the project was very much a product of its times; ten years earlier, New Don Pedro would probably have faced no major opposition, and ten years later, it would have met growing environmental activism and a host of new laws and regulations. As it was, New Don Pedro was not an environmental issue in the sense of that term that became common after 1970. No one quoted Muir or Thoreau on the philosophic value of wild places or wild creatures as a justification for saving the salmon, instead testimony dealt with the dollar value of the commercial catch or the number of recreational fishermen. What's more, relatively little attention was paid to the habitats that would be innundated or even the historic and prehistoric sites in the reservoir area. The New Don Pedro controversy, however, was one indication of developing trends in conservation policy. The recognition that fish and wildlife, water quality maintenance or recreation, should be major goals of water development projects dated back at least to the federal Fish and Wildlife Coordination Act of 1934, and over the years additional legislation increased the authority of what might be termed "environmental" agencies, like the U.S. Fish and Wildlife Service or the California Department of Fish and Game. These agencies became increasingly active as the detrimental side-effects of some water projects, like the demise of the San Joaquin River salmon fishery, became obvious.

Had New Don Pedro been a federal project, the environmental elements of its design and operation would have been negotiated largely within the federal bureaucracy, in consultation with counterpart state agencies. That was exactly what happened in the case of the New Melones project, authorized by Congress in 1962 for construction on the Stanislaus River. New Melones was similar in size and location to New Don Pedro, and involved somewhat similar fish and wildlife problems. There were controversies, but they were all solved within the framework of the administrative agencies involved.[55] The Turlock and Modesto irrigation districts were not, of course, covered by the same requirements that governed federal construction agencies, but through the licensing authority of the Federal Power Commission, they could be made to adhere to the same kind of environmental criteria. Lacking the process of interagency negotiations that applied to federal undertakings, the FPC became the arbiter between the environmental agencies and the districts, and its judgement was then tested and confirmed in the courts. Although the procedures followed were different, both New Melones and New Don Pedro demonstrated the increasing influence of environmental factors in water project design and operation.

The advocates for the Tuolumne salmon included not only the professional biologists and bureaucrats of the Department of Fish and Game and U.S. Fish and Wildlife Service, but also a variety of sport and commercial fishing organizations. The relationship between these groups and the DFG is not entirely clear. It seems reasonable to assume that they received information from the agency, or its employees, concerning fishery issues, and they were the DFG's natural constituency, but they were not reluctant to attack the department's policies when they did not go far enough to satisfy them. The groups played a pivotal role in New Don Pedro, for without their hue and cry, the negotiated compromise would probably have been adopted by the Fish and Game Commission and included in the FPC license. They were active not only in the New Don Pedro matter, but made their influence felt in other water development issues as well. One of the most notable examples of their activity came shortly after New Don Pedro, when alternative plans for a Delta water facility were being considered. The fishermen joined the DFG in supporting a Peripheral Canal around the Delta, and commenting on a 1964 hearing, the chairman of the California Water Commission said:

> An interesting aspect of the hearing was that, for the first time in our memory, large groups of fish and wildlife and recreation interests supported, almost without qualifications, a proposed water project. In fact, the entire San Francisco Bay fishing fleet declared a holiday so that the skippers and their families could be present at the hearing.[56]

The decision to go forward with the Peripheral Canal, which later came under fire from environmentalists, demonstrated not only the influence and activism of the sport and commercial fishing groups, but also the growing significance of "intangible" (non-economic or unquantifiable) environment benefits and detriments associated with water projects. Both facts were also amply demonstrated at New Don Pedro.

Although the districts may have thought the existence of their project was being threatened, there was never really any question of whether or not the dam should be built, only how it would be operated. In 1965, the year the courts finished with New Don Pedro, Congress passed a law directing the Secretary of the Interior to study anadromous fisheries and cooperate with the states to conserve them, and two years later, the Supreme Court used that law to, in effect, stop the proposed High Mountain Sheep dam on the Snake River.[57] The legal and administrative policies that had mandated a measure of environmental mitigation at New Melones and New Don Pedro had now stopped a project where adequate mitigation was impossible. The decade of the 1960s was a complex and critical time in the history of American conservation. The wilderness movement was growing toward the activism that characterized it in the environmental era of the 1970s. On a parallel, or perhaps converging, track, attitudes on conservation and resource development were changing in executive agencies, Congress and the courts. New Don Pedro was part of that process at work.

Even after the license survived all the legal assaults on it, there was no assurance the project would actually be built. Despite its massive size, New Don Pedro offered the TID little in the way of additional irrigation benefits. Nearly all the land in the district that could use water was already using it, and dry years were being handled reasonably well with drainage pumps and private wells to supplement the supply from the river. According to one estimate, the average amount of "new" water created by the project would be only 34,000 acre feet annually.[58] Though that figure understated the value of holding water from normal or wet years for use in drier ones, or the importance of additional storage to offset increased diversions by San Francisco, there was little question that the increased irrigation supply alone could not justify the expenditure of millions of dollars. On the other hand, the larger dam could produce nearly five times the peak power of the old, and two and a half times as much energy on an annual basis. For the districts, then, the project's feasibility was linked to how much electricity it could produce and the cost of that power compared to alternative sources. When the agreement with the DFG broke down in 1962, Meikle and the TID board began to worry that the project would not remain viable, especially if fish releases reduced its ability to make power and delays pushed the cost of construction higher. In August1962 the board asked its financial consultants, Stone and Youngberg, to ascertain the potential capital value of New Don Pedro to the Turlock District. The result was a figure of approximately $28 million.[59] As early as October 1962, even before the FPC hearings, Meikle was thinking of abandoning the project if it appeared that its financial feasibilty was destroyed.[60]

The Modesto Irrigation District was always more inclined to push for early action on New Don Pedro, presumably because they needed more power sooner than the TID. In November 1962 the MID was urging immediate negotiations with San Francisco on a Fourth Agreement to cover in detail the allocation of New Don Pedro costs. Meikle commented:

> In my opinion this would be a waste of time because we have nothing definite to talk about. When T. I. D. starts to talk 4th Agreement, we should know what the project is worth to us and so state.
>
> The City has been somewhat indifferent concerning New Don Pedro. The T. I. D. can get along with present project for 10 or even

> 50 years, while the City must have more storage soon. For Districts to start negotiation on 4th Agreement would accomplish nothing and indicate weakness.[61]

In response to further prodding from the MID's Plummer a few months later, the TID chief engineer noted that, "the City would have to find out by itself how badly it needs New Don Pedro Project and then provide the financing to acquire it."[62]

The legal challenges to the project produced long delays during which nothing useful could be accomplished. The possibility of licensing conditions so restrictive that the project might have to be dropped forced the TID to consider alternatives to New Don Pedro. The power that would be produced by the new dam could be generated in other ways, and preliminary consideration was given to the installation of gas turbine generators. The supply of irrigation water from the old dam posed no immediate problems, but as San Francisco's diversions increased, it would be necessary to conserve as much as possible of the water reaching Don Pedro. To that end, plans originally prepared by Burton Smith half a century earlier for a reservoir at Dickinson Lake, in the foothills just west of Turlock Lake (the name given Owens Reservoir when a state park was established there), were dusted off and updated. A 100,000 acre-foot reservoir there could capture winter power releases from Don Pedro for use during the following irrigation season.[63] Contingency plans were also made to overhaul the old dam. By repairing the existing dam, perhaps building a relatively inexpensive foothill reservoir and finding other sources of power, the TID could forego New Don Pedro and remain debt free. It would also escape any efforts to require fishery releases or otherwise interfere with its water rights at least until 1980, when the FPC minor part license for the old dam expired. New Don Pedro was still seen as a good project, but it was not the only course of action open to the district.

Drafts of a Fourth Agreement were circulating by September 1963 but Meikle reiterated the district's position that it could not pay more for the project than it was worth to the TID. A September 5, 1963, memorandum by Meikle said in part:

> Developments over the past 10 years including inflation, State and Federal demands, and possible contingencies have caused the estimated cost of the New Don Pedro Project to increase more than 28%.

> A financial analysis of the present situation as it affects Turlock Irrigation District indicates that the District can afford to pay only a fixed maximum amount as its contribution toward the new project, due to the fact that the present debt free Don Pedro Project yields a profit of better than one million dollars annually.
>
> It would appear that a new start will have to be made on the project with all interested parties including the Government paying for benefits received.[64]

The last sentence was another way of saying that if the state and federal governments wanted benefits in the form of fish releases or other functions beyond the original purpose of the project, they should be prepared to pay for them. Of course, they were not willing to do so, which meant that the project's partners would have to decide who would pay the added costs mandated by the license. In the division of responsibilities outlined in the Third Agreement, the financial impact of the license would fall wholly on the districts. If the TID's demand that its share of costs not exceed its estimate of the project's value was to be met, some means would have to be found for dealing with expenses and obligations unanticipated at the time of the Third Agreement.

The Supreme Court's refusal to review the lower court decision upholding the FPC license forced negotiations between the city and the districts into high gear because the districts would soon have to either accept or reject the license. At the end of 1965 and through the early months of 1966, the future of New Don Pedro hung in the balance.[65] The Turlock District insisted that it would pay no more than $28 million as its share, and since the ratio of ownership between the irrigation districts was fixed in the same proportion used at the old dam, any concessions would have to come from San Francisco. Meanwhile, the possibility that New Don Pedro might be abandoned if its partners could not reach agreement was attracting interest in other quarters. On February 1, 1966, Meikle had lunch with A. Gomez of the Corps of Engineers, who made it plain that if the districts declined to accept the FPC license, the Corps would like to take over the project for itself. Gomez added that the Bureau of Reclamation also wanted New Don Pedro, but he noted that the Corps had more money at that time.[66] Although they had been effectively barred from the Tuolumne for over twenty years, the federal construction agencies were still ready, even eager, to develop the river if the cooperative plan faltered. The TID did not want to abandon New Don Pedro, which, after all, had wide public support, but the district's principal negotiators, Meikle and board chairman Abner Crowell, remained unbending in their demands and circumspect about their ultimate intentions as they played what amounted to a high stakes poker game over New Don Pedro.[67]

The Fourth Agreement was completed in May 1966 on terms satisfactory to the intransigent TID. In it, San Francisco became responsible for over half the costs of roads, fishery research and recreation added since the Third Agreement. After deducting an expected state grant of $7.5 million for recreation and fisheries enhancement under the Davis-Grunsky program, the Turlock District's share of the $105 million project came to $28,216,904. A provision was inserted in the agreement at the TID's insistence that if estimated costs rose above that figure, the district could declare that its costs exceeded its benefits and withdraw from the project. Similar provisions protected the MID and San Francisco, but since they had not publicly set such stringent limits on their participation, they were much less likely to walk away from New Don Pedro. The city also agreed that if the water releases for salmon infringed on the districts' recognized entitlements, it would bear over 51 percent of the burden of meeting the fishery requirements. By threatening to scuttle the project if their terms were not met, Meikle and Crowell had won important concessions insuring that the project would be profitable for the districts. The agreement was promptly signed by all parties, and the districts then accepted the FPC license, clearing the way for the construction of New Don Pedro.[68]

New Don Pedro was designed as an earth-and rock-fill dam, 580 feet high from its foundation and 2,800 feet thick at its base. It was something like building a wide, gently sloped mountain in the bed of the Tuolumne River. The relatively narrow core of the mountain would be compacted clay, held in place and protected by the massive rock shell on each side. Material for the dam was to come from near La Grange. To get enough clay soil, the districts bought the old olive orchard near La Grange owned by the La Grange Gold Dredging Company, only to discover another, more favorable deposit on the Brescia property.[69] Twelve million of the over sixteen million cubic yards of material used in the dam would be rock. Years before, big gold dredgers had scooped up the river bottom, sifted through it for gold and dumped the rock, which was perfectly suited for this type of dam construction, in long, unsightly piles.

Building the dam was a relatively simple job of hauling the materials into place. Two tunnels had to be bored in the hills to the left of the dam. One would be a diversion tunnel to carry the river around the construction site. Later, it would be fitted with three massive gates to serve as a flood control outlet. The other, smaller tunnel was the intake to the powerhouse. Three generating units, each capable of supplying 50,000 kilowatts, were to be installed in an outdoor powerhouse at the foot of the dam. While the old dam had filled almost every year and water commonly went down its spillway, New Don Pedro was so large, and the capacity of its diversion and power tunnels so great, that its spillway was expected to be used rarely, if at all. Still, the spillway was on the same massive scale as the rest of the project. Three radial gates, each forty-five by thirty feet in size, controlled an operating spillway capable of passing 78,000 second-feet when the reservoir was at its normal maximum elevation of 830 feet above sea level. A 995-foot wall to one side of the operating spillway was built as an emergency, uncontrolled spillway. Any water going over the spillways would first destroy a road leading to the dam and then cascade down natural ravines to the river some distance below the powerhouse.[70]

The districts called for the first bids, covering turbines for the powerhouse, in September 1966, but Bechtel did not have the final specifications for the dam, powerhouse, switchyard and other works ready until early 1967. On June 22, 1967, an overflow crowd gathered at the TID office for the opening of bids on the dam and powerhouse. Happily, two of the seven proposals for the main construction job were under the engineers' estimates. The low bidder was the Guy F. Atkinson Company of South San Francisco, whose bid of $49,693,960 was about $3.6 million below the estimate. Bids on powerhouse equipment were also lower than expected.[71] The city and the districts all sold their bonds for the project on the same day, August 1, 1967, and on August 18, Thomas B. White of the San Francisco Public Utilities Commission handed the presidents of the two districts a check for $45 million for the city's share of the project.[72] The Don Pedro Board of Review, a policy and management committee made up of representatives of the three partners, appointed long-time MID irrigation engineer Charles D. Crawford to the post of project coordinator to oversee the acquisition of property, accounting and the many other details involved in a project of that magnitude. Atkinson's crews began work within a month after receiving the formal contract in August, and on October 6,

1967, TID board president Everett Tomlinson joined representatives of the MID and San Francisco in setting off three dynamite blasts in the canyon to officially break ground for the project and begin work on the diversion tunnel.[73]

The diversion tunnel, running 3,400 feet through solid rock, took almost a year to build. Crews worked around the clock, drilling and blasting out eleven foot sections at a time. The drills were mounted on specially built trucks and 1,500 pounds of explosives were used on each section. The whole thirty-foot-wide tunnel was given a concrete lining at least fifteen inches thick. Rock plugs kept water out of the tunnel until it was finished, and on September 5, 1968, R. V. Meikle touched off the explosion that shattered the plug. Once the debris was cleared, the river could flow through the tunnel and allow work to begin on the dam itself.[74] The haul road from the river near La Grange to the dam site was already under construction. It was a virtual superhighway, sixty feet wide with uniform grades. The road network ran almost twenty miles and included four bridges, one of which carried a county road across the haul road. To keep slow moving sprinkler trucks from interfering with the rock and dirt haulers, eight miles of the road were given a sprinkler system to keep dust down.

The trucks that used the roads were one of the most notable features of the construction project. The KW-Dart bottom-dump haulers could move seventy-five cubic yards of material at a speed of up to fifteen miles per hour and could go back for another load at up to forty-five miles per hour. Their 700 horsepower engines were matched to automatic transmissions built specifically for the grades and conditions found at New Don Pedro. Fully loaded, each truck weighed just under 200 tons. The Atkinson company bought forty of the giant machines, and five fifteen cubic yard loaders to fill them. The trucks ran twenty-four hours a day, five days a week, and on weekends they were serviced and the haul road regraded. Refueling stops resembled those of an auto racer. The big haulers first went through an automatic wash rack and then into the fueling station, where the truck got a thirty-point maintenance check and 350 gallons of diesel fuel in ten minutes. Every eighth stop, it got an oil change that added only five minutes. In case of breakdowns, the trucks featured a modular design so that a whole new engine or transmission could be installed in a matter of hours and the defective part sent back to San Francisco for repairs. With an average of thirty-seven trucks on the job, Atkinson was able to place 60,000

Big equipment for a big job: KW Dart bottom-dump truck and loader.
Source: Bechtel Corporation.

cubic yards of material on the dam each day.[75] When estimates on the quantity of dredger tailings in the original borrow area proved to be in error, the haul road was simply extended to other piles further down the river.[76] The last load of dirt, with Roy Meikle joining the driver in the cab, was deposited after a brief ceremony on May 28, 1970.[77]

While final work went forward on the powerhouse and other parts of the project, plans were made to begin storing water behind New Don Pedro in the fall of 1970, after the irrigation season ended. The schedule was threatened briefly in early September when state highway engineers announced that three of the massive girders in a bridge being built to carry Highway 120 across an arm of the new reservoir were defective and would have to be reinforced or replaced. Two of the girders were already in place, but the third, measuring ten feet high and a hundred feet long, still rested on the canyon floor. After discussions with state officials and the steelmaker, the contractor promised to lift the girder out of the streambed by December 15. The likelihood that completion of the highway would be delayed by remedial bridge work created yet

Building the New Don Pedro powerhouse.
Source: Bechtel Corporation.

another concern, but after reviewing historic stream-flow data, the Board of Review judged that there was little danger of flooding the old road before the May 1, 1971, deadline and decided to proceed as planned.[78]

The birth of New Don Pedro reservoir meant the death of old Don Pedro, the dam that had supplied water and power to the districts for forty-seven years. In early October, one of the five generators in the old powerhouse was turned off, and the job of dismantling and salvage began. The rest of the powerhouse was shut down forever on October 15. The TID hoped to be able to save the three older 5,000 kilowatt units, and possibly the other two as well, if there was enough time before the gates of the old dam were opened on November 2. Since the powerhouse was built around the generators, the downstream end of the building had to be knocked out to remove them.[79] At 10:00 a.m. on November 2, 1970, a crowd of over a thousand people gathered at the old dam to witness the final opening of its valves and the transfer of storage from the old reservoir to the new. Don Pedro resident engineer Charles P. Arnold II opened the first valve, one of the lower irrigation outlets in the face of the dam. He had been born at Don Pedro, where his father had been the first superintendent. Other long-time district employees opened the remaining valves until water was thundering out of all twelve ports. Mist filled the canyon below the dam where last minute salvage operations were continuing as water poured into the space between the two Don Pedros and rose fifteen feet in the first hour. The heavy mist made movement of the last turbine from the old powerhouse more difficult, and spectators watched the truck struggle up the hill on the makeshift road built for the dismantling. Around four o'clock that afternoon the powerhouse was finally submerged. By early the following morning, the level stabilized and bulldozers went to work destroying the old employees' village and schoolhouse that, like the dam itself, would soon be under water.[80] The transfer of storage was a milestone in the history of the New Don Pedro project and of the Turlock Irrigation District, but the man most responsible for it all stayed away from Don Pedro that day. Roy V. Meikle told his associates simply that he did not go to funerals. Although not given to emotion, Meikle was deeply saddened by the end of the dam he had built and operated for so many years. In an interview a week after the transfer of storage, his voice broke when he acknowledged his sadness at the loss of the old dam.[81]

Two days after the first water reached the new dam, releases were begun to help migrating salmon. It took another month to raise the water level enough to begin testing the first generating unit, and as the lake rose in early 1971 the other two units were started.[82] Dedication ceremonies and a barbecue attended by 3,000 people were held in May 1971 for a project that had gone smoothly once construction was finally underway.

Unfortunately, two issues – fish and recreation – that bedeviled the districts during the licensing process continued to create problems once the dam was in operation. The FPC license required only that the districts make specified releases and cooperate in a research program to clarify the kind of environment needed by salmon. The districts, however, went a step further and applied for a state Davis-Grunsky grant, administered by the Department of Water Resources, to enhance spawning conditions. About $2 million was allocated to rehabilitate and maintain the gravel where salmon made their nests and where the newly-hatched young remained until they started their journey to the ocean in the spring. As part of the work, the Department of Fish and Game informed the districts in 1969 that they would have to acquire access to property along a fifteen mile stretch of the river beginning a short distance below La Grange. The result was a controversy between the districts, the state and the affected landowners.[83] The squabble was complicated and, in the long run, unimportant, but it seemed to illustrate how essentially peripheral issues associated with New Don Pedro took up an inordinate amount of time and energy. In 1971 the districts were finally able to hire a contractor to improve about a million square feet of gravel, reshaping the river channel in accordance with DFG instructions. That year, 21,900 salmon returned to the Tuolumne River, but most refused to use the reworked gravel.[84] Additional work was done to revised specifications in 1971, but the results were no more encouraging, and the following year the state suspended the gravel rehabilitation program.[85] Exactly why the improved gravel did not attract salmon is unknown, but as time altered the reworked gravels to a closer approximation of a natural streambed, the fish began to use those areas once more.[86]

The gravel rehabilitation program was not the only thing that went wrong with the Department of Fish and Game's plan to save the salmon. The various release schedules proposed in the early 1960s and the one adopted by the Federal Power Commission

The New Don Pedro Dam and Powerhouse.

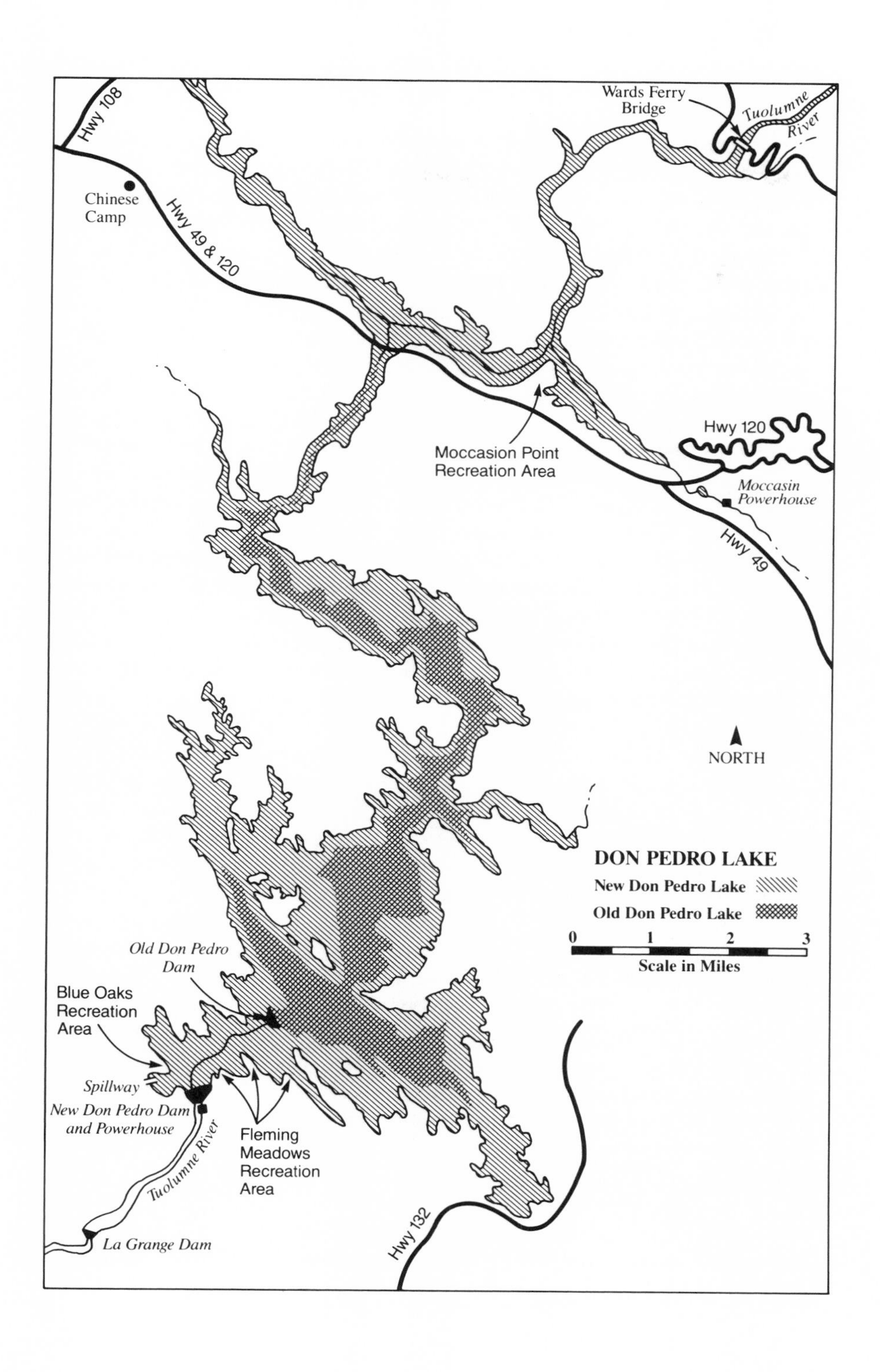

Hwy 108
Chinese Camp
Hwy 49 & 120
Wards Ferry Bridge
Tuolumne River
Moccasion Point Recreation Area
Hwy 120
Moccasin Powerhouse
Hwy 49
NORTH
DON PEDRO LAKE
New Don Pedro Lake
Old Don Pedro Lake
0
1
2
3
Scale in Miles
Old Don Pedro Dam
Blue Oaks Recreation Area
Spillway
New Don Pedro Dam and Powerhouse
Fleming Meadows Recreation Area
Tuolumne River
La Grange Dam
Hwy 132

were all concerned primarily with the condition of the river in the fall and winter, when the salmon ran upstream to spawn. The dam had only been in operation a short time when the DFG reported that scheduled releases during the spring were not large enough to move the fingerlings out of the river.[87] A brief flushing flow was tried in May 1973 which succeeded in moving most of the young salmon out of the Tuolumne.[88] The experiment helped convince the DFG that inadequate spring outflow, rather than insufficient spawning gravel, was the limiting factor for Tuolumne River salmon, and that conclusion influenced the final decision to halt the gravel rehabilitation program. On the assumption that the total amount of water designated for fish releases was established but that the timing of the releases could be changed, the DFG asked in October 1974 that that year's fall flushing flow be withheld and other flow levels reduced slightly to make water available for higher releases the following May.[89] The districts agreed to these requests, at least temporarily, since the overall amount of water remained the same. However, in some circumstances, holding water back in the fall for spring fishery releases could adversely affect power generation.[90] After having been almost lost in the early 1960s, the salmon run rose to over 32,000 fish in 1969 and then fell to an average of about 1,200 to 1,400 fish a year from 1974 to 1980. The number climbed to 14,000 in 1981 and has remained near that figure.

Recreation on and around New Don Pedro proved nearly as troublesome as fishery issues. At the FPC hearings in 1962, Tuolumne County, where the reservoir would be located, took the districts to task for not giving more thought to recreation. Fishermen had come to the old reservoir, but there were no boat ramps or campgrounds for them. There had been a few houseboats, but Meikle had considered them just unsightly and unsanitary floating shacks. When confronted with the recreation issue, the districts had tried to argue that they could not legally be in the recreation business, but the FPC believed otherwise and required them to construct or provide for the construction of recreational facilities for boating, swimming and camping as a condition of their license. The Fourth Agreement made San Francisco responsible for about half of these additional costs, but there was hope that the districts could get a Davis-Grunsky grant to offset some of the initial expense. In 1965 the legislature authorized a grant of not over $7 million for recreation and fish and wildlife enhancement (the gravel program) in connection with New Don Pedro, plus an additional sum for water supply and sanitary facilities at the lake. The districts made an application for a total of $8.6 million in 1967 and received state funds after revising their application in 1969.[91]

Construction went forward as the dam neared completion, but there was still uncertainty over who would operate the recreational features of the project. Tuolumne County made it clear that it would be willing to do the job, and that it would pattern its approach after Napa County's operation of Lake Berryesa. That was not something the directors of the two districts wanted to hear, since a delegation that visited Berryesa came away unimpressed. At that reservoir, the county acted only as an overseer for private operators, and the result, in the districts' view, was an emphasis on mobile homes along the shore and short shrift to picnickers and campers.[92] The districts tried to interest the National Park Service, Forest Service, Bureau of Land Management or the state park system in taking the job, but only the Bureau of Land Management expressed even passing interest.[93] In April 1970 the districts decided to hire a recreation director themselves to supervise the construction and operation of the facilities at least temporarily.[94] Former Forest Service employee George S. James was named director of the Don Pedro Recreation Agency when it was formally created by the districts and San Francisco later that year. The agency was thought of as a temporary measure, to last only until a permanent operator could be found, but none ever appeared. The recreation agency has continued to manage the campgrounds, boat launching ramps and picnic sites at the lake, while private concessionaires operate the marinas at Fleming Meadows and Moccasin Point.

The districts, and the recreation agency they formed, were responsible for recreation within the project boundaries, which, except for a larger area around the dam, generally extended above the high water line at elevation 830 feet to the 845-foot contour. Although there would be no privately owned lake frontage, real estate developers looked forward to subdividing private lands near the lake for vacation homes that would be readily accessible to water-oriented recreation. Pacific-Cascade Land Development Company, a subsidiary of Idaho-based Boise-Cascade, bought 16,000 acres east and west of the new lake and announced plans for subdivision. The company, however, had no domestic water supply for their proposed communities except the reservoir itself. After an initial rebuff by the districts in late 1967, the company decided to work through Tuolumne County Water District No. 2 in negotiations for as much as 20,000 acre-feet a year from New Don Pedro. The Turlock board told representatives of the company, "that Turlock was sympathetic to the plan subject to an investigation by the District's legal counsel."[95] The districts made no promises, but Pacific-Cascade nevertheless pressed ahead with plans for a sales office complex to be known as Hacienda de Don Pedro, and the first lots went on sale in the summer of 1968.[96] The company was soon advertising its five acre "rancheros," three acre "ranchitos" and smaller homesites in national magazines and offering charter flights from southern California for prospective buyers. One advertisement, complete with pictures of serene foothill landscapes, promised that, "Intelligent private developers are working with State and Federal master planners to preserve the gently rolling land around Don Pedro's lake for all the future generations of vacationers, fishermen, water skiers, and nature-lovers."[97]

The "intelligent private developers," however, made a serious error in dealing with their potential water suppliers. In buying land for their subdivisions, they raised the value of land the districts still needed to buy in the project area. By July 1968 just as land sales were beginning, Meikle informed TID secretary Reynold Tillner that the districts could pump enough irrigation water to compensate for the use of domestic water from the lake, but "now is not the time to admit it."[98] The next month, the board ordered its attorney to tell the company and Tuolumne County that a water supply contract could not be considered until land acquisition for the project was complete.[99] Matters remained stalemated until early 1970, when Boise-Cascade announced it would not sell the 321 acres it owned within the project boundaries unless a water supply agreement was considered at the same time. Modesto director Thomas Beard responded that, "We can condemn their land but they can't condemn our water."[100] The districts did condemn the land they needed and took no further interest in providing water for the subdivisions. The company got water for its property southeast of the reservoir from the Merced Irrigation District's Lake McClure, but the Hacienda and the roads that lead nowhere across the hills are all that remain of the dry subdivisions west of the lake.

In November 1971 the name of the New Don Pedro Project was officially changed to Don Pedro Dam.[101] Dropping the word "new" from the title may have been done for simplicity, but it was also appropriate if New Don Pedro was viewed simply as a bigger version of an exisiting project. The original dam had given the Turlock Irrigation District a full irrigation season and its own electrical utility, and the new one succeeded it without any visible effect on the farms and towns below. There were, however, some important differences in the purpose of the two projects. The new dam symbolized cooperation among the river's developers, and it served the needs of San Francisco as much as it did those of the irrigation districs. New Don Pedro was also conceived, at least in part, as a instrument of "home rule," intended to prevent federal or state agencies from staking their own claims on the river. Here, the result was ironic. Flood control, fishery and recreation requirements all infringed somewhat on the independence the developers had sought to retain. The Tuolumne was still under local control, but it did not, and could not, exist in a vaccum, and, as the New Don Pedro project showed, it was not beyond the influence of state or national trends in water policy and envoirnmental management.

CHAPTER 13 – FOOTNOTES

[1] Board minutes, vol. 30, p. 417 (Dec. 18, 1950).
[2] Tuolumne County Water District. No. 2, "Protection and Development of Water Resources of Tuolumne County," (no date), in Meikle files, vol. 21.
[3] Board minutes, vol. 33, p. 161 (Mar. 14, 1955).
[4] *Turlock Tribune,* Apr. 7, 1955.
[5] R. V. Meikle to B. F. Sisk, Apr. 27, 1955, in Meikle files, vol. 21.
[6] *San Francisco Examiner,* June 2, 12, 1956.
[7] A general discussion of the Engle bill is found in Stephen P. Sayles, "Hetch Hetchy Reversed: A Rural-Urban Struggle for Power," *California History,* vol. LXIV, no. 4 (Fall 1985), pp. 254-263.
[8] Applications to Appropriate Water, Nos. 14126, 14127 (both Jan. 16, 1951). The former covered the use of water for irrigation, the later covered its use for power generation.
[9] "Progress Report, Jan. 1, 1957, New Don Pedro Project," in Meikle files, vol. 28, item 37.
[10] Memorandum on New Don Pedro Project, Oct. 12, 1959, in Meikle files, vol. 30, item 92.
[11] Memorandum by R. V. Meikle, July 20, 1960, in Meikle files, vol. 33, item 21; *Modesto Bee,* May 1, 1961; Board minutes, vol. 37, p. 50 (May 1, 1961).
[12] See Fourth Agreement, June, 1966, Appendix A, p. 3.
[13] Stone & Youngberg and Blythe & Co., *Financing Report: Turlock Irrigation District, New Don Pedro Project,* (July, 1961); *Modesto Bee,* June 13, 1961.
[14] *Modesto Bee,* Nov. 8, 1961.
[15] *Modesto Herald,* Dec. 27, 1877.
[16] *Modesto Daily Evening News,* June 6, 7, 1886, Oct. 24, 28, 1889.

[17] *Stanislaus County Weekly News,* Dec. 18, 1903, Dec. 2, 1904.
[18] "Fall-Run Chinook Salmon Stocks in the Tuolumne River, 1940-," (ca. 1970), in Meikle files, vol. 1970, item 56.
[19] Author's interview with Tim Ford, June 24, 1985.
[20] U. S. Dept. of the Inter., *Central Valley Basin,* Sen. Doc. 113, 81st. Cong., 1st. Sess. (1949), p. 413.
[21] "Rebutal Testimony, William R. Gianelli," before Federal Power Commission, 1962, pp. 1-2.
[22] "Time Table, New Don Pedro, TID and MID and State DFG," in Meikle files, vol. 36, item 7. [Hereafter cited as "Time Table."]
[23] "Time Table" (unpaginated).
[24] Summary of demands for fish releases, Sept. 2, 1970, in Meikle files vol. 1970, item 44.
[25] "Time Table."
[26] *Modesto Bee,* Nov. 8, 1961.
[27] Testimony of Milo C. Bell before Federal Power Commission in *Joint Appendix, State of California, Turlock Irrigation District, Modesto Irrigation District and United States of America on relation of Stuart L. Udall, Petitioners v. Federal Power Commission, Respondent, in the United States Court of Appeals for the Ninth Circuit,*vol. I, pp. 112-115. [Hereafter cited as *Joint Appendix.*]
[28] "Time Table."
[29] *Modesto Bee,* May 15, 1962.
[30] "Time Table."
[31] *Idem.; Modesto Bee,* June 25, 1962.
[32] *Modesto Bee,* June 29, 1962.
[33] *Modesto Bee,* July 1, 1962.
[34] Board minutes, vol. 38, p. 63 (July 2, 1962).
[35] *Modesto Bee,* July 17, 1962.
[36] *Ibid.,* July 22, 1962.
[37] *Modesto Bee,* Oct. 16, 1962.
[38] "Testimony of Milo C. Bell," before Federal Power Commission, 1962, p. 30.
[39] *Joint Appendix* vol. I, pp. 25-26.
[40] "Rebuttal Testimony of Maurice L. Dickenson," before Federal Power Commission, 1962, pp. 1-5.
[41] *Joint Appendix* vol. II, p. 429.
[42] *Ibid.,* pp. 431-432.
[43] California Dept. of Water Resources, Dept. of Fish and Game, Central Valley Regional Water Pollution Control Board, *Problems of the Lower San Joaquin River Influencing the 1963 Salmon Run* (Jan. 15, 1964), pp. 1-3.
[44] "Exceptions of Commission Staff Counsel to Findings and Conditions of Fact and Law contained in the Decision of the Presiding Examiner," July 24, 1963, in *Joint Appendix,* vol. II, pp. 628-635.
[45] *Modesto Bee,* Feb. 28, 1963.
[46] "Opinion No. 420, Opinion and Order Issuing License," Mar. 10, 1964, in *Joint Appendix,* vol. II, p. 661.
[47] *Ibid.,* p. 681.
[48] *Ibid.,* pp. 662-663.
[49] *Ibid.,* p. 666.
[50] Memorandum by R. V. Meikle, June 15, 1964, in Meikle files, vol. 37, item 47.
[51] Board minutes, vol. 39, p. 13 (Apr. 4, 1964).
[52] *Modesto Bee,* July 7, 1964.
[53] *Ibid.,* June 8, 1965.
[54] R. V. Meikle, "The Turlock Irrigation District," 1971, p. 15, in Meikle files, vol. 44, item 38.
[55] W. Turrentine Jackson and Stephen D. Mikesell, *The Stanislaus River Drainage Basin,* pp. 82-95.

[56] Quoted in W. Turrentine Jackson and Alan M. Paterson, *The Sacramento-San Joaquin Delta: The Evolution and Implementation of Water Policy: An Historical Perspective,* (Davis, Calif., 1977), p. 98.
[57] William A. Hillhouse, II, "The Federal Law of Water Resource Development," in Erica L. Dolgin and Thomas T. P. Gilbert, editors, *Federal Environmental Law,* (St. Paul, 1974), pp. 891-892.
[58] TID and MID, *New Don Pedro Project: Application Report for Davis-Grunsky Grant,* (July 1967), vol. I, pp. 5-15.
[59] Meikle, "The Turlock Irrigation District," p. 14.
[60] Memorandum by R. V. Meikle, Oct. 12, 1962, in Meikle files, vol. 34, item 12.
[61] Memorandum by R. V. Meikle, Nov. 14, 1962, in Meikle files, vol. 33, item 71.
[62] Notes by R. V. Meikle, Jan. 25, 1963, in Meikle files, vol. 34, item 23.
[63] Memorandum on alternatives to New Don Pedro, Sept. 25, 1964, in Meikle files, vol. 37, item 82.
[64] Memorandum by R. V. Meikle, Sept. 5, 1963, in Meikle files, vol. 34, item 63.
[65] *Modesto Bee,* Dec. 17, 21, 1965.
[66] Notes on lunch with A. Gomez, Corps of Engineers, Feb. 1, 1966, in Meikle files, vol. 39, item 29.
[67] Author's interview with Beth Crowell Hollingsworth, Feb. 22, 1983; author's interview with William R. Fernandes, May 4, 1983.
[68] Board minutes, vol. 40, pp. 114-120 (May 16, 1966), p. 124 (May 23, 1966); *Modesto Bee,* May 11, 17, 1966.
[69] "Information for the Press from New Don Pedro Board of Review," Oct. 23, 1967, in Meikle files, vol. 40, item 12.
[70] New Don Pedro Project brochure, ca. 1969.
[71] Board minutes, vol. 41, p. 72 (June 22, 1967); *Modesto Bee,* June 23, 1967.
[72] *Modesto Bee,* August 20, 1967. [73] *Ibid,* Oct. 6, 1967.
[74] *Modesto Bee,* Sept. 4, 6, 1968, *Western Construction,* (Aug. 1968), pp. 44-45, 48-49.
[75] *Modesto Bee,* Sept. 29, 1968; *Engineering News-Record,* (July 9, 1969), pp. 34-35.
[76] *Modesto Bee,* July 2, 1969. [77] *Ibid.,* May 29, 1970.
[78] *Modesto Bee,* Sept. 20, 24, 1970. [79] *Ibid.,* Oct. 11, 1970.
[80] *Modesto Bee,* Nov. 3, 1970; *Turlock Journal,* Nov. 3, 1970; Reynold Tillner interview.
[81] Norman Boberg interview; R. V. Meikle interviewed by Gertrude Vasche, Nov. 9, 1970.
[82] TID, *Annual Report,* 1970, p. 12; *Modesto Bee,* Jan. 29, 1971.
[83] *Modesto Bee,* June 4, 1969, Feb. 13, 25, 1970. [84] *Ibid.,* Dec. 12, 1971.
[85] Board minutes, vol. 43, p. 263 (Aug. 28, 1972); P. A. Towner to Ernest Geddes, July 10, 1973. [86] Ford interview.
[87] E. C. Fullerton to W. R. Gianelli, Mar. 12, 1973.
[88] *Modesto Bee,* May 11, 14, 29, 1973. [89] *Ibid.,* Oct. 9, 1974.
[90] Norman Boberg to Ernest Geddes, May 29, 1973.
[91] Dept. of Water Resources, *Transactions under the Davis-Grunsky Act: 1974 Report to the Legislature,* (Jan. 1975), pp. 66-67.

[92] *Modesto Bee,* Aug. 22, 1969, Mar. 6, 1970. [93] *Ibid.,* May 1, 1970.
[94] Board minutes, vol. 42, p. 289 (Apr. 27, 1970).
[95] *Ibid.,* vol. 41, p. 307 (Feb. 13, 1968).
[96] *Sonora Daily Union Democrat,* Apr. 9, June 27, 1968.
[97] Advertisement (undated) from *Life* magazine.
[98] R. V. Meikle to R. S. Tillner, July 12, 1968, in Meikle files, vol. 43, item 8.
[99] Board minutes, vol. 41, pp. 439-440 (Aug. 5, 1968).
[100] *Modesto Bee,* Feb. 12, 1970.
[101] Board minutes, vol. 43, p. 118 (Nov. 8, 1971).

14

A Changing District

Although most residents of the Turlock District could scarcely notice the change from old Don Pedro to New Don Pedro, the new project was one of a series of events that made the early 1970's a turning point in the district's history. Another such occasion came in July 1971 when, with the new powerhouse in operation, Roy V. Meikle relinquished his job as Don Pedro engineer, a position he had held since 1929. The following month, his official duties were reduced to serving as an alternate member of the Don Pedro Board of Review and an advisor to the district.[1] The man who had guided the TID for nearly six decades was stepping down, and on December 1, 1971, at age eighty-seven, Meikle formally retired.[2] It had been fifty-nine years since he had arrived in 1912 to help prepare the case against Hetch Hetchy and fifty-seven years since he had become chief engineer. Meikle had been ready to retire years before, but the board convinced him to remain on the job to oversee the New Don Pedro project and protect the district's interests in negotiations with San Francisco.[3] During his last years with the district, he still maintained a full and active schedule, insisting that he should put in a day's work for a day's pay despite his advancing age. About the only sign of old age came when he declined to accompany groups over the rough terrain at Don Pedro to see what, as he said, he had seen before.[4] Meikle's long tenure as chief engineer and de facto manager had given the TID a remarkable continuity in policy and outlook and a sense of stability. Roy Meikle died at his Turlock home on June 15, 1975.

Others who had helped run the district for many years also retired at about the same time as the chief engineer. Meikle's assistant in the irrigation department since 1926, Reinhold Schmidt, left in 1969. Hale Parker, who had joined the district as a draftsman in 1922 and became its second electrical engineer after R.W. Shoemaker departed in 1929, retired in 1967. Another long-time employee, Reynold S. Tillner, had been named general manager in 1968 after serving as auditor and secretary. He planned to retire in

1975. Even the Board of Directors had remained fairly static until the late 1960s. Between 1930 and 1970 only twelve men had occupied the five seats on the board, and from the 1951 to 1967 there were no changes at all. In 1967 Abner Crowell, who had joined the board in 1945 and been its president continuously since 1947 died, leaving a vacuum of sorts in the leadership of the board.[5] In the four years after his death, three other seats changed hands, so that by 1971 only Everett Tomlinson, first elected in 1947, remained of the "old" board. As the 1970s began, the men who had managed the affairs of the district for a generation or more were all passing from the scene. Passing, too, was the kind of world they had known.

On the eve of his retirement in 1975, Reynold Tillner told a reporter that labor relations and the search for additional electric power, two issues that had not been pressing ten years earlier, had become the dominant concerns of district management.[6] The labor problem was not so much a crisis as it was a basic change in the way things were done. Except for the strike threats in 1942 and a short-lived organizing effort by the International Brotherhood of Electrical Workers just after World War II, labor relations at the TID had been uneventful. Annual salary increases were handled with the utmost simplicity; Electrical Superintendent "Rosie" Wallin would talk the matter over with Meikle, who made the final decision.[7] It was a system that obviously left no room for participation by the employees. In the late 1960's, a Turlock Irrigation District Employees Association (TIDEA) was formed in an effort to give workers a voice in the heretofore private business of negotiating their wages. At the same time, the IBEW resumed its long dormant attempt to become the bargaining agent for district employees. A representation election was held in 1969 that resulted in a victory by the local employees association.[8] In February 1970 the board adopted rules governing employee negotiations that provided for talks with whatever organization represented a majority of the employees and for a memorandum of agreement on wages and related matters when negotiations were completed.[9] Despite official recognition of the legitimacy of bargaining and the establishment of a mechanism to provide for it annually, the ultimate authority remained with the district's Board of Directors. In 1973, for example, management insisted that it could pay no more than it had offered and when no agreement was reached with the TIDEA, the board unilaterally adopted the management's schedule of wage increases.[10] The next year, 1974, the association hired a professional negotiator for the first time, but faced with a district offer that was judged unacceptable, they appealed directly to the board and won a compromise of sorts.[11] Perhaps because of the outcome of these negotiations, the IBEW became the majority representative in early 1975, but retained that position for only a year. The changes brought about by the TIDEA and a policy of annual negotiations were scarcely revolutionary and soon became part of the district's routine, but they were one measure of the kind of changes that had taken place within the district organization itself.

The other issue Tillner mentioned in 1975 was power. The story of the search for adequate and affordable electric resources is told in the following chapter, but behind it was a significant change in the Turlock region – a trend toward urbanization and population growth. In the West, irrigated agriculture can be a logical precursor to non-agricultural development. Level land, an established population, market towns and agriculturally-related industries found in irrigated areas all help create the potential for further development. There are numerous places where irrigated farms were succeeded by cities and industries. Phoenix, Arizona, was one such place, and California's Santa Clara Valley was another. It was a major fruit-growing region, but the growth of the San Francisco Bay area turned it into the high-tech industrial suburb known as Silicon Valley. In southern California, where the irrigation ideal of small farms and intensive agriculture was perhaps best exemplified, images of sunshine and oranges encouraged tourism and migration until the orange groves were mostly replaced by housing tracts, and by the freeways that became the new symbol of the area.

The Santa Clara Valley and the Los Angeles basin were transformed by California's explosive and sustained post-World War II growth, and both were relatively close to long established urban centers. The Turlock and Modesto irrigation districts were not directly adjacent to any major cities, but they were still affected by the state's general pattern of development. Modesto was clearly the focus of growth in the local area. It had always had the largest population, and after the war it began to add new industries, primarily in food processing and in related businesses like packaging. Modesto's growth spilled southward across the Tuolumne River into the Turlock District in Ceres and south Modesto. Ceres got its first subdivision, the Caswell tract, in 1946, and others followed. Turlock grew gradually, adding a new industry, mainly in agricultural processing, or a few new houses at a time. The town's business district looked in 1960 remarkably like it did in 1940.[12] In 1959 it was announced that a new state college would be located on Monte Vista Avenue, which was expected to bring new business, prestige and population to the town. The prophecy was accurate, even though Stanislaus State College grew more slowly than first expected. Turlock expanded northward toward the college, and Geer Road was soon lined with shopping centers, offices and fast food franchises. Denair, Hilmar and Hughson all attracted new residents willing to exchange a short commute for the chance to live in a small town. Ironically, as these small towns grew in population, their merchants often found it impossible to compete with the newer and larger stores opening in Turlock, Ceres or Modesto, where many of their new residents worked, and often shopped.

Change was a gradual process, but its cumulative impact on the landscape was unmistakable. Denair and Turlock grew closer together, and from Ceres north across the Tuolumne as far as Salida, a solid metropolitan area emerged with only islands of agriculture, often only neglected remnants waiting for a developer's bulldozer, left to remind an observer of the area's past. One measure of these changes that had particular relevance to the Turlock Irrigation District was the number of electric meters in service. A meter could be connected to a home or a store or an industry, so another meter could mean more residents or more business, but whatever the use of the power, it all meant growth. Between 1936 and 1945 the number of meters rose from 7,130 to 9,950, with little expansion during the war due to critical shortages of wiring and other supplies. Pent-up demand, postwar prosperity and the beginning of suburban expansion led to a rapid increase in customers between 1945 and 1949, and at the end of the period the TID had almost 15,000 meters. The rate of increase slowed between 1950 and 1968, the numbers rising from 15,900 meters in 1950 to 26,103 in 1968. Thereafter, growth accelerated rapidly, passing 35,000 meters in 1974 and reaching 46,028 in 1984 as the rate of increase slowed only slightly.[13] These figures indicate that in the Turlock District, growth, though ever-present, leaped dramatically in the late 1960s and early 1970s.

The rising population meant more electrical customers and the need to acquire the power they would use. It also meant the abandonment of some community ditches when the farms they served were subdivided. Vandalism of canal gates increased, and where open canals passed through new housing tracts the risk that unwary children or swimmers could become drowning victims increased, too.

Despite the obvious extent and impact of suburban growth, the Turlock District was still predominantly an irrigation empire. The irrigated acreage in the district peaked at almost 173,000 acres in 1969, at the beginning of the suburban boom. Thereafter, it declined gradually to 162,000 acres in 1985. The acreage devoted to pastures and clover fell steadily from the early 1950s, and was overtaken by trees and vines in 1960. In 1985 orchards and vineyards covered over 40 percent of the Turlock District's farm land, followed in importance by corn, grain, pasture and alfalfa.[14] Even the rural areas, however, felt the upsurge in population. Small farms, which were found primarily on the east side of the district and near Hilmar and Delhi, were sometimes divided into suburban-style hobby farms or "ranchettes." Each had a house and a few acres, seldom more than five, with room for horses or cows or a few trees. They offered not an economically viable farm, but rather a chance for some urbanites to get "back to the land" by living, though not working, in a rural environment. Besides altering the character of some neighborhoods, the proliferation of ranchettes could also affect TID water service. The new irrigators were often inexperienced and their tiny farms were often difficult to irrigate with the district's standard fifteen second-foot head of water. Scheduling could be a problem, too, since most ranchette owners wanted water only on their days off. The overall impact on the Turlock District of these rural "suburbs" was relatively minor but they were one more indication of the changes occurring in the Turlock region.

One of the most serious potential problems created by suburbanization was the possibility of rural-urban conflict over district governance. It was made especially likely because, unlike some irrigation agencies, cities were included in the district, and because the TID was also an electric utility, giving urban power consumers a definite stake in district policy. In general, the internal politics of local water agencies are uncontroversial, marked by a broad consensus on basic issues, unanimity within the governing board and low voter participation.[15] Once it passed its first decade and a half of irrigation, the Turlock District conformed to that model, and continued to do so even after it entered the electrical business. For at least half a century, electoral controversies were rare, and when a contest did develop over a directorship, personalities more than issues divided the candidates. Even with the beginning of suburban development after World War II, the district's population retained an agricultural orientation and, perhaps more importantly, no issues arose that could threaten its harmony. When a controversy did finally arise, it was soon translated into a political struggle with rural-urban overtones.

The price of power established when the district went into the electrical business in 1923 had remained substantially unchanged over the years. There had been some adjustments during the first decade primarily to simplify the rate structure and eliminate inequities, and although some bills went up in the process, there was no real attempt to raise the price of electricity. In 1955 an interim agreement between the districts and San Francisco covering power purchased for the districts' use from PG&E at times when their demands exceeded the Hetch Hetchy supply carried an impact warning of higher power costs in the future, and prompted a reexamination of TID rates. Of all the various classes of service, only one, the commercial-industrial rate, was found to be in need of change because it produced a much smaller margin of profit than the domestic rate. The TID staff developed a new commercial-industrial rate and R.W. Shoemaker, who had designed much of the rate structure in the first place, was hired to review it. His studies produced a rate proposal very similar to the staff's, and in 1959 the board approved the first major electrical rate increase in over thirty years.[16]

Leading industrial power users vigorously protested the rate increase, but to no avail.[17] Largely as a result of the dispute over the commercial-industrial rate, two of the largest customers, W.O. Thompson of Turlock Refrigeration and Thornton Snider of Snider Lumber Products, sued the district not over the rate itself but over the way the board members, who held the ultimate rate-making authority, were selected. The Irrigation District Act required that each director's division be as nearly equal in size as possible, but Thompson and Snider argued that the existing geographical division violated the concept of one man, one vote and allowed a minority of rural voters to choose the directors. After a prolonged court battle, the district was ordered to redistrict with reference to both population and the geographic area of each division, but the result had no perceptible impact on board elections or on the kind of power rates adopted by the district.[18]

The changing environment within which the district operated and the increasing importance and complexity of its electric utility function were reflected in the selection of a successor to General Manager Reynold Tillner in 1975. For over sixty years, the TID had been managed along the conservative guidelines laid down by Roy Meikle. The board sensed the need for a new management style and, accordingly, they sought a manager from outside the ranks of district personnel, advertising for someone with academic training in public or business administration and at least ten years experience.[19] In January 1975 the board announced the selection of Leroy J. Louchart, the manager of the customer service department of the Sacramento Municipal Utility District, another public power agency.[20] Louchart took charge of the TID on Tillner's retirement at the beginning of April 1975. Exactly what went wrong after that is difficult to determine. Clearly, the staff and the new manager soon came to view one another with a measure of suspicion. Less than a year after Louchart's arrival, Tillner was relaying complaints about his performance to board president Lloyd Starn, who assembled the directors informally to discuss the situation.[21] On March 16, 1976, the board was ready to take action. It was reported that TID general counsel Ernest Geddes asked Louchart, "what it would take to get you to resign."[22] When he replied that he would not resign, he was called into an executive session of the board and read a brief notice of dismissal. Louchart began an ultimately futile legal action to recover his job, and a week later, the board appointed attorney Geddes the acting general manager, based on his familiarity with the district.[23] Several months later, after a number of executive sessions on district management, the board asked Geddes to remain in his new post. Since he had worked closely with the district staff and dealt with policy issues for four years, his appointment as general manager practically amounted to a return to a policy of continuity in district management.

Whatever the changes occuring in its service area or management, the Turlock Irrigation District was still basically a water agency, depending on its dams, and on Mother Nature, for the water that drove its generators and filled its canals. In 1976 and 1977 the water that it needed ran short and the district faced its greatest modern crisis. The problem began over the Pacific. When the rainy season should have begun in late 1975, a high pressure ridge positioned itself in the atmosphere between California and Hawaii. The ridge proved unusually persistent, pushing most

Pacific storms far to the north, away from the Turlock region and the Tuolumne watershed. Dry conditions forced the district to begin the 1976 irrigation season on February 1, the earliest start since at least 1923. The 624,000 acre-feet that finally flowed down the Tuolumne River that year was only 35 percent of normal and just slightly more than the record dry year of 1924.[24] Private pumps were turned on to supplement the Don Pedro supply, but the district chose not to reduce the 1976 water entitlement below the usual four acre-feet per irrigated acre. The long season meant heavy water use, despite several unusually heavy summer tropical storms. In all, the TID delivered 941,000 acre-feet for irrigation in 1976, including 164,000 acre-feet from groundwater.[25] Although it was not full when the drought began, New Don Pedro allowed the district to limit the impact of the dry year on irrigation, and as the 1976 season ended, it seemed that the district, and the new reservoir, had weathered a major crisis. But only if winter brought a return to more normal precipitation.

The winter of 1976-1977 was even worse than the preceding one, bringing only 25 percent of normal runoff from the Tuolumne watershed. Anxious farmers watched the sky for any hint of rain through a winter that saw some owners of private pumps begin irrigating their orchards in January. At the beginning of 1977, the district owned a little less than 100,000 acre-feet in Don Pedro reservoir, and with little prospect of getting more, it was estimated that instead of the usual four acre-feet per acre no more than eight acre-inches (two-thirds of an acre-foot) would be available from the Tuolumne River. The district conducted a census of private irrigation pumps and found nearly 800 of them which would become, along with the TID's drainage pumps, the main source of supply during the drought. Owners of private wells were asked, in effect, to turn them over to the district, which would pay $6.00 per acre-foot for pumped water and pass the charges on to the irrigators who used it. This was a departure from previous practice, which had treated the sale of pump water as a private arrangement between individual irrigators. The severity of the water famine was such that the district itself took control of the groundwater available from private wells and distributed it as needed throughout the system. Not all private well owners elected to participate in the district program, preferring to use their water themselves and forego the use of any district water. The price paid pump owners was raised to $8.00 per acre-foot, and when rising power costs threatened

Low water in San Francisco's Lake Eleanor and Cherry reservoirs reflected the severity of the 1977 drought.

to make that figure uneconomical, the district decided to provide pumping power without charge and pay $4.00 per acre-foot for the water itself.[26]

Understandably, there was a scramble for new wells. In order to help farmers get water as soon as possible, the district bought a well developer, a machine used to clear sand out of newly drilled wells and establish the flow. Some small well drillers did not have developers of their own, and all drillers could move on to the next job faster if someone else developed the well. The district's developer ran two shifts, seven days a week, and another, smaller developer was soon added. The rule became, in the words of Director Tilden Genzoli, "Sacrifice everything to find water," and other construction work was curtailed to keep the well developers running. As further encouragement for new wells, powerline extensions to them were provided free of charge up to nearly a mile from existing TID lines.[27] Probably the most effective way the district could promote well drilling was through the formation of improvement districts to help farmers finance the work. About 125 improvement district wells were put in during 1977, along with perhaps sixty new private wells. For many years, the district had purchased improvement district warrants itself, but with $3 million worth of work being done and its budget strained by power purchases, it did not have the money to buy them, and $2 million worth had to be sold to investors.[28] With many pumps running continuously, the water table beneath the district began to drop rapidly, and some domestic and irrigation wells had to be deepened to remain in operation.

The first release of the district's scarce reservoir supply came in March, when about 8,000 acre-feet were parcelled out primarily to trees, vines and young plantings. Other brief releases were made in April and May. Meanwhile, the district was doing all it could to get more surface water. It applied for 25,000 acre-feet of the water made available when the Metropolitan Water District of Southern California turned back 400,000 acre-feet of its entitlement to the State Water Project. For the district to have gotten the water, a complicated exchange would have had to take place, with San Francisco releasing some of its Hetch Hetchy water to the district and being compensated with state water from the South Bay Aqueduct. Competition was intense for any share in the state's extra water, and the Turlock District was passed over in the allocation. A brief effort to win federal emergency financing for new wells

Pumps kept water in the TID canals during 1977.

was abandoned becasue it would have involved unacceptable delays and red-tape.[29] The last apparent opportunity to add to the irrigation supply was to use some of the 309,000 acre-feet of "dead storage" in the minimum pool behind Don Pedro. The minimum pool, below which the reservoir was not supposed to be lowered, was originally established to maintain an adequate head on the powerplant. To deplete that "untouchable" water required the approval of the Modesto Irrigation District, the Federal Power Commission and the Department of Fish and Game. Permission was granted by the middle of June, and as the draw-down proceeded, the top of old Don Pedro Dam surfaced in August. When the districts stopped taking water from Don Pedro in mid-September, the TID had used a little less than 60,000 acre-feet from the minimum pool.[30]

Between the use of over 300,000 acre-feet of pumped water and 123,000 acre feet from Don Pedro, and with extremely cautious use by irrigators, the district managed to avoid an agricultural disaster. It was, of course, a tense and worrisome season, and pointed questions were asked about why the TID had so little water in Don Pedro, especially since the Modesto District planned to finish 1977 with 40,000 acre-feet still in storage. Criticism centered on the release of water during the two winters prior to the drought for power generation rather than keeping it in storage. The explanation offered by TID officials was that the district had decided to make power in the winter of 1974-1975 to bolster its sagging finances. The next year found the reservoir at an even higher elevation, so the same generation schedule was adopted. The decision on power generation made each fall could not be easily altered since constant releases for salmon spawning were required from November through most of December, with no more than a four-inch reduction in river level permissable thereafter until the end of March. By the time it became apparent that the winter of 1975-1976 would be extremely dry, the district had already committed itself to a release schedule that sent 365,000 acre-feet down the river before the 1976 irrigation season started. The problem was compounded by the decision not to reduce the irrigation allotment for 1976.[31]

Criticism of the district's water management policies and the electric rate structure adopted to offset large purchases of outside power was reflected in the most conroversial election campaign in decades. Four of the five board members, including one appointed that summer after long-time director Everett Tomlinson died, had to face the voters in November 1977. One group of vocal dissidents, calling itself VOTEMOUT (Victims of Turlock's Energy Mismanagement On Utility Troubles), attacked the all-farmer board as being unrepresentative of the electric rate payers and levelled other charges of mismanagement at the incumbents. Further allegations were made that district's energy conservation efforts were too little and too late, that a series of district newspaper advertisements explaining water and power policies were essentially political ads for the incumbents and that the appointment of Phillip Short to fill the vacancy created by Tomlinson's death was improper because his house, though not his legal residence, was just outside district boundaries.[32] Fourteen candidates, including ex-General Manager Louchart, competed for the four seats. Only one of the incumbents was defeated, although in two other divisions, the directors received less than a majority of the votes cast.[33] The election did not alter the make-up or outlook of the board to any significant extent, but the controversy surrounding it was one indication of the severity of the district's worst water and power crisis since 1914. Although the issue of rural versus urban control was raised in 1977, it did not, in the short run at least, become a major point of contention. Several contested elections occurred after 1977, including one in 1983 that drew a large number of candidates critical of district power policies. However, voter turnout remained low and no incumbent has been turned out of office since 1977.

Happily, the rain and snow Mother Nature had been withholding for two years arrived the next winter. Although forecasters had predicted that it would take at least two years to recover from the drought, Don Pedro returned to normal levels in a single season. Irrigation resumed its usual routine quickly after 1977, although the board was naturally more cautious about husbanding the district's water supplies when it considered each winter's power generation program. If it had done nothing else, the drought had refocussed attention on irrigation at a time when the TID and the Turlock region were undergoing some of the most important changes since the arrival of water three-quarters of a century earlier.

CHAPTER 14 – FOOTNOTES

[1] Board minutes, vol. 43, p. 32 (July 12, 1971), vol. 43, p. 54 (Aug. 9, 1971).
[2] Board minutes, vol. 43, p. 119 (Nov. 15, 1971).
[3] William R. Fernandes interview.
[4] Author's interview with Charles D. Crawford, Mar. 7, 1983.
[5] Reynold Tillner interview.
[6] *Modesto Bee,* March 23, 1975.
[7] Richard Vollrath interview.
[8] Board minutes, vol. 42, p. 178 (Sept. 15, 1969).
[9] *Ibid.,* pp. 262-265 (Feb. 24, 1970).
[10] *Modesto Bee,* Dec. 6, 17, 1973.
[11] *Ibid.,* Jan. 7, 1975; Board minutes, vol. 44, p. 369 (Jan. 6, 1975).
[12] *Streams in a Thirsty Land,* pp. 156, 231.
[13] The number of electric meters can be found in the district's *Annual Report.*
[14] TID, "Comparison of Crops" (annual).
[15] See James Jamieson, Sidney Sonenblum, Werner Z. Hirsch, Merrill R. Goodall and Harold Jaffee, *Some Political and Economic Aspects of Managing California Water Districts,* Pub. No. 190, Institute of Government and Public Affairs, (Los Angeles, 1974).
[16] Memorandum to the Board of Directors on commercial-industrial power rate, Sept. 28, 1959, in Meikle files, vol. 30, item 69; Board minutes, vol. 35, pp. 390-391 (Mar. 6, 1959).
[17] Board minutes, vol. 35, p. 405 (Mar. 23, 1959).
[18] *Modesto Bee,* Aug. 8, 1967; David Lincoln Martin, "California Water Politics: The Search for Local Control," (Ph.D. thesis, Claremont Graduate School, 1973), pp. 42-48.
[19] Board minutes, vol. 44, p. 293 (Sept. 30, 1974).
[20] *Ibid.,* p. 374 (Jan. 20, 1975).
[21] *Modesto Bee,* Oct. 22, 1976.
[22] *Ibid.,* May 18, 1978.
[23] *Ibid.,* Mar. 14, 1976.
[24] California Dept. of Water Resources, *The California Drought, 1977: An Update,* (Feb. 15, 1977), p. 5.
[25] *Ibid.,* p. 65.
[26] *Modesto Bee,* Feb. 8, 15, March 24, 1977.
[27] *Ibid.,* Mar. 22, April 8, May 17, 1977.
[28] *Ibid.,* Apr. 21, Oct. 4, 1977; *Turlock Daily Journal,* Sept. 28, 1977.
[29] *Modesto Bee,* Feb. 24, Apr. 21, 1977.
[30] *Ibid.,* May 27, June 17, Aug. 3, Sept. 12, 1977.
[31] *Ibid.,* Oct. 16, 1977.
[32] *Ibid.,* Sept. 5, 13, 15, Oct. 4, Nov. 4, 1977.
[33] *Ibid.,* Oct. 18, Nov. 9, 1977.

15

The Search for Power

Electricity was originally thought of as a by-product of irrigation; a profitable "cash crop" for an irrigation district. In the years following the Second World War, rapid increases in the amount of power demanded by the district's customers gradually turned power, especially the question of finding enough of it, into the district's principal preoccupation. The reasons why an energy surplus became an energy deficit are not hard to find. Population growth was an obvious factor. Somewhat less conspicuous was a pattern of rising consumption, as each of the district's customers, on the average, used more and more energy. That fact is indicated by a comparison of the number of meters with the amount of energy delivered by the distribution system. Such a comparison does not take into account the effects of large industrial users or agricultural pumps, but it can still establish a general trend in demand. In 1950 the system handled about 6,634 kilowatt-hours (kwh) annually per meter, but in 1960, the number was up to over 10,000 kwh, and in 1970, over 17,000 kwh.[1] In 1975 the average residential customer used over 10,000 kwh per year.[2]

One of the principal causes of rising energy use was the increasing use of air conditioners. Early settlers had only shade trees, once they grew, to combat hot summer temperatures. Later, homemade evaporator or "swamp" coolers were built. These utilized a fan to blow air through sacks or fibers kept damp, and therefore cooler, by dripping water. Manufactured evaporator coolers were available after World War II, but they were less effective than refrigerated air conditioning. Unfortunately, air conditioners gulped huge amounts of energy. In his testimony to the Federal Power Commission in 1962, the MID's Clifford Plummer commented that in earlier years, "our peak demand was experienced in September and October due to large requirements by the wineries and continued requirements by the canneries. As will be noted this peak shifted in 1959 to August. The greater utilization of air conditioning throughout the region is responsible for this shift in peak."[3]

Until the end of World War II, the only sources of power available to the TID were its Don Pedro and La Grange plants, with emergency back-up service from the San Joaquin Light and Power Company, and its successor, PG&E. Except in the critically dry year of 1931, the district had had enough power to meet its own needs as well as the requirements of its wholesale contract. The 1945 agreement with San Francisco to take Hetch Hetchy power came fortuitously at a time when the district's demand was finally beginning to exceed Don Pedro's capacity. Although the TID was able to continue wholesaling power at times when Don Pedro's generation, which was dictated by irrigation and flood control requirements, exceeded what its consumers needed, it had to begin buying power in 1946. Until 1947 Don Pedro had enough capacity to supply all the power needed at the time of peak demand on the system, but thereafter, additional power had to be purchased to meet the peak load.[4]

The district's first connection to Hetch Hetchy was indirect. The city's transmission line crossed the Modesto District, so the MID was able to tie its system directly to San Francisco's. Modesto purchased whatever power the two districts needed, and then resold Turlock's share to the TID over a line running to the TID's Ceres substation. That practice continued until 1956, when the TID built its own transmission line from a Hetch Hetchy substation south of Oakdale to the district's Tuolumne substation.

The districts' contract with the city committed San Francisco to supply whatever energy the two districts required, even if it exceeded the amount available from Hetch Hetchy facilities. To make up any shortfall, the city bought power from PG&E and resold it to the districts, passing on the higher costs of such supplemental purchases. Further development of the upper Tuolumne watershed increased the supply of Hetch Hetchy power. Cherry powerhouse, with a capacity of 135,000 kilowatts (KW), went into operation in 1961, and a few years later, the Kirkwood powerhouse at Early Intake added another 67,000 KW to the system. New Don Pedro dramatically increased the TID's own generation, but by the time its powerhouse was in full production, it could not meet more than two-thirds of the Turlock District's peak demand nor more than the same proportion of the district's annual energy demands.[5] The situation was even less favorable in the Modesto District, which had a greater population but a smaller share of Don Pedro.

The Hetch Hetchy contract guaranteed the sale of whatever

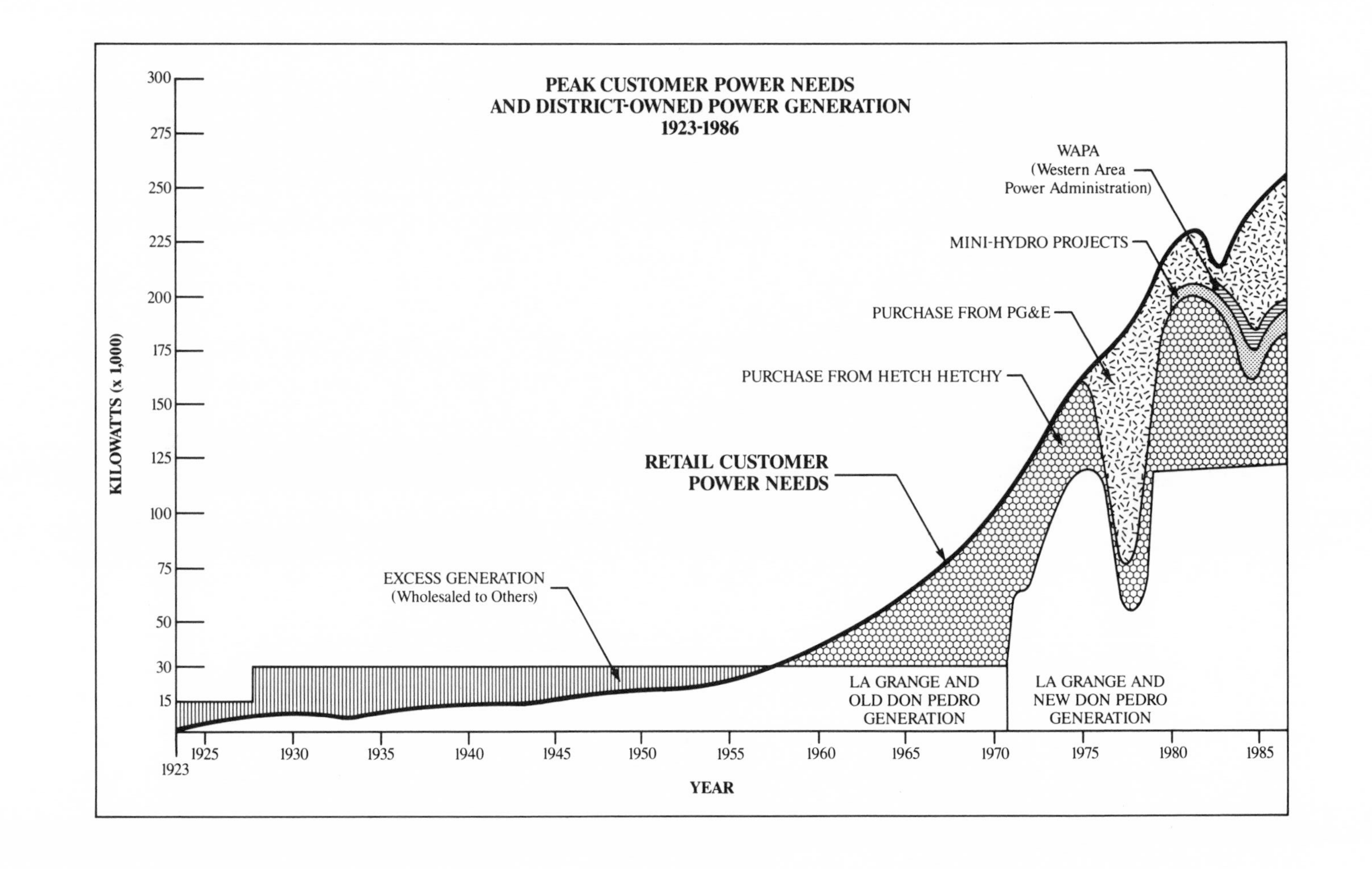
PEAK CUSTOMER POWER NEEDS
AND DISTRICT-OWNED POWER GENERATION
1923-1986
KILOWATTS (x 1,000)
300
275
250
225
200
175
150
125
100
75
50
30
15
WAPA
(Western Area
Power Administration)
MINI-HYDRO PROJECTS
PURCHASE FROM PG&E
PURCHASE FROM HETCH HETCHY
RETAIL CUSTOMER
POWER NEEDS
EXCESS GENERATION
(Wholesaled to Others)
LA GRANGE AND
OLD DON PEDRO
GENERATION
LA GRANGE AND
NEW DON PEDRO
GENERATION
1923
1925
1930
1935
1940
1945
1950
1955
1960
1965
1970
1975
1980
1985
YEAR

additional energy the districts needed, but it did not guarantee that the districts would be able to physically link adequate outside power sources to their own distribution systems. Projections made in 1973 showed that as early as the summer of 1976 the transmission lines carrying Don Pedro and Hetch Hetchy power, including the supplemental power from PG&E, would be inadequate to meet the combined demands of the two irrigation districts. Rolling blackouts were thought to be possible if the districts did not get access to more power.

The solution to the problem of transmission capacity was a new 230 kilovolt line connecting the districts to PG&E's main north-south transmission system.[6] The new intertie would be a joint project of the Turlock and Modesto districts running from the PG&E lines on the west side of the valley through the Turlock District to a point where the line would divide, with branches running to the MID and to the TID's Walnut substation, west of Turlock. The cost of the joint portion of the intertie was to be borne equally by the two partners, and each would pay the full cost of its own branch. The project encountered several objections to the location of the big powerline, especially to the section of the Modesto branch that followed the TID's Ceres Main Canal, but a lawsuit by the City of Ceres failed to halt it.[7] The pre-assembled steel poles that carried the line were set in place by helicopters in 1975, and by the summer of 1976, it was ready to carry power.

The intertie permitted more energy to flow into the districts, but when it came from PG&E, it carried a high price tag. Unless the districts found less expensive alternative sources, they were going to have to buy more and more of the private utility's costly power to supplement Don Pedro and the Hetch Hetchy system. To make matters worse, San Francisco was expected to gradually increase its own use of Hetch Hetchy power for municipal purposes, leaving less to sell to the irrigation districts.[8] With forecasts of a rapidly growing gap between the supply of reasonably priced power and consumer demand, the Turlock and Modesto districts were receptive to any opportunity to increase their own power resources, for it was generally cheaper to own and operate power plants than to buy wholesale power from outside suppliers.

One potential opportunity to develop additional power came in 1972. In January of that year officials from both districts met quietly with PG&E at the company's request, to discuss the joint construction of a nuclear powerplant in the foothills of eastern

Stanislaus County.[9] Nuclear plants require large amounts of cooling water, and for that reason, most had been located along the coast. The passage of California's coastal protection initiative in 1971 made powerplant construction there more difficult, forcing utilities to look inland for plant sites. It was water that drew PG&E to the Tuolumne River irrigation districts. A plant with two 1,100,000 KW units would need 50,000 acre feet of water annually, and it was hoped that the districts would be willing to supply it.

The question, of course, was where they would get that much water. The Tuolumne River was the most obvious source, and was undoubtedly what PG&E had in mind when talks started.[10] However, the districts were soon warned by their attorneys and by their consultant, former Department of Water Resources director William Gianelli, that their existing water rights did not cover the use of water for powerplant cooling. Amending their rights or filing for new ones would almost certainly lead to another controversy over the fishery or other environmental effects, so the use of river water from New Don Pedro was never seriously pursued, even though it continued to receive limited study.[11] As alternatives, Gianelli suggested that either deep or shallow wells on the west side of the districts could yield the needed water, or perhaps some means could be found to collect water routinely spilled into the rivers at the ends of the districts' laterals.[12] The engineering firm of Bookman, Edmonston was hired to evaluate these possibilities, but even before their report was completed in late 1975, groundwater from wells near the mouth of the Tuolumne had become the focus of discussion.[13]

In 1974 the districts signed a letter of intent with PG&E committing them to supply water to the proposed plant, but the terms under which that would be done remained hazy. Despite these considerable uncertainties, the districts proceeded with plans for a cooling water supply. In July 1975 they formally agreed to share the costs and benefits of supplying water to the plant on an equal (50/50) basis. The districts had hired Arthur D. Little, Inc. to evaluate their future needs and the best ways to meet them. That study, completed in April 1975, endorsed participation in a nuclear plant as the best means of complementing the districts' hydroelectric resources.[14] Participation, in the form of co-ownership of the plant, was what the districts had had in mind when negotiations with PG&E first began. They had hoped to use their water as a bargaining tool to win a share in the project for themselves. PG&E,

on the other hand, finally made it known that it wanted only a water service contract that would repay the districts for their investment in wells and pipelines, and reimburse them for the water delivered at the same rate they charged farmers for irrigation water.[15] The company did not rule out letting the districts share in the plant, but it was to be treated as a separate issue. That was hardly what the districts had in mind. Negotiations on all aspects of the project were continuing in the summer of 1977 when URS Research, hired in 1976 to study half a dozen alternative water supplies, and select the best, submitted its report. To no one's surprise, they chose deep wells in the Jennings Road area west of Keyes. The water would be carried in a pipeline eastward to the foothills and across the Tuolumne to the site on the Barnett ranch, near Roberts Ferry, where the reactors would be located.[16] PG&E used that information when it submitted its formal Notice of Intent to the California Energy Commission in August 1977. The commission rejected the application the following month.[17] Although PG&E kept the project alive for several more years and waged various legal battles on its behalf, the California Energy Commission's rejection of the proposal essentially finished the East Stanislaus Nuclear Project. No further discussions on water supply or participation were held.

In the irrigation districts as elsewhere around the country, sentiment was turning against nuclear powerplants, and concerns over safety and the high cost of construction would certainly have made the project even more controversial than it already was in the local area. Although nothing concrete came out of half a dozen years of negotiations and studies, the nuclear project was a milestone in the history of the Turlock Irrigation District. It focussed the district's attention on its long-term energy future, and for the first time, the district looked beyond the hydroelectric resources of the Tuolumne River for a major project.

Rising electrical demand and the costly steps taken to meet it led to increased rates for district customers. Although the district's financing consultants had promised in 1961 that rates would not have to go up to pay for New Don Pedro, by the time the new powerplant was in operation in 1971, it was clear that steeper than anticipated increases in power use and the high cost of supplemental PG&E power made a general rate increase unavoidable. Rate consultant Roy Wehe was hired in 1971 to make the first thorough review and revision of the district's rates in nearly sixty years of power distribution. Wehe's recommendations, which included an average increase of 16 percent in domestic rates and slightly lower increases in the lighting and commercial-industrial catagories, were adopted by the board in April 1973.[18]

Six months later, it appeared that rates would have to be raised again to pay for the planned intertie line, and Wehe was asked to suggest further revisions.[19] He had no sooner been hired to plan the next round of rate increases than district officials began discussing a temporary surcharge to quickly raise more money while the rate study was underway. For three consecutive years, the district had used its reserves to fund budget deficits, and by early 1974, its already anemic finances were weakened further by the effects of the Arab oil embargo and the nationwide energy crisis. The cost of PG&E's oil-fired power soared, and with it, the cost of supplemental energy purchased by the districts. At the same time, the emphasis on energy conservation reduced consumption and cut the district's revenues. The combination of higher costs and lower income threatened to undermine the TID's standing in the financial marketplace, which could in turn affect the sale of the warrants needed to pay for the intertie. On April 1, 1974, the board approved a 13 percent surcharge on all electric bills and doubled the irrigation water charge, first imposed in 1972, to $2.00 per irrigated acre.[20] The surcharge was replaced in December by a new rate increase of about 14 percent, but in March 1975 another surcharge, this time for 7 percent, was added to electrical bills and the water charge was advanced to $2.50 per acre. The latest surcharge came at the behest of Stone and Youngberg to help market unsold New Don Pedro bonds needed to cover additional expenses on that project.[21] Everywhere the district turned, it seemed to encounter higher costs. Unfortunately, the period of rapid growth in population and electrical demands coincided with a trend toward scarcer and more expensive energy. Still, the various increases and surcharges sound more devastating than they were. The district's rates before 1973 had been so low that even with two general rate increases and two surcharges within two years, they remained well below those charged by private utilities.

As the cost of PG&E power escalated, the TID could take comfort in its Hetch Hetchy contract, which provided a large quantity of low-cost power from San Francisco's hydroelectric plants. All that changed suddenly in June 1976 when Hetch Hetchy manager Oral Moore informed the two districts that the city could no longer sell them so-called Class 3 power at the price provided for in their existing contracts. The Class 3 category covered 90 percent of the power the districts purchased from Hetch Hetchy. The remainder was power for agricultural pumping and municipal uses that, under the Raker Act, had to be sold to the districts at cost, and was now referred to as Class 1 power. Class 2 power had been a temporary share of the output of the Cherry powerhouse granted by the Third Agreement and had expired when New Don Pedro was completed. Moore acknowledged that the city had an obligation to supply power to the districts until 1985 and he admitted the price of Class 3 power was fixed by the contract, but he insisted there was nothing to keep the city from selling its Hetch Hetchy output directly to industrial customers assigned to it by PG&E, at PG&E rates, rather than to the districts. In the event it did so, it would have no Class 3 power to sell to the irrigation districts and would supply them entirely with costly PG&E power. The alternative was for the districts to agree to rates "more in keeping with market values," which the city thought meant what PG&E was getting for its power. Until they came to terms, Moore announced, the districts would get only PG&E power.[22] San Francisco also raised the rates it charged the airlines at San Francisco International Airport and Norris Industries, the operator of the old aluminum plant- turned-arsenal in Riverbank.

The districts asked the secretary of the interior to intervene, contending that the Raker Act required him to approve Hetch Hetchy rates. In March 1977 the districts sued San Francisco to halt its alleged violation of the Raker Act, and in May, Secretary of the Interior Cecil Andrus joined in the suit on grounds that he had to approve the rates and that they should be based on the cost of production rather than on what the market would bear.[23] The city based its counter-argument on a novel interpretation of the language of the Raker Act. Section 9 (o) required that rates "conform to the laws of the State of California," or in the absence of appropriate state regulation, be approved by the secretary of the interior. The California Public Utilities Commission could not legally exercise control over Hetch Hetchy so the secretary had always been viewed as the final authority on power rates. The city ingeniously contended that its city charter was a state law, and that since its own Public Utilities Commission, which was also responsible for Hetch Hetchy management, had rate-making authority, it was also entitled to decide what rates were appropriate.[24]

The U.S. District Court in Sacramento that heard the districts' suit ruled in February 1978 that Hetch Hetchy power could indeed be sold only at rates approved by the secretary of the interior, though it did not necessarily have to be sold to the irrigation districts. The ruling was a victory for the districts, but a similar suit brought by the airlines hit with sudden rate increases had an altogether different outcome, as the federal court in San Francisco decided in favor of the city.[25] Two different federal district courts had given contrary opinions in essentially the same case. The whole matter then went to the Ninth Circuit Court of Appeals, which upheld the city's point of view. The districts appealed to the U.S. Supreme Court, but even as they did so, their cause was being undermined by the secretary of the interior. In a January 3, 1980, letter, San Francisco mayor Diane Feinstein asked Secretary Andrus not to appeal the circuit court's decision, and on January 22 she met with him in Washington. At that point, Andrus agreed to reverse himself and drop the case. Speculation as to the reason for the abrupt about-face turned to politics. Andrus's boss, President Carter, had supported Feinstein in a recent election and the mayor was in turn actively backing the president. Some called the decision a political wedding gift to the newly married Feinstein.[26] Whatever his motives may have been, Andrus had repudiated the position assumed by other interior secretaries, going back to Harold Ickes and before, that San Francisco had to be held accountable to the terms, and the intent, of the Raker Act. Solicitor General Wade McCree, Jr. wrote the Supreme Court in April 1980 that, "the United States agrees with petitioners (TID and MID) that the Court of Appeals erred" because the decision did not recognize the congressional intent to provide an independent review of rates and low-priced public power. Despite that fact, McCree concluded that the case lacked enough "general importance" to warrant a review by the Supreme Court.[27] A month later the Supreme Court declined to hear the districts' appeal.

At the outset of the controversy, the city began billing the districts at PG&E rates for all the power it delivered. The TID recomputed each bill and paid only the contractural price of Class 3 power. In 1978 San Francisco returned to billings based on existing Class 3 rates, but until 1981 included an "advisory" billing showing what the billing would have been with all energy at PG&E rates. The billing dispute remained unresolved, but it appeared that for all practical purposes, Class 3 power continued to flow at the old rates, and would do so until the expiration of the contract in

1985.[28] At that time, if the districts remained customers of the Hetch Hetchy system, it would be at drastically higher rates. The last source of inexpensive outside power was disappearing.

San Francisco's attempt to raise Hetch Hetchy rates was not the only bad news that came the district's way in 1976. That was the first of the back-to-back drought years, and although irrigation deliveries were maintained at regular levels, electrical generation at Don Pedro fell off dramatically. In 1974 and 1975, the TID had been able to meet about two-thirds of its annual energy demand from its own sources, but in 1976 Don Pedro and La Grange provided only 42 percent of what was needed.[29] Purchasing more outside power meant another rate increase to cover the additional costs, this time a 10 percent hike that went into effect on July 1, 1976.

By December 1976 the financial ravages of one dry year and the increasing likelihood that another one was beginning prompted yet another electric rate increase. The board raised rates 10 percent and added a further flexible charge that would vary depending on the amount of power the district had to buy from Hetch Hetchy and PG&E. The rate increase was designed to carry the district through another year like 1976, but it was soon clear that there would be even less water, and less hydroelectric power, in 1977. By May the district was faced with the grim prospect of spending $5 million more than the budget adopted only a few months earlier had called for. To deal with the emergency a new rate schedule went into effect on June 1 which adjusted the way consumers were billed for the extra power the TID had to buy. The district committed $666,666 a month to purchase outside power and any amount over that would be passed on directly to the rate-payers in the form of a separate "purchase power charge" on their bills.[30] The additional charge on top of the district's regular rates caused some confusion and even anger from customers, but it did remind residents just how dependent the TID was on outside generation, especially PG&E's fuel-fired plants since Hetch Hetchy power production was also curtailed by the drought. In fact, during 1977 Don Pedro produced less than 7 percent of the district's energy.[31] With its Don Pedro transmission lines carrying very little power, the value of the year-old intertie was confirmed.[32]

Ironically, as the drought was beginning in 1976, the Turlock Irrigation District began to examine additional hydroelectric facilities. New Don Pedro was the last major storage reservoir contemplated for the Tuolumne River, but even though little more could be done to add to the availability of water for irrigation or domestic use, the untapped power potential of the river was enormous. The district's Groveland project and the competing plans of the Yosemite Power Company were the first attempts to exploit the fall of the river and its tributaries between the Hetch Hetchy system and Don Pedro. These projects, with the exception of Meikle's plans for a dam near Wards Ferry, emphasized the development of the tributary forks rather than the main stem of the river. San Francisco's ability to divert so much of the summer flow into its Mountain Tunnel at Early Intake made it unlikely that a project of any size on the main river between that point and Moccasin Creek would have enough water, except in flood seasons or wet years. All that changed when the city modified its plans for the Cherry and Eleanor reservoirs to have them pour their waters into the river below Early Intake. A map of future Hetch Hetchy construction by Max Bartell, drawn in 1937 and revised in April 1941, showed a proposed 50,000 acre-foot regulating reservoir near the confluence of Cherry Creek and the Tuolumne River, and a further revision in December 1941 added a 500 second-foot diversion from that reservoir to a powerhouse further down the river.[33] The details were sketchy, but it was the sort of project that Meikle and others undoubtedly had in mind when they added Article 19 to the Third Agreement to cover the division of any project to develop power between Cherry powerhouse and New Don Pedro.

Less than a year after the Third Agreement was signed, R. V. Meikle applied for water rights on behalf of the Turlock and Modesto districts for a power project like that outlined on San Francisco's 1941 map. The application described a diversion dam just below Cherry Creek and a ninteen mile, ten-foot-diameter tunnel to a powerhouse near Wards Ferry, which would be at the upper end of New Don Pedro reservoir.[34] Although there were no immediate plans to construct the project, estimated in 1954 at 60,000 KW, the districts kept the water rights permit granted in 1953 alive.[35]

In 1968 San Francisco had R.W. Beck and Associates draw up a long-term plan for additions to the Hetch Hetchy system. A three-stage proposal was presented that included, as its first stage, a power project on the same stretch of the river covered by the districts' permits. What came to be known as the Clavey-Wards Ferry project called for a regulating reservoir below Cherry Creek to control releases from upstream Hetch Hetchy facilities and divert them into a tunnel leading to a reservoir on the Clavey River, a tributary flowing into the Tuolumne from the north. From the reservoir on the Clavey, water would be released into a tunnel leading to a powerhouse at the mouth of the Clavey capable of generating 300,000 KW. A dam at Wards Ferry would back up water almost to the Clavey powerhouse and would be able to produce an additional 100,000 KW.[36] The Clavey-Wards Ferry project was designed primarily to provide peaking power by releasing large amounts of water through its turbines in a short period of time to meet daily peak electrical demands. Because the facilities used to produce peaking power sit idle or run at only partial capacity most of the time, the price of each kilowatt-hour of energy they generate must be priced relatively high to recover the cost of the plant. That makes it the most expensive kind of power to buy, and the most valuable to own or sell. However, in 1968 when the Beck report was prepared, the power the Clavey-Wards Ferry project could produce was still too expensive to justify construction.

Engineers were not the only ones interested in the Tuolumne River below the Cherry powerhouse. In late 1968 Gerald Meral and Richard Sunderland, both members of the Sierra Club and avid whitewater kayakers, floated down the river from Lumsden Bridge, about five miles below Cherry Creek, to Wards Ferry. Several trips with kayaks and whitewater canoes were made in 1969, establishing a new recreational use for the river.[37] Although at high flows its rapids could be more challenging than those found on other popular whitewater streams, by 1972 commercial rafters were ready to begin carrying passengers from Lumsden to Wards Ferry for two or three days of wilderness adventure. The Forest Service conducted an environmental assessment of the impact of rafting in 1972 and the following year prepared to issue permits for the activity.[38] The interest in recreational use of the river coincided with and complemented efforts to place the Tuolumne in the wild and scenic river system authorized by Congress in 1968. Rivers, or sections of rivers, could be designated wild or scenic to protect them from dams or other developments that could alter them. Suggestions that the Tuolumne should be protected came as early as 1970.[39] In 1974 legislation, supported by both California senators and local Congressman John McFall, providing for the studies needed to justify wild and scenic designation for the Tuolumne passed Congress, and it was signed into law in 1975.

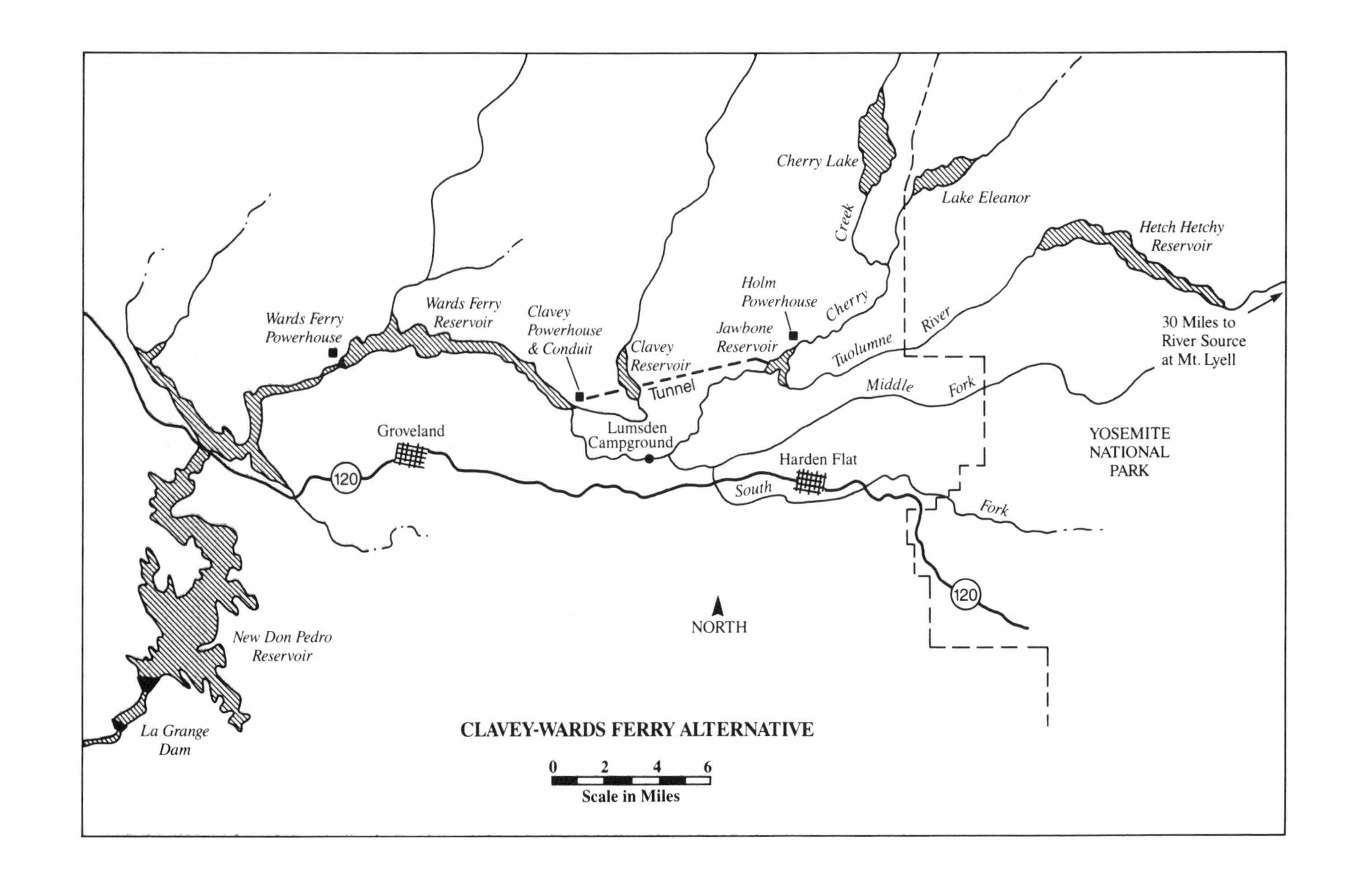

CLAVEY-WARDS FERRY ALTERNATIVE

As the study began, the districts and San Francisco hired R.W. Beck to update its 1968 report to give the federal study team accurate information on the hydroelectric potential of the river and on plans to develop it.[40] Beck reported in early 1976 that the economic feasibility of the project had vastly improved since 1968 because the cost of alternative power sources had gone up more rapidly than the projected cost of the Clavey-Wards Ferry scheme. As depicted in the new Beck report, the project appeared to be a good source of future power, so in April 1976 the Turlock and Modesto districts requested a Federal Power Commission preliminary permit for it.[41] If granted the permit would give the districts a priority over any other developers and allow them three years to study the physical and economic feasibility of the project and its environmental effects. The districts were now committed to pursuing the new project, and the stage was set for a confrontation between protecting the river as it was or developing the last available section of its main stem.

Since the wild and scenic study process was already underway, the districts, which had been joined by San Francisco under its Third Agreement rights, faced an uphill battle in their quest for a preliminary permit. Supporters of a wild and scenic river worried that even though such a permit was no more than a right to make studies, its issuance would legitimize arguments for more dams on the river and possibly diminish the chances of winning federal protection. By July 1977 the Department of Fish and Game and the California Department of Navigation and Ocean Development, Tuolumne County and the Sierra Club had all filed petitions with the FPC to intervene in the permit process and, they hoped, prevent any action on the Clavey-Wards Ferry project.[42] Over a year later the commission, now renamed the Federal Energy Regulatory Commission (FERC), finally decided to withhold a permit until the districts and San Francisco submitted a detailed work plan for review by federal and state agencies and the Sierra Club.[43] The work plan requirement grew out of a fear that geologic and other on-site studies conducted under the preliminary permit would disturb the environment of the Tuolumne canyon at the very time its protection was being considered. In response, Beck drew up a proposal to use helicopters rather than roads and to restore work areas. An informal hearing, required by the FERC order, drew the predictable criticism of the work plan from the agencies and organizations already opposed to the preliminary permit on any terms.[44] The whole issue of a work plan was put to rest a few week later when the Forest Service announced that, regardless of any FERC action, no rights-of-way or special use permits would be issued for the Clavey-Wards Ferry project.[45]

Faced with such determined and effective opposition from federal and state administrative agencies and environmental activists, the districts fought back as best they could. The Turlock District took the initiative by hiring attorney Lee White, a former FPC commissioner, to lobby Congress and federal executive agencies on behalf of the project and against a wild and scenic designation.[46] The study team's draft report issued in June 1979 stopped just short of endorsing wild and scenic status, instead calling that the "preferred alternative" for the main stem of the Tuolumne from its headwaters to Don Pedro reservoir except for the short sections already occupied by Hetch Hetchy reservoir and the Early Intake complex. Hearings were held in Columbia, Modesto, Oakdale and San Francisco during the ninety-day period provided for public review and comment on the draft report. Wild and scenic status was supported by the State of California through its Resources Agency, and by environmental organizations and whitewater rafting companies. Friends of the River, an environmentalist group founded to oppose the innundation of the popular Stanislaus River whitewater run by New Melones Dam, presented a petition with over 18,000 signatures. The districts enlisted support from other water agencies, Chambers of Commerce and labor unions. San Francisco took an official position of neutrality on the wild and scenic question, despite its participation in the preliminary permit application.[47] Meanwhile, the Tuolumne was receiving national attention, thanks to an article in the September 1979 *Readers Digest* reprinted from the *New York Times* that carried an emotionally charged appeal to readers to help save the river from further development.[48] On October 2, 1979, President Carter, acting on advice from the secretaries of Interior and Agriculture, asked Congress to make the Tuolumne River part of the federal wild and scenic system, even though the formal study process was not quite finished.[49] The districts had lost the first round in their battle for the Clavey-Wards Ferry project; the next move was up to Congress.

At the other end of the hydroelectric spectrum from the massive Clavey-Wards Ferry project was an ambitious program to develop the power potential of irrigation canals. The idea of putting small powerplants in the canal system was not new. The earliest plans for electrical development by Burton Smith and R.V. Meikle both called for a plant at Hickman drop, where the main canal fell about eighteen feet. In southern California, the Imperial Irrigation District was already producing power from its canals. Although the energy potential of falling water in the canal system had long been recognized, it was not economically worthwhile to develop it until the price of oil and of alternative power sources soared in the early 1970s. The district's own engineering staff began studying the feasibility of installing small generators at three sites in 1976. The following year, the decision was made to build plants at two of the sites, Hickman drop and the outlet from Turlock Lake.[50] Rather than hiring an outside contractor, the district built the two plants with its own labor force. This was done in part because the TID already had the skilled personnel and most of the equipment needed, and in part because it could maintain full control of its canal system when it managed all facets of construction itself. The 1,100 KW plant at Hickman drop was built first, and it went into operation in October 1979. The Hickman plant and the other canal generators could only operate when water was in the canal, but the long irrigation season and the fact that it would be running during the summer, when peak electrical demand occured, made it a valuable new power resource. Work on the 3,300 KW Turlock Lake plant adjacent to the old outlet gate began in 1979, with completion the following summer.[51]

The district did not limit its search for suitable sites for the so-called "mini-hydros" to its own canals. Application was made for a government grant for help pay for a study of canals in the Merced, South San Joaquin and Oakdale irrigation districts as well as the Turlock District, and in 1978, funding was arranged and the study began. Fluid Energy Systems, Inc. reported in early 1979 that of nineteen sites originally considered, eight proved uneconomical, but the remaining eleven could produce benefits exceeding their costs.[52] The Oakdale District was found to have no usable mini-hydro locations. The TID and the other two districts developed plans for a cooperative construction and operation program to be paid for and managed by the TID, which would have the right to purchase the output of the plants on favorable terms. The Turlock District proposed an $80 million revenue bond issue to raise money for the mini-hydros, and in April 1980 district voters gave their overwhelming approval. Agreements signed with the South San Joaquin and Merced districts in June 1980 gave them ownership of the plants on their canals and provided for the purchase of

Hickman powerhouse was the first TID mini-hydro project.

Operators at TID's Broadway control room oversee the district's electrical system and operate the Don Pedro and mini-hydro power plants by remote control.

power by the TID.[53] Subsequent changes in federal law, however, resulted in a new arrangement. Instead of connecting TID transmission lines to each of the new powerplants, it became more lucrative to sell their output to PG&E and divide the profits, giving the TID more cash to pay for power purchased from San Francisco or PG&E.[54]

A total of eight sites in the three districts were selected for initial development. Besides the plants at Hickman drop and Turlock Lake, Dawson Lake on the main canal near La Grange was scheduled to receive a 4,000 KW plant, and two plants were scheduled for construction in the South San Joaquin District and three in the Merced District.[55] The network of mini-hydros, finished in 1984, made the TID a leader in that aspect of energy development. Together the plants could turn out over 20,000 KW, almost as much as Turlock's share of the old Don Pedro powerhouse. All eight powerhouses were operated by remote control from the TID's Broadway control room.

Only the highest drops could turn the conventional turbines used in the district's mini-hydros. The other, lower drops also had a power potential but its development would require a different technology. A device that looked like an oval waterwheel with hydrofoil vanes was billed by its inventor, Texas physician Daniel J. Schneider, as an answer to the problem of very low head power generation. District engineers happened to hear of the "hydrodynamic power generator" while doing research on mini-hydros, and in 1977 the TID agreed to participate in experiments of the invention.[56] With a U.S. Department of Energy grant to cover much of the cost, testing a model of the machine at the University of California, Davis, was to be followed by the installation of a prototype at Drop 6 on the main canal above Hickman. Delay after delay plagued the project, but the unit was finally installed just before the start of the 1983 irrigation season. The clanking, chain-driven machinery was run at reduced speed all during its first year. In the summer of 1984, it was finally brought up to full load, but after only six days of promising operation, one of the vanes broke loose and jammed the machine. Despite growing skepticism on the part of district officials, Schneider was still hopeful the problem could be remedied, but even if it could, it would be years before his invention would be ready for widespread use.[57]

The search for new sources of power was accompanied by an increased interest in energy conservation. At one time, of course, the TID had done all it could to encourage the use of electricity by consumers. The retail store was the symbol of those halcyon days of plentiful and inexpensive power. Long after other appliance stores had opened in the Turlock area, the TID store continued to do a substantial business based on its competitive prices and what was reputed to be unparalleled service.[58] By the 1960s, however, sales were down to the point that when the chief store clerk announced that he was leaving to open a Westinghouse store of his own, the district decided to end its appliance sales. Repair work and the sale of small electrical parts continued until the end of 1975, when the store closed completely.[59] The final demise of the store came at a time when conservation rather than consumption had become the byword. The oil shortage of 1973-1974 had made Americans energy conscious. Apart from the immediate crisis, saving energy was increasingly viewed by environmentalists and policymakers as a viable alternative to the construction of costly new powerplants. Residents of the Turlock Irrigation District were no less familiar with the trend toward conservation than electric users elsewhere, and several sharp rate increases made it obvious that electricity would cost more in the years ahead. However, even with the rate increases of the mid-1970s, TID rates were only about 40 percent of what PG&E charged its customers, due primarily to the district's ownership of powerhouses at Don Pedro and La Grange and the availability of low-cost Hetch Hetchy power. The district was slow to initiate conservation programs because its low-cost energy provided less of an incentive to conserve. The first concrete step toward reducing electric use came in the rate increase that took effect in March 1981. Until then the district used so-called "declining block rates," which charged less per kilowatt-hour as usage increased. Such rates were an obvious disincentive to conservation. They were replaced by a flat rate per kilowatt-hour for the winter months and a two-tier summer rate that increased the price on consumption over 850 kilowatt-hours per month. The summer rate penalty was designed to encourage savings when use was highest and the district had to buy the most expensive outside power. The district also launched an advertising campaign with hints on how to save energy. More active conservation measures adopted in the following years included home energy audits and a program to move the operation of swimming pool pumps to off-peak hours. It was a comparatively modest start, but the rising costs of additional power suggested that further efforts in that direction would soon become more cost effective.

Energy conservation had the potential to reduce the rate at which energy demands grew, and indeed, average residential use did stabilize and even decline slightly between 1979 and 1984.[60] However, as long as the district's population continued to rise or new industries arrived, the imbalance between the TID's needs and its existing resources could only grow worse. The mini-hydros were helpful, but they could not close the gap. The district could get no more Hetch Hetchy power than it was already purchasing, and as San Francisco's own use of its power increased, there would be less for sale to the districts. In addition, the city's legal victory on Hetch Hetchy rates guaranteed that whatever energy the districts did purchase could be almost as expensive as PG&E power. Like any utility, and especially a publicly-owned one, the TID had a

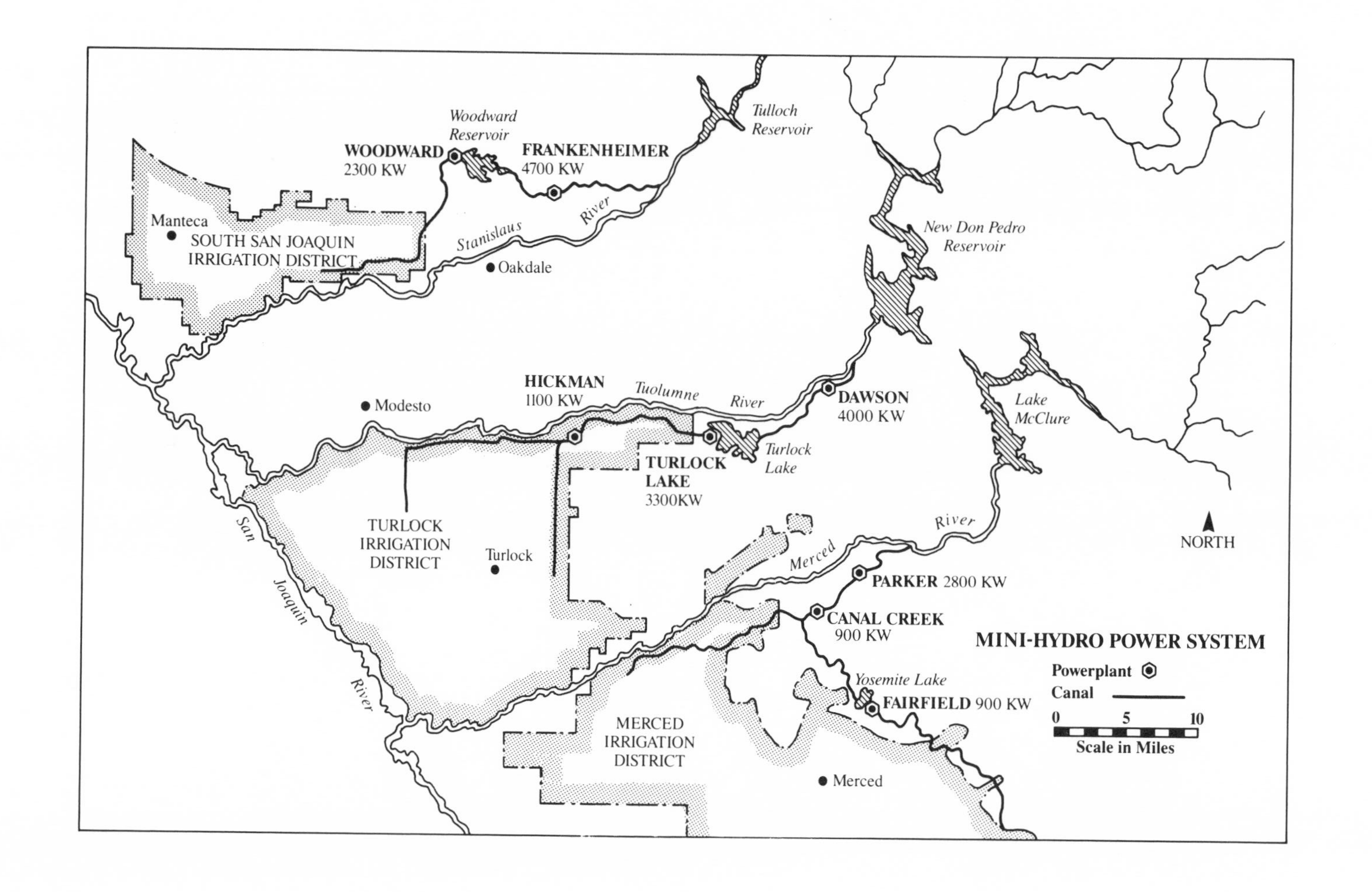
MINI-HYDRO POWER SYSTEM
Powerplant
Canal
0 5 10
Scale in Miles
NORTH
WOODWARD
2300 KW
Woodward
Reservoir
FRANKENHEIMER
4700 KW
Tulloch
Reservoir
Manteca
SOUTH SAN JOAQUIN
IRRIGATION DISTRICT
Stanislaus
River
Oakdale
New Don Pedro
Reservoir
HICKMAN
1100 KW
Tuolumne
River
Modesto
DAWSON
4000 KW
TURLOCK
LAKE
3300KW
Turlock
Lake
Lake
McClure
San
Joaquin
River
TURLOCK
IRRIGATION
DISTRICT
Turlock
Merced
River
PARKER 2800 KW
CANAL CREEK
900 KW
Yosemite Lake
FAIRFIELD 900 KW
MERCED
IRRIGATION
DISTRICT
Merced

responsibility to supply its rate-payers with the most economical power possible, but unless it found a way to acquire new generating facilities of its own, it would become more and more dependent on PG&E for its power, and its rates would approach PG&E's retail prices. For that reason, the TID did whatever it could to find new sources of power, and in the early 1980s the focus of its efforts became the Clavey-Wards Ferry project.

The president's recommendation in 1979 that the Tuolumne River be placed under the protection of the wild and scenic river system triggered an automatic three-year moratorium on development to give Congress time to consider the question. In 1980 Senator Henry M. Jackson of Washington introduced the legislation needed to make the formal wild and scenic designation, but it died in committee.[61] While the TID kept a watchful but unobtrusive eye on the situation, nothing further was done until the moratorium was near the end. In July 1982 Oakland congressman Ronald Dellums introduced legislation to protect the Tuolumne, largely because he believed it would also protect the tributaries from development. San Francisco had applied for preliminary permits to study several power projects on the South and Middle forks of the river, and some of the proposed dams would have affected Berkeley's city-owned camp used by Dellums' constituents.[62] Environmentalists, meanwhile, had been working with Senator S.I. Hayakawa on a bill to merely extend the moratorium one more year. Hayakawa's bill was introduced in August and was soon cosponsored by the state's other senator, Alan Cranston.[63] Even though the moratorium expired in October without congressional action on the bills, the Federal Energy Regulatory Commission promised not to issue a preliminary permit for the Clavey-Wards Ferry project until after the end of the congressional session in December.[64]

As plans for the big hydroelectric project emerged from the twilight of the moratorium, its partners were far from united. The TID was always the most enthusiastic backer of the project. Since shortly after the moratorium went into effect, it had continued to pay for Lee White's lobbying and Beck's limited engineering work, and as activity in Congress increased in late 1982, the district fought almost alone behind the scenes to keep the project alive. The Modesto District had shared the costs of earlier studies, but after 1980, it scaled back its financial support for a project that seemed on the verge of extinction, and took only a limited hand in the effort to avert last-minute congressional action.[65] San Francisco's position was the most complex. Mayor Feinstein was publicly in favor of wild and scenic designation, and the city's Board of Supervisors voted in November 1982 to support a continuation of the moratorium on development for another three years.[66] On the other hand, the city refused to formally abandon its interest in the Clavey-Wards Ferry project and was still studying alternative plans, including a second Mountain Tunnel to Moccasin powerhouse that would have had much the same effect on the river.[67]

Congress did not act in the closing days of the 1982 session and the way was cleared for the issuance of a preliminary permit, which was finally approved in March 1983. Three months later, the districts hired R.W. Beck for the $15 million, three-year study of the design, feasibility and environmental impact of the Clavey-Wards Ferry project.[68] As plans were made to begin work the political sky over the project began to darken. San Francisco strengthened its stand in favor of wild and scenic status and several bills were presented in Washington that would protect the Tuolumne. One measure of the city's dedication to stopping the project it nevertheless refused to abandon, was its effort to win support in Tuolumne County. It offered more and cheaper water for Groveland from its Mountain Tunnel as well as a share of any future non-Raker Act project in exchange for the county's support of wild and scenic status. On the other hand, the irrigation districts were also talking to county officials about a partnership in the Clavey-Wards Ferry project, which could provide the county with some much needed domestic water. The county chose to throw in its lot with the districts and eventually won the right to participate in both the water and power benefits of the project.[69]

Tuolumne County was one of several battlefields, but the war for the future of the river could only be won or lost in Congress. Senator Cranston had introduced a wild and scenic bill in January 1983, and just over a year later California's junior senator, Pete Wilson, announced a package that tied preservation of the Tuolumne to the designation of 1.7 million acres of California as wilderness. Wilson hoped to use the Tuolumne to win over critics who wanted to add a greater acreage to the wilderness system.[70] Since Cranston and Wilson did not agree on how much land should become wilderness, the linkage established by Wilson between the wilderness bill and the Tuolumne meant that as long as an impasse continued on the wilderness issue, it was unlikely that any action would be taken on the river.

To offset criticism that the Clavey-Wards Ferry project would dry up one part of the river and flood another with the Wards Ferry reservoir, R.W. Beck engineers drew up plans for a somewhat more benign project known as the Ponderosa alternative, which was unveiled in January 1984. The Jawbone diversion dam was little changed, but a longer tunnel was proposed to carry water under the Clavey River to a dam and forebay that would regulate the flow into penstocks leading to the 390,000 KW Ponderosa powerhouse, just north of the river three miles above Don Pedro reservoir. Another reservoir, Golden Rock, would be built on the South Fork. Releases from Golden Rock would replace whitewater rafting flows diverted into the tunnel. Since summertime rafting depended on releases from Cherry powerhouse, whitewater recreation was tied to the city's power generation schedules, which curtailed releases on Saturdays and eliminated them on Sundays. Proponents claimed Golden Rock releases could be managed in whatever way was best suited to the rafters. In addition, the Ponderosa alternative eliminated the Wards Ferry dam and reservoir and the Hunter Point dam on the Clavey River, so the impact of the project below the Jawbone diversion dam would be limited primarily to changes in streamflow. As originally designed, the Ponderosa plan could generate slightly less power than the Clavey-Wards Ferry project configuration, but planners later enlarged the prospective size of the Golden Rock reservoir and added a 50,000 KW powerhouse where its releases were discharged into the river near Early Intake. As a result, the new proposal would have produced more power than the old one. The Ponderosa alternative was clearly so much more attractive politically than the Clavey-Wards Ferry plan that by the spring of 1984 it had become the preferred alternative.[71]

It is hard to tell how much impact the Ponderosa alternative had on public opinion. Congressman Ed Zshau of Los Altos, whose district included part of Stanislaus County, had publicly endorsed wild and scenic status in November 1983 but in April 1984, he indicated that the emergence of the Ponderosa alternative had caused him to rethink his position.[72] "I haven't made up my mind on the Ponderosa project," he said, "I think the irrigation districts and the customers made a real good-faith effort to meet the concerns of people worried about losing rafting and the rugged inner canyon wilderness . . . I agree there's not enough information to tell if the environmentalists' concerns are still valid."[73] Zshau's comment that more information was needed was exactly what the

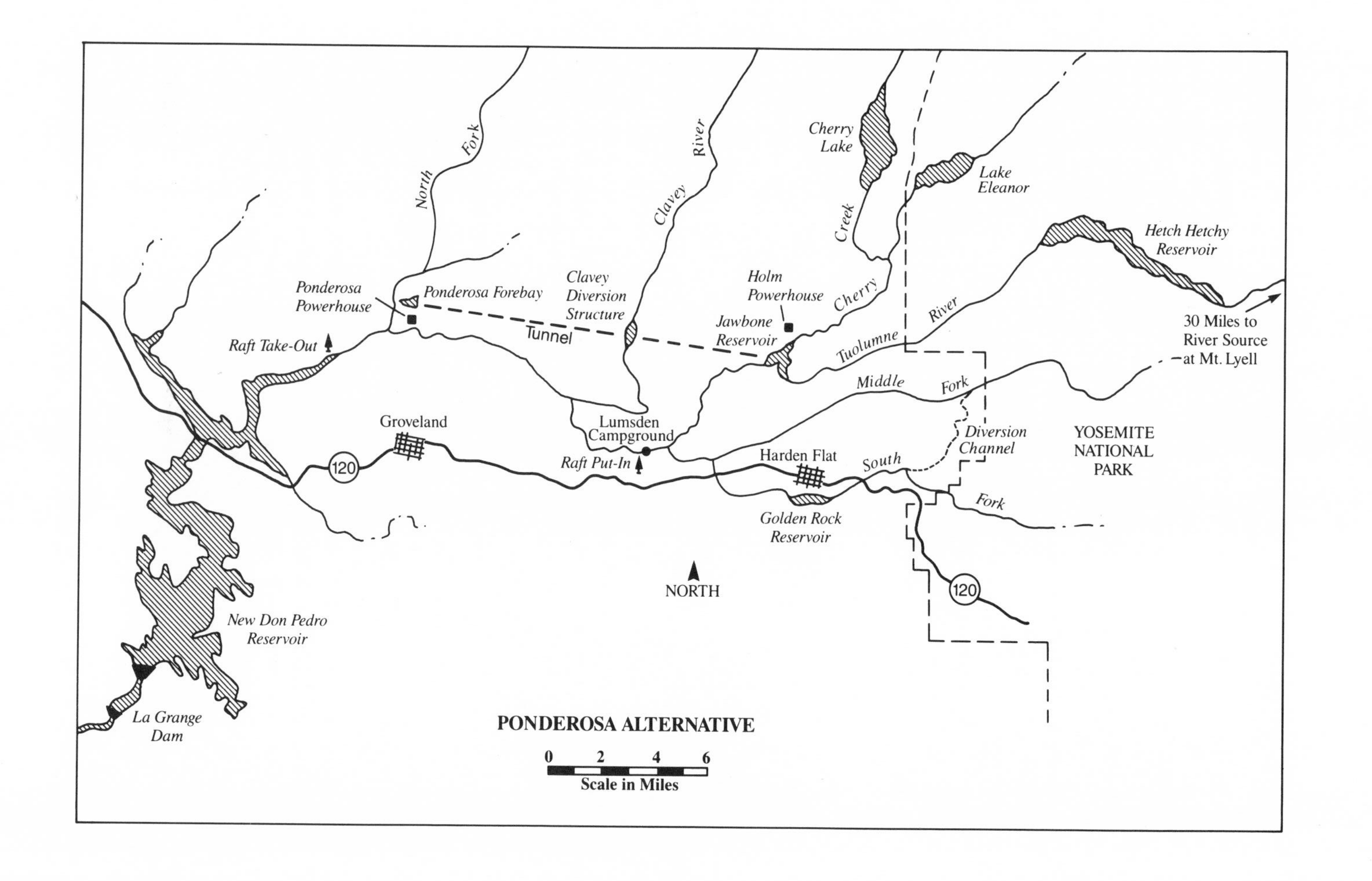

PONDEROSA ALTERNATIVE

districts wanted to hear. The project needed time more than anything else; time to complete the studies that would hopefully show it was a feasible development that would not unduly disrupt the environment of the Tuolumne canyon.

The Ponderosa alternative did nothing to defuse vocal opposition to any further development of the river. The new proposal, with its publicized provisions for water releases for the benefit of whitewater sports, implied that rafting was the primary obstacle to development. A group of civic and business leaders organized as PACE (Public Affordable Clean Energy) in early 1984 sounded the same theme in a newspaper advertisement.

> 9 to 5 . . . and never on Sundays.
> Water releases from the Holm Powerhouse on the Tuolumne River are timed during daylight hours, on weekdays only, to provide power for high daytime energy demands. It is these *controlled* releases which create the white water which had made commercial rafting possible since 1969. It is *a coalition of rafters and others who have proposed legislation which would stop all studies* of the potential for additional hydro-electric developments on the river.[74]

PACE and other backers of the project consistently identified rafting with environmental opposition, in effect merging the two while emphasizing only the former. Environmentalists objected to the suggestion that wild and scenic status was being sought primarily for the protection of whitewater recreation. One letter appearing in the *Modesto Bee* summarized the environmentalist viewpoint.

> I get the impression from the articles it is thought the key reason for saving the Tuolumne is white water rafting.
>
> This is very misleading. The primary goal of a wild and scenic designation is to preserve a natural wonder that symbolizes freedom, beauty and quality of life . . . The Stanislaus River was lost to this destruction. We cannot let it happen again.[75]

The reference to the Stanislaus River is significant. There, on a watershed adjacent to the Tuolumne, a major battle had been fought over the filling of New Melones reservoir, which flooded a popular rafting run. There, too, the issue, at least to environmentalists, was not only rafting but the flowing river and the wild canyon. Rafting had made it possible to visit and experience the river wilderness, and in turn became the focal point of a broader struggle.[76] Had the central issue been the recreational use of the

Tuolumne, some form of compromise might have been conceivable, but it was really a question of preservation versus development; a further replaying of the themes that had emerged decades earlier in the battle over Hetch Hetchy.

The debate was not waged exclusively over the virtues of wilderness, but also over the economic necessity or wisdom of the project. Its critics, led by the Tuolumne River Preservation Trust, Stanislaus County League of Conservation Voters and the Sierra Club, were basically motivated by their environmental outlook; an outlook that embraced not only an appreciation of wild and unspoiled places but a skepticism about growth and traditional patterns of energy consumption and development. They criticized the high cost of the project, which was expected to approach one billion dollars, and suggested that energy conservation and smaller-scale alternative energy sources were preferable from both a philosophic and economic point of view. On the other hand, PACE and the irrigation districts pointed to the future requirements of the Modesto-Turlock area and, in the words of Modesto councilwoman Carol Whiteside, argued that, "we would be allowing ourselves and future generations to be held hostage by unaffordable energy."[77] Proponents also pointed out that hydroelectricity was "clean" power when compared to the emissions from coal- or oil-fired plants or the radiation risks of nuclear power.

The publicity campaign continued to grow warmer on both sides as the preliminary permit studies neared the end of their first year, but the turning point had already been passed. When Senator Wilson announced his wilderness package including a wild and scenic Tuolumne, almost all that stood between the districts and the end of their project was the disagreement between Wilson and Cranston over the amount of land that should be declared wilderness. When they reached agreement on a compromise in June 1984, a wild and scenic future for the river seemed assured. Virtually the only important congressional opposition came from Congressman Tony Coelho of Merced, who had tried unsuccessfully to push a bill of his own to allow the Ponderosa studies to continue with provisions for further congressional review before construction. By August the wilderness bill had passed the Senate, and Coelho conceded that there was no real hope of stopping it in the House. By making the Tuolumne a part of the wilderness package, the merits of the Ponderosa project versus a wild and scenic river were never debated. As lobbyist Lee White said, "No one wants to be blamed for stopping the California Wilderness Bill and everybody's leery and suspicious of us."[78] The bill was passed by the House and signed into law by President Reagan in September 1984.

Within a few days, the field studies, which had already been curtailed, were terminated and the Ponderosa project was over. Congress, however, specifically provided for develpment of the river's forks and other tributaries so long as it did not adversely affect the protected main stem. The basic purpose of the unusual provision was to allow Tuolumne County to develop domestic water from small projects, but in March 1985 a preliminary permit was issued to the Turlock and Modesto irrigation districts for the Golden Rock reservoir and powerhouse on the South Fork. The districts hired R.W. Beck to analyze data from the Ponderosa studies to see if any satisfactory tributary projects were still possible.[79] Beck reported that the most feasible new project was not Golden Rock but an 80,000 KW development on the Clavey River. Predictably, environmentalists and rafters were quick to condemn the still sketchy plans as incompatible with the wild and scenic status of the main river.[80]

The same month that witnessed the final defeat of the TID's plans for the Ponderosa project also saw a dramatic victory of sorts for the district. In a surprise move planned by General Manager Ernest Geddes and attorneys for the irrigation districts, a five-line amendment backed by Congressman Coelho and other Central Valley congressmen was added to a budget bill working its way through Congress. If enacted, it would have required San Francisco to sell its Hetch Hetchy power at cost and made the secretary of the interior rather than the city's Public Utilities Commission responsible for reviewing rates. The amendment would have effectively rolled back the Court of Appeals ruling that gave San Francisco the unlimited right to raise its rates. The city had opened negotiations with the districts on a new contract to take effect when the old one expired in 1985, while simultaneously inviting other agencies to bid for Hetch Hetchy power. None did and the city had little realistic alternative but to deal with the irrigation districts. The talks had made it clear that the city expected to get 75 to 85 percent as much as PG&E's wholesale rates, which would have increased the cost of Hetch Hetchy power by up to 800 percent. The funding bill carrying the amendment was being rushed to meet the start of the new federal fiscal year on October 1, so any action had to come within a matter of days. Officials of both districts acknowledged that the TID was far more enthusiastic about the legislative approach, and more aggressive in its negotiations, than the Modesto District, which still feared the city might market its power elsewhere.[81]

As soon as word of the amendment reached San Francisco, officials there started lining up their forces for a showdown, aided by Senator Wilson, fresh from his victory over the districts' hydroelectric plans, who vowed to halt any attempt to restrict the city's rate-making authority.[82] Negotiations which would avoid the all-or-nothing risk of the amendment were soon suggested. Senator Cranston met with both sides and Coelho announced that he preferred a "fair and honorable compromise" to another bitter battle in Congress.[83] Several days of intensive negotiations involving Coelho, Cranston's staff, the districts and San Francisco resulted in an eleventh hour compromise. Both sides consented to a thirty-year contract fixing Hetch Hetchy power prices at 60 to 65 percent of PG&E's rates. The agreement was an odd sort of victory; the 500 percent increase in Hetch Hetchy rates hardly seemed to fit the public power pretensions of the Raker Act, but it was a better price than the districts could have negotiated without the threat posed by Coelho's amendment, which was then withdrawn.[84] Details of the proposed contract were still not finalized when the old contract expired in June 1985, so an interim, two-and-a-half-year contract was agreed upon while negotiations continued.[85] At the same time, the price paid for PG&E standby service to carry the districts through the loss of Don Pedro or Hetch Hetchy generation went up dramatically. The TID had been preparing for 1985 since it had failed to win a hearing before the U.S. Supreme Court and had been raising its rates annually to accumulate a reserve to cushion the "rate shock" that would occur if its customers suddenly found their bills had skyrocketed.

Rate increases were also necessary to underwrite the development of alternatives to high-priced Hetch Hetchy and PG&E purchase power. During the watershed month of September 1984 two such projects, a geothermal partnership and a gas-turbine peak power plant, were given approval by the Board of Directors, completing the redirection of the district's power program. The two projects met different needs. Geothermal power was developed from steam produced deep in the earth beneath a mountainous region known as The Geysers in Lake County. With a temperature of up to 400 degrees Fahrenheit, this natural, pressurized steam could be tapped by deep wells and harnessed to spin the turbines of electrical generators. The TID had been approached by the Northern California Power Agency (NCPA), a group of a dozen publicly-owned electric systems, with an offer to sell surplus power from NCPA Geothermal Project No. 3, a 110,000 KW powerplant already under construction. While receptive to the idea, the district preferred to own additional generating sources whenever possible. Negotiations resulted in a compromise, with several NCPA members surrendering part of their ownership shares to the TID, so that in all the district would own 10,000 KW, or not quite 10 percent of the plant's capacity. At least until the end of the century, the TID also agreed to purchase specific amounts of additional energy, up to 20,000 KW. The deal was particularly favorable because it required no immediate cash investment since construction financing was already arranged. Instead, the TID simply had to pay its share as it received power. The NCPA partnership provided the district with a source of base load power, used to meet the basic energy consumption level maintained throughout the year.[86]

While one of the great virtues of the geothermal plant was its constant base load energy output day-in and day-out, the whole point of the gas-turbine plant was that it would run only for short periods. It was a peaking plant, designed for operation only when power consumption was at its highest. On summer days when the system peaks usually occurred, the district's own resources and its share of Hetch Hetchy were inadequate to meet all the demands placed on the system by consumers, forcing a reliance on PG&E power to meet the peak. That was a costly practice since the higher the demand the TID placed on PG&E, the higher the cost of each kilowatt-hour the district had to buy from the private utility. The gas-turbine plant, with two 24,900 KW generating units, was expected to reduce the district's dependence on PG&E for peaking power, and in turn lower the cost of outside power. Compared to other power sources, gas-turbines are relatively inefficient in their use of fuel but inexpensive to build. When used only for peak power, the high cost of fuel for such plants is offset by their low capital cost.[87] The gas-turbine plant, located at the district's Walnut substation west of Turlock, was expected to cost $19.2 million and, like the geothermal plant, be in operation in 1986.

Another project with peaking power potential undertaken in the mid-1980s was a proposal to expand the Don Pedro powerplant. As wet winters sent water pouring out of the dam's flood control valves, studies were made on adding a fourth generator so that water bypassed around the main powerhouse could still be utilized to make electricity. A new 28,000 KW generator was expected to enhance the peaking capacity of Don Pedro and, because it could operate more efficiently at low loads than the existing units, its use could reduce wear and tear on the larger generators. The Turlock and Modesto districts moved to amend their FERC license to permit the fourth unit and in June 1985 they opened bids on the generating equipment. At the same time, the Sierra Club filed a protest against the project, requesting FERC to reject it unless new environmental conditions were attached to the Don Pedro license. The Sierra Club's motion was in some ways symbolic of the changes in environmental attitudes and activism that had occured since the initial dispute ended in 1966. The club complained that the fourth unit would increase fluctuations harmful to the salmon and would somehow even be detrimental to the wild and scenic river above Don Pedro reservoir. The motion also alleged that the districts had excess water, the use of which should be investigated by the commission, and requested that the license be amended to require the release of additional water into the lower Tuolumne, the removal of a logjam where the river entered the reservoir and the installation of restrooms at Wards Ferry Bridge. The various assertions and requests, many of which had no connection to the fourth unit, were an indication of the kind of scrutiny and demand for environmental mitigation and recreational enhancement that any modern water and power project could expect to encounter, and in fact, extended beyond the specific project to broader issues of water management.[88]

In terms of the number and magnitude of the projects considered, no period in the Turlock Irrigation District's history could equal the years from 1972 to 1985. Virtually no stone was left unturned in the search for additional power. Besides the projects discussed above, the district won the right to buy a small amount of power from the federal Western Area Power Administration, investigated participation in coal-fired plants, discussed co-generation and joined a consortium of other utilities in an effort to increase the capacity of transmission lines needed to bring surplus power from the Pacific Northwest to California. Dominated by electrical utility problems, the character of the 1970s and 1980s was, for the district, so distinctive that those years were undoubtedly a turning point in the district's history. While the changes were obvious, their full meaning for the district and its irrigation empire will unfold during the Turlock Irrigation District's second century.

CHAPTER 15 – FOOTNOTES

[1] TID, *Annual Reports;* "Historical Electrical Load Data."

[2] TID, *Annual Report,* 1978, p. 21.

[3] "Testimony of C. E. Plummer," before Federal Power Commission, 1962, p. 22.

[4] "Historical Electrical Load Data." [5] *Idem.*

[6] *Modesto Bee,* Mar. 7, 1973 [7] *Ibid.,* June 26, 1974, Apr. 23, 1975.

[8] Arthur D. Little, Inc., *An Analysis of Future Electric Power Needs and Sources for the Turlock and Modesto Irrigation Districts,* (Apr. 1975), vol. 1, pp. II-10 – II-12.

[9] Draft letter, J. D. Worthington to TID and MID, Aug. 29, 1973.

[10] *Modesto Bee,* Sept. 9, 1973. [11] *Ibid.,* Oct. 10, 1973.

[12] William R. Gianelli, *Investigation of Drainage Water and Well Supplies for Nuclear Power Plant Cooling Water, Stanislaus County,* (July 1974).

[13] See "Memorandum of Understanding between Turlock and Modesto Irrigation Districts Re: Proposed East Stanislaus Nuclear Power Plant," July 22, 1975.

[14] Arthur D. Little, Inc., vol. 1, pp. II-18 – II-24.

[15] *Modesto Bee,* June 2, 1977. [16] *Ibid.,* July 19, 1977.

[17] *Ibid.,* Sept. 11, 29, 1977.

[18] Board minutes, vol. 43, p. 452 (Apr. 23, 1973); *Modesto Bee,* Apr. 19, 1973.

[19] Board minutes, vol. 44, p. 156 (Feb. 19, 1974); *Modesto Bee,* Nov. 30, 1973.

[20] *Modesto Bee,* Mar. 29, Apr. 2, 1974. [21] *Ibid.,* Jan. 21, 1975.

[22] *Modesto Bee,* July 15, 18, 1976. [23] *Ibid.,* May 5, 1977.

[24] *Ibid.,* Feb. 23, 1978. [25] *Idem.*

[26] *Ibid.,* Jan. 23, 24, April 27, 1980. [27] *Ibid.,* Apr. 30, 1980.

[28] See "Notes to Financial Statements," TID, *1984 Annual Report,* p. 24.

[29] "Historical Electrical Load Data." [30] *Modesto Bee,* May 17, 1977.

[31] *Ibid.,* July 30, 1977; "Historical Electrical Load Data."

[32] *Modesto Bee,* Sept. 7, 1977.

[33] Maps of "Present and Proposed Future Construction, Hetch Hetchy Project," in Meikle files, vol. 14, item 1.

[34] Application to Appropriate Water, No. 13604 (Feb. 27, 1950).

[35] R. V. Meikle and J. M. Grigsby to State Water Rights Board, Aug. 31, 1966.

[36] R. W. Beck and Associates, *Appraisal Report: Hetch Hetchy Water and Power Stage I Additions,* (Jan. 1976).

[37] Sierra Club, *Tuolumne River: A Report on Conflicting Goals with Emphasis on the Middle River,* (Modesto, 1970), pp. 19-22.

[38] *Modesto Bee,* July 11, 1973.

[39] R. V. Meikle to Robert McCarty, Oct. 30, 1970.

[40] *Modesto Bee,* Aug. 27, 1975, Board minutes, vol. 45, p. 38 (Sept. 22, 1975).
[41] *Appraisal Report; Modesto Bee,* Feb. 11, Apr. 27, 1976; Board minutes, vol. 45, p. 161 (Apr. 26, 1976).
[42] *Modesto Bee,* July 21, 1977. [43] *Ibid.,* Oct. 12, 1978.
[44] *Ibid.,* Dec. 10, 16, 1978. [45] *Ibid.,* Dec. 27, 1978.
[46] *Ibid.,* Apr. 11, 1979.
[47] U. S. Forest Service and National Park Service, *Tuolumne Wild and Scenic River Study: Final Environmental Statement and Study Report,* (Oct. 1979), pp. 68-98.
[48] *Modesto Bee,* Sept. 2, 1979. [49] *Ibid.,* Oct. 3, 1979.
[50] *Ibid.,* Jan. 27, 1976, Board minutes, vol. 45, p. 372, (Dec. 5, 1977).
[51] U.S. Dept. of Energy, *Small-Scale Hydroelectric Power Demonstration Project, Turlock Irrigation District Drop No. 1 Power Plant: Final Technical and Construction Cost Report,* (Jan. 1981), pp. 2-3, 5-8.
[52] *Modesto Bee,* Apr. 25, 1978, Mar. 8, 1979.
[53] TID, *Official Statement – Electric Revenue Bonds, Series A,* July 29, 1980, p. 16.
[54] *Modesto Bee,* Feb. 24, 1982.
[55] *Official Statement – Electric Revenue Bonds,* pp. A10-A13.
[56] *Modesto Bee,* Aug. 16, 1977.
[57] *Ibid.,* Jan. 26, Mar. 8, 1983, Sept. 19, 1983, Aug. 9, 1984.
[58] Carl Muller interview; author's interview with George Yuge, Mar. 2, 1983.
[59] Board minutes, vol. 38, p. 323 (July 22, 1963), vol. 45, p. 69 (Oct. 14, 1975).
[60] TID, *1983 Annual Report,* p. 12; TID, *1984 Annual Report, p. 14.*
[61] *Modesto Bee,* Apr. 15, 1980. [62] *Ibid.,* July 27, 1982.
[63] *Ibid.,* Aug. 6, 1982. [64] *Ibid.,* Oct., 16, 1982.
[65] *Ibid.,* Dec. 22, 1982; "Allocation of Clavey-Wards Ferry Project Sunk Costs," (computer printout), July 14, 1983. [66] *Modesto Bee,* Nov. 24, 1982.
[67] *Modesto Bee,* July 22, 1981, Mar. 24, 1982. [68] *Ibid.,* June 1, 1983.
[69] *Ibid.,* Sept. 21, Nov. 1, 9, 1983 [70] *Ibid.,* Feb. 10, 1984.
[71] R. W. Beck and Associates, *Clavey-Wards Ferry Project, FERC Project No. 2774, Phase II – Feasibility Evaluation, Presentation on Project Alternatives,* (Jan. 12, 1984), Exhibits III-1, III-2, III-3, pp. IV-2 – IV-3.
[72] *Modesto Bee,* Nov. 22, 1983, Apr. 1, 1984 [73] *Ibid.,* Apr. 1, 1984.
[74] *Ibid.,* Apr. 25, 1984 [75] *Ibid.,* Feb. 5, 1984.
[76] On the Stanislaus River see W. Turrentine Jackson and Stephen D. Mikesell, *The Stanislaus River Drainage;* and Tim Palmer, *Stanislaus: The Struggle for a River,* (Berkeley, 1982).
[77] *Modesto Bee,* Dec. 29, 1983 [78] *Ibid.,* Sept. 6, 1984.
[79] *Ibid.,* Mar. 6, 1985 [80] *Ibid.,* June. 4, 1985.
[81] *Ibid.,* Sept. 20,26, 28, 1984 [82] *Ibid.,* Sept. 25, 1984.
[83] *Ibid.,* Sept. 26, 1984 [84] *Ibid.,* Sept. 28, 1984.
[85] *Ibid.,* June 5, 1985 [86] *Ibid.,* Sept. 29, 1984.
[87] *Ibid.,* Sept. 12, 1984; also see Arthur D. Little, Inc., vol. 1, pp. II-18 – II-19.
[88] *Modesto Bee,* Jan. 9, 1984, June 26, 1985; *Protest and Motion to Intervene of the Sierra Club,* before Federal Energy Regulatory Commission, June 10, 1985.

Epilog

An Approaching Centennial

In 1937, when the Turlock Irrigation District turned fifty years old, W.H. Lockwood wrote:

> Spirits from the other world were walking at Don Pedro. The spirits of Judge Waymire, Roger Williams, Ed Kiernan, Pat Griffin, Albert Chatom, and other old timers, met on the hills above the powerhouse; and there in the light of their spiritual sight they reviewed their work, including the shimmering web of copper that carries the juice to light and serve the thousands of homes in the contented and happy valley of the Tuolumne. They voted it worth all their efforts, and declared it worthy of a Golden Jubilee.[1]

Nearly fifty years later, the spirits Lockwood described would have to walk a higher hill above a larger Don Pedro, and their number would have swelled to include Roy Meikle and the scores of other employees and directors who had continued the job. Lockwood was expressing what a great many people felt, but local pride and civic boosterism aside, there was good reason to think that if those spirits were to gather they would indeed pronounce the result of their labors worthwhile. The district had done what its founders and their successors had hoped it would do; turn a vast expanse of grain fields into a thriving community of small farms and growing towns.

In 1937, as it had been since the turn-of-the-century, people in the Turlock region identified strongly with their irrigation district. Its water was the source of their wealth and its power had made rural living easier and more pleasent. Nearly fifty years later, some 70 percent of the electric users polled in the district's first public opinion survey held the TID in high regard and only 7 percent were actively hostile in their opinions, but less than half knew it was a public agency and only one in five could name a director.[2] The local pride was still there, but the sense of involvement had diminished. Irrigation did not directly affect as many residents as it once had, and although everyone was an electric user, electricity had become something usually taken for granted. Most district residents now know the TID primarily as an electric utility, and the emphasis in the district's planning reflects the struggle to keep pace with the power demands of a burgeoning population. The TID retains the form of an irrigation institution, but its function has evolved into that of a complex water and power agency. There is a certain tension, not always apparent on the surface, between its two purposes and even between their seemingly distinct constituencies.

Population growth and all it entails may be the central fact of the district's future. As its centennial approached, the Turlock region was once again beginning to change. Since 1970 a fourth epoch in local history appeared to be taking shape, one of suburbanization. Its coming was more subtle and gradual than the change from grazing to grain or from grain to irrigated farming. In the mid-1980s the Turlock region was still very much an irrigation empire, but the inroads of housing tracts, shopping centers and rural homes were unmistakable. Forecasts of the future are not necessarily a part of history, but if the projections are correct and the population of Stanislaus County increases two-, or three-, or four-fold in the coming decades, it will mean that the irrigation empire is giving way to some other, as yet undefined, kind of place. There are those who fear that the region's eventual fate will resemble that of the Santa Clara Valley, where agriculture was shouldered aside and productive farm land was paved over to create a metropolitan area. If that prophecy were to come true, the decade of the 1970s will probably be seen as the beginning of the transition for the Turlock region, just as it was a time of important transition for the Turlock Irrigation District. Whether fifty or a hundred years in the future, another, much larger group of spirits will gather above Don Pedro to praise the value of their work will depend on how well the district responds to the changes in the community it helped create.

EPILOG – FOOTNOTES

[1] W. H. Lockwood, "The Story of an Idea," *Turlock Tribune,* Souvenir Edition, June 11, 1937.

[2] *Modesto Bee,* May 15, 1985.

Notes on the Sources

A glance at the footnotes will reveal a heavy reliance on unpublished Turlock Irrigation District records and newspapers. Under the circumstances a standard bibliography did not seem appropriate. As an alternative, these notes were prepared to provide readers with a brief commentary on the most useful sources.

Of course, no sources were more basic than the TID's own records. The minutes of the Board of Directors and a limited number of legal and financial documents are the district's only record of its operations during most of its first twenty years. For the period from 1900 to about 1915, the minutes remain the best single reference, but some correspondence, engineering data, maps and photographs have survived to help flesh out the story. After about 1915 the documentation improves, due in large part to the efforts of Roy V. Meikle, chief engineer from 1914 to 1971, who kept not only his correspondence but also copies of important reports and memoranda. As one would expect, information on the district's recent activities is plentiful, but it is often so scattered between various files and reports that it can be surprisingly difficult to collect.

The local newspapers were almost as important as the district's records. Beginning with the *Tuolumne City News* in 1868, the papers consulted included its successors, *Stanislaus County News, Stanislaus County Weekly News,* and *Modesto Daily Evening News,* as well as the *Modesto Herald, Modesto Morning Herald, Modesto News-Herald,* and *Modesto Bee,* all of Modesto; and the *Turlock Journal* (or *Turlock Daily Journal*) and *Turlock Tribune.* These newspapers are available on microfilm at California State University, Stanislaus, the Stanislaus County Library and at the *Turlock Journal* office.

A great deal of valuable information came from interviews with district residents and past and present TID employees. Whenever possible, interviews were tape-recorded, but informal conversations and even telephone interviews also proved useful. Those interviewed included (in alphabetical order):

Arlene Abel
Edna Anderson
Jim Arnold
Michael Berryhill
Norman Boberg
Earl Caswell
Christine Chance
Frank Clark
Richard Clauss
Charles Crawford
Dalton Christman
Lester Domecq
Clarice Espinola
William R. Fernandes
Forest Fiorini

Tim Ford	Isabelle Martin	Ted Schuld
Ron Hawkins	Don McDonald	Phillip Short
Beth Crowell Hollingsworth	Norman Moore	Arthur Starn
	Carl Muller	Lloyd Starn
Leroy Kennedy	Velma Openshaw	William Tell
Ray Knowles	Grant Paterson	Paul Terrell
Samuel Kronberg	Charles O. Read	Reynold Tillner
Grant Lucas	Robert Reuther	William Trent
Mildred Lucas	Charles Rose	Steven Vilas
Arvid E. Lundell	H.R. Sahlstrom	Richard Vollrath
David Lundquist	Henry Schendel	George Yuge

In addition, interviews with Hughson-area residents conducted by Margaret Sturtevant, William Hurd and others between 1967 and 1971 were very helpful, as was an interview Gertrude Vasche taped with Chief Engineer Meikle in 1970.

Although not used specifically to any great extent in writing the history of the Turlock Irrigation District, readers wishing to know more about the general development of water in California might consult the following sources; William Kahrl, et. al., *The California Water Atlas* (Sacramento, 1978); S.T. Harding, *Water in California* (Palo Alto, 1960); and Erwin Cooper, *Aqueduct Empire* (Glendale, Calif. 1968). For a perceptive look at geographic, economic and social changes in another part of the San Joaquin Valley where irrigation is important, see William L. Preston, *Vanishing Landscapes: Land and Life in the Tulare Lake Basin* (Berkeley, 1981).

The San Joaquin Valley and its native grasslands are described in Elna Bakker, *An Island Called California* (Berkeley, 1972) and L.T. Burcham, *California Range Land: An Historico-Ecological Study of the Range Resources of California* (Sacramento, 1957). U.S. Department of Agriculture, *Soil Survey of the Modesto-Turlock Area,* by A.T. Sweet, J.F. Warner and L.C. Holmes (1909) describes the area's general soil characteristics, while two University of California, Agricultural Experiment Station publications by Rodney J. Arkley, *Soils of Eastern Stanislaus County, California,* Soil Survey No. 13 (1959), and *Soils of Eastern Merced County, California,* Soil Survey No. 11 (1954) provide both basic information and a series of detailed soil maps.

The best general histories of the Turlock area are Helen Hohenthal, "A History of the Turlock Region," (M.A. thesis, University of California, Berkeley, 1930) and Helen Alma Hohenthal and others, *Streams in a Thirsty Land: A History of the Turlock Region* (Turlock, 1972), which includes portions of her thesis. The "Supplementary Material" attached to Hohenthal's thesis, which includes letters and interviews with old-time residents, is also a valuable source. I.N. Brotherton, *Annals of Stanislaus County: River Towns and Ferries* (Santa Cruz, 1982) deals with aspects of early settlement. Sol Elias, *Stories of Stanislaus* (Modesto, 1924) describes colorful episodes of the county's early days as well as giving many useful details on the grain era and the struggle for irrigation. The *History of Stanislaus County* (Elliot and Moore, publishers, San Francisco, 1881) and George Tinkham, *A History of Stanislaus County* (Los Angeles, 1921) are both "mug books" combining a brief but sometimes useful history of the county with complimentary profiles of leading citizens. Mildred Lucas, *From Amber Grain . . . to Fruited Plain: A History of Ceres, California, and its Surroundings* (Ceres, 1976) is also worthwhile.

While there are a number of studies dealing with the disposal of public lands, Richard Allen Eigenheer, "Early Perceptions of Agricultural Resources in the Central Valley of California," (Ph.D. dissertation, University of California, Davis 1976) provides a detailed look at land sales in the San Joaquin Valley. Khaled Bloom, "Pioneer Land Speculation in California's San Joaquin Valley," *Agricultural History,* 57 (July 1983) is also helpful, as is Hohenthal's work.

The wheat era in California deserves more careful study. The local histories cited above all deal with the subject. Wallace Smith, *Garden of the Sun* (Fresno, 1939) covers farm practices and agricultural technology. Rodman Paul, "The Wheat Trade Between California and the United Kingdom," *Mississippi Valley Historical Review,* XLV (December 1958) provides a discussion of the Liverpool market, and his article "The Great California Grain War: The Grangers Challenge the Wheat King," *Pacific Historical Review,* XXVII (November 1958) deals with some of the efforts to overcome the speculators who preyed on the grain growers. Charles Nordhoff's description of the wheat growers he encountered in *California for Health, Pleasure and Residence: A Book for Travellers and Settlers* (New York, 1873) seems to parallel the viewpoint taken by novelist Frank Norris in *The Octopus* (New York, 1901). How close these vivid impressions were to the truth needs further study.

Donald J. Pisani, *From Family Farm to Agribusiness: The Irrigation Crusade in California and the West, 1850-1931* (Berkeley, 1984) is an excellent, comprehensive account of irrigation in California. Thomas E. Malone, "The California Irrigation Crisis of 1886: Origins of the Wright Act," (Ph.D. dissertation, Stanford University, 1964) actually covers much more than the events of 1886 and can be read along with Pisani for an understanding of the search for effective irrigation institutions. Elwood Mead, *Irrigation Institutions* (New York, 1903) is an important early survey of legal and institutional problems. Richard J. Hinton, *A Report on Irrigation,* Senate Executive Document 41, Part 1, 52nd Congress, 1st Session (1893) records some of C.C. Wright's own thoughts on his irrigation district plan. Arthur Maass and Raymond L. Anderson, *. . . and the Desert Shall Rejoice: Conflict, Growth and Justice in Arid Environments* (Cambridge, Mass., 1978) describes the development of irrigation in the Fresno and Hanford areas in California, and provides useful comparisons with irrigation systems in Utah, Colorado and Spain.

Irrigation in the Turlock region is treated most thoroughly in Hohenthal's thesis and in Benjamin F. Rhodes, Jr., "Thirsty Land: The Modesto Irrigation District: A Case Study in Irrigation Under the Wright Act," (Ph.D. dissertation, University of California, Berkeley, 1943). Neither of these, however, provides a complete history of irrigation plans south of the Tuolumne. Three other histories of the Modesto Irrigation District that deserve mention are Robert M. Graham, "An Epic of Water and Power: A History of the Modesto Irrigation District," (M.A. thesis, College of the Pacific, 1946), Albert Peter Beltrami, "The Modesto Irrigation District: A Study in Local Resources Administration," (M.A. thesis, University of California, 1957) and Dwight H. Barnes, *The Greening of Paradise Valley* (Modesto, 1987).

Two reports by Frank Adams, *Irrigation Districts in California, 1887-1915,* California Department of Engineering, Bulletin No. 2 (1916), and *Irrigation Districts of California,* California Department of Public Works, Division of Engineering and Irrigation, Bulletin No. 21 (1930), are basic sources of information about the evolution of California irrigation district law and the histories of individual districts.

The construction of La Grange Dam is discussed by Sol Elias, *Stories of Stanislaus,* James D. Schuyler, *Reservoirs for Irrigation, Water Power and Domestic Water Supply* (New York, 1901), C.E. Grunsky, *Irrigation Near Merced, California,* U.S. Geological Survey, Water Supply and Irrigation Paper No. 19 (1899), and Herbert Wilson, *Irrigation Engineering* (6th ed., New York, 1909). The significance of the 1890 agreement between the Turlock and Modesto districts is readily apparent when comparisons are made with other areas. Maass and Anderson, . . . *and the Desert Shall Rejoice* describes the water rights problem on the Kings River and U.S. Department of Agriculture, Office of Experiment Stations, *Report on Irrigation Investigations in California,* Bulletin No. 100 (1901) deals with the same problem on the Kings River and several other California streams.

The goals of the national irrigation crusade, which were similar to those of the local movement, are eloquently stated by one of reclamation's greatest publicists, William E. Smythe in *The Conquest of Arid America* (New York, 1907), and Pisani and others cited above deal with the same question. Although Kevin Starr devotes only a few pages in *Americans and the California Dream, 1850-1915* (New York, 1973), and *Inventing the Dream: California Through the Progressive Era* (New York, 1985) to irrigation, his comments are well worth reading.

Most accounts of settlement and agriculture in the Turlock District after irrigation were frankly promotional in purpose, but they often contained specific information about the area. Among them were James J. Rhea, *The Turlock District, Stanislaus County, California* (San Francisco,

1912), R. L. Adams and W. W. Bedford, *The Marvel of Irrigation: A Record of a Quarter Century in the Turlock and Modesto Districts, Califorina* (San Francisco, 1921), John T. Bramhall, *The Story of Stanislaus* (Modesto, 1914), National Irrigation Congress, *Modesto and Turlock Districts, San Joaquin Valley, California* (Bulletin No. 3, Publicity Series, 1907), and the articles in the Santa Fe Railway publication *The Earth* (February 1910). The same kind of writing for California as a whole can be found in A. J. Wells, *California for the Settler: The Natural Advantages of the Golden State for the Present Day Farmer* (San Francisco, 1914) and R. E. Hodges, *Farming in California* (San Francisco, 1923). The *Modesto Morning Herald* "Irrigation Jubilee Supplement," April 22, 1904, is a useful look at the area and its hopes at the beginning of the irrigation era. E.J. Cadwallader, "History of Turlock" (typescript, 1945) is an excellent first person account of real estate work and community growth. Individual farms were the focus of U.S. Department of Agriculture, Office of Experiment Stations, Irrigation Investigations, *The Settlement of Small Farms in Modesto and Turlock Irrigation Districts,* by Wells Hutchins, (Berkeley, 1915).

How farmers chose to irrigate their fields is a subject that deserves more study. General treatises like Lute Wilcox, *Irrigation Farming: A Handbook for the Practical Application of Water in the Production of Crops* (New York, 1895), or Henry Stewart, *Irrigation for Farm, Garden and Orchard* (New York, 1886) are of interest, but they deal primarily with methods more commonly used elsewhere. Perhaps the most complete discussion of irrigation practices relevant to the Turlock area is in Elwood Mead, *Textbook for Reading Course in Irrigation Practice* (Berkeley, 1906). Frank Adams, "The Distribution and Use of Water in Modesto and Turlock Irrigation Districts, California," in U.S. Department of Agriculture, Office of Experiment Stations, *Annual Report of Irrigation and Drainage Investigations, 1904* (Separate No. 3, 1905) describes the design and construction of irrigation systems in the Turlock area. Adams' informative report also dealt with the districts' early history and the operation of their irrigation works. B.A. Etcheverry, *Irrigation Practice and Engineering,* volume 1 (New York, 1915) is also useful.

The California Irrigation District Bond Commission, *Report on Turlock Irrigation District* (1914) summarizes the TID's finances and engineering works in early 1914. S.T. Harding, *The Operation and Maintenance of Irrigation Systems* (New York, 1917) describes many of the practical problems of operating irrigation canals. On the important matter of drainage, J.H. Dockweiler, *Water Needs of the Turlock and Modesto Irrigation Districts and the Quantity of Water Remaining for Storage and Diversion to the Cities Around San Francisco Bay* (San Francisco, 1912) gave detailed information on the lakes and other evidence of waterlogging. U.S. Department of Agriculture, Office of Experiment Stations, *Drainage of Irrigated*

Lands in the San Joaquin Valley, California, by Samuel Fortier and Victor M. Cone, Bulletin No. 217 (1909) dealt with efforts to solve the problem in the Turlock and Fresno areas.

A great deal has been written about the Hetch Hetchy controversy, although little of it deals with the role of the irrigation districts. Ray W. Taylor, *Hetch Hetchy: The Story of San Francisco's Struggle to Provide a Water Supply for Her Future Needs* (San Francisco, 1926) gives a basic journalistic account of the project, while City Engineer M. M. O'Shaughnessy added his own perspectives in *Hetch Hetchy: Its Origins and History* (San Francisco, 1934). John Muir set the tone for the nature-lovers' fight to save the valley with writings like *The Yosemite* (New York, 1912). Roderick Nash, *Wilderness and the American Mind* (revised edition, New Haven, 1973), and Holway Jones, *John Muir and the Sierra Club* (San Francisco, 1965) deal with Hetch Hetchy as a wilderness issue. Elmo Richardson, *The Politics of Conservation: Crusades and Controversies, 1897-1913* (Berkeley, 1962), Kendrick A. Clements, "Politics and the Park: San Francisco's Fight for Hetch Hetchy, 1908-1913," *Pacific Historical Review,* 48 (May 1979), and Samuel P. Hays, *Conservation and the Gospel of Efficiency: The Progressive Conservation Movement, 1890-1920* (Cambridge, Mass., 1959) discuss the political aspects of the struggle. Walton Bean, *Boss Ruef's San Francisco* (Berkeley, 1952) tells the story of the unsuccessful attempt to sell San Francisco another water supply. Essential to any understanding of the events leading to the Raker Act is John R. Freeman, *On the Proposed Use of a Portion of the Hetch Hetchy, Eleanor and Cherry Valleys . . . for the Water Supply of San Francisco* (San Francisco, 1912). The irrigation districts replied with the *Report of H.S. Crowe, Engineer of the Modesto Irrigation District, and Burton Smith, Engineer of the Turlock Irrigation District, in Answer to Reports on Behalf of San Francisco . . . ,* volume 1, (Turlock, 1912). *Hetch Hetchy Valley: Report of the Advisory Board of Army Engineers,* House Document No. 54, 63rd. Congress, 1st. Session (1913), and *Hetch Hetchy Dam Site: Hearing Before the Committee on Public Lands, House of Representatives, 63rd. Congress, 1st. Session on H.R. 6381,* June 25, 1913, volumne 1, are also helpful. Despite the attention given to the decision to build Hetch Hetchy, there remains no comprehensive history of the project. The important controversy over the disposal of Hetch Hetchy power is perhaps best summarized in a short document released by the city in 1944 titled "A History of Efforts to Dispose of Electricity Generated on Hetch Hetchy Project." Stephen P. Sayles, "Hetch Hetchy Reversed: A Rural-Urban Struggle for Power," *California History,* LXIV (Fall 1985) deals with the question of the Early Intake power site.

The TID has extensive records on the construction of Don Pedro Dam. A. J. Wiley, *Report on Don Pedro Reservoir for the Turlock and Modesto Irrigation Districts* (1918) gives the consulting engineer's description of the dam's design and purpose. R. W. Shoemaker, *The Don Pedro Project: Proposed Power Development Covering the Generation, Distribution and Sale of Electrical Energy by the Turlock Irrigation District* (1922) became the blueprint for the district's entry into the electric utility business.

Plans for the development of the Groveland project are outlined in water rights applications no. 5094 (July 12, 1926) and no. 5266 (November 12, 1926), and in *Amended Application for Preliminary Permit for Groveland Project Submitted to Federal Power Commission by Turlock Irrigation District and Modesto Irrigation District* (1928), and the districts' *Application for License: Project No. 761 – California* (1932). John C. Beebe, *Report to Federal Power Commission on Application for License, Project No. 761* (San Francisco, 1932) is interesting because it questions certain parts of the districts' proposal.

The story of the abortive state attempt to establish a model farm community at Delhi has several sources. The rationale behind the state land settlement concept is discussed in Paul Conkin, "The Vision of Elwood Mead," *Agricultural History,* 34 (April 1960), and also in Kevin Starr, *Inventing the Dream.* Frank Adams, *Report on Water Supply for the Wilson Tract* (prepared for State Land Settlement Board, 1918), and the transcript of Adams' oral history interview in the Bancroft Library help explain the selection of the Delhi site. The reports of the State Land Settlement Board dated June 30, 1918, and September 30, 1920, are of considerable interest. Roy J. Smith, "The California State Land Settlements at Durham and Delhi," *Hilgardia,* 15 (October 1943) is a standard treatment of the subject. Also useful are Tony Kocolas, "California's Experiment at Delhi: Evaluation of the Delhi Land Settlement Project of 1920 to 1931," (graduate paper, Stanislaus State College, 1971), and an interview of former Delhi superintendent J. Winter Smith by J. Carlyle Parker in the California State University, Stanislaus, archives.

Comparative information on California irrigation districts during the Depression can be found in Harding, *Water in California* and in the "Conclusions and Recommendations of the Report of the California Irrigation and Reclamation Financing and Refinancing Commission," December 1, 1930, in *Appendix to the Journals of the Senate and Assembly, 49th session (1931),* volume 5. Kenneth R. McSwain, *History of the Merced Irrigation District* (Merced, 1978) describes the effects of the Depression on another irrigation district close to the TID.

Documentation on the post-1913 relationship between the TID and San Francisco can be found primarily in TID files. Reports by B.A. Etcheverry and Thomas H. Means, including *Report on Use of Water from Tuolumne River by Turlock Irrigation District* (1915), and *Report on Use of Water from Tuolumne River by Modesto Irrigation District and Turlock Irrigation District* (1920) were part of an effort to prepare for eventual litigation. They can be found at the California Water Resources Center Archives in Berkeley. California Division of Water Resources, *San Joa-*

quin River Basin, Bulletin No. 29 (1931) describes tentative state plans for the Tuolumne. The Max Bartell collection at the Water Resources Center Archives contains the memoranda leading to the Second Agreement.

Among documents of interest in the planning and construction of the New Don Pedro project are Stone and Youngberg, Blythe & Co. *Financing Report: Turlock Irrigation District, New Don Pedro Project* (1961), California Department of Water Resources, *Review of Report on Tuolumne River Development Proposed by Modesto and Turlock Irrigation Districts* (prepared for California Districts Securities Commission, 1961), and TID and MID, *New Don Pedro Project: Application Report for Davis-Grunsky Grant, Volume I – Feasibility Report,* by Angus Norman Murray (1967). The lengthy transcripts of the hearings conducted by the Federal Power Commission in 1962, together with prepared statements and exhibits submitted by the districts, are important sources. Portions of the transcripts are contained in *Joint Appendix: State of California, Turlock Irrigation District, Modesto Irrigation District and United States of America on Relation of Stuart L. Udall, Petitioners v. Federal Power Commission, Respondent, in the United States Court of Appeals for the Ninth Circuit,* which also includes petitions, FPC staff reports and the Commission's Opinion and Order Issuing License.

On the salmon controversy, comments by the state's Department of Fish and Game in U.S. Department of the Interior, *Central Valley Basin,* Senate Document 113, 81st. Congress, 1st. Session (1949) give the first clue that the fishery could become an issue. TID documents, the FPC hearing transcript and the *Joint Appendix* detail fishery negotiations and the ultimate outcome. More work needs to be done on the overall history of environmental management and attitudes in this period, but comparisons with New Don Pedro, and other useful information, can be derived from, among others, W. Turrentine Jackson and Stephen D. Mikesell, *The Stanislaus River Drainage Basin and the New Melones Dam: Historical Evolution of Water Use Priorities* (Davis, Calif., 1979), W. Turrentine Jackson and Alan M. Paterson, *The Sacramento-San Joaquin Delta: The Evolution and Implementation of Water Policy: An Historical Perspective* (Davis, Calif., 1977), William A. Hillhouse II, "The Federal Law of Water Resource Development," in Erica L. Dolgin and Thomas T.P. Gilbert, editors, *Federal Environmental Policy* (St. Paul, 1974), and William Ashworth, *Hells Canyon: The Deepest Gorge on Earth* (New York, 1977).

On the question of modern irrigation district governance, see James Jamieson, Sidney Sonenblum, Werner Z. Hirsch, Merrill R. Goodall and Harold Jaffee, *Some Political and Economic Aspects of Managing California Water Districts* (Los Angeles, 1974), and David Lincoln Martin, "California Water Politics: The Search for Local Control," (Ph.D. disser tation, Claremont Graduate School, 1973).

The California drought of 1976-1977 is covered by California Department of Water Resources, *The California Drought, 1977: An Update* (February 1977) and by the same agency's, *The 1976-1977 California Drought: A Review* (May 1978).

The TID's electric power needs and options were evaluated in the Arthur D. Little, Inc. study, *An Analysis of Future Electric Needs and Sources for the Turlock and Modesto Irrigation Districts,* 2 volumes, (1975). Nuclear power plant cooling water was the subject of William R. Gianelli, *Investigation of Drainage Water and Well Supplies for Nuclear Power Plant Cooling Water, Stanislaus County* (1974), and URS Company, *Project Report on the Water Supply System for the Stanislaus Nuclear Project* (1977).

On the Clavey-Wards Ferry and Ponderosa projects, the best general descriptions are found in two reports by R.W. Beck and Associates; *Appraisal Report: Hetch Hetchy Water and Power Stage I Additions* (1976), and *Clavey-Wards Ferry Project, FERC Project No. 2774, Phase II – Feasibility Evaluation, Presentation on Project Alternatives* (1984). The Sierra Club, *Tuolumne River: A Report on Conflicting Goals with Emphasis on the Middle River* (Modesto, 1970) describes the beginning of recreational use of the river above Don Pedro. U.S. Forest Service and National Park Service, *Tuolumne Wild and Scenic River Study: Final Environmental Statement and Study Report* (1979) summarizes reaction to the wild and scenic proposal. The Stanislaus River and the filling of New Melones reservoir provide an interesting comparison to the Tuolumne River and the Clavey-Wards Ferry project in terms of environmental opposition. On that subject, see Tim Palmer, *Stanislaus: The Struggle for a River* (Berkeley, 1982), and the 1979 study by Jackson and Mikesell cited above.

The TID's mini-hydro project is described in the district's *Official Statement – Electric Revenue Bonds, Series A* (1980), and in U.S. Department of Energy, *Small Scale Hydroelectric Power Demonstration Project, Turlock Irrigation District Drop No. 1 Power Plant: Final Technical and Construction Cost Report* (1981).

Index